DYNAMICS OF MECHANICAL AND ELECTROMECHANICAL SYSTEMS

DYNAMICS OF MECHANICAL
AND ELECTROMECHANICAL SYSTEMS

STEPHEN H. CRANDALL
DEAN C. KARNOPP
EDWARD F. KURTZ, JR.
DAVID C. PRIDMORE-BROWN

DEPARTMENT OF MECHANICAL ENGINEERING
MASSACHUSETTS INSTITUTE OF TECHNOLOGY

EDITED BY
STEPHEN H. CRANDALL

KRIEGER PUBLISHING COMPANY
MALABAR, FLORIDA

Original Edition 1968
Reprint with corrections 1982

Printed and Published by
ROBERT E. KRIEGER PUBLISHING COMPANY, INC.
KRIEGER DRIVE
MALABAR, FLORIDA 32950

Copyright © 1968, by McGraw-Hill Inc.
Reprinted by arrangement

Library of Congress Cataloging in Publication Data

Main entry under title:

Dynamics of mechanical and electromechanical
    systems.

    Reprint. Originally published: New York: McGraw-
Hill, 1968.
    Includes index.
    1. Dynamics.   I. Crandall, Stephen H.
TA352.D96   1982   620.1'04   82-9890
ISBN 0-89874-529-2   AACR2
10 9 8

PREFACE

This book has been written to serve as a text for an intermediate course in engineering dynamics. At M.I.T. this course is the third in a sequence of mechanics courses for engineers. The preceding courses provide an introduction to the mechanics of deformable media and an introduction to dynamic systems. The present text has been designed to reinforce the reader's knowledge of the fundamental principles of dynamics and to extend his experience by considering more complex systems and more complex dynamic phenomena.

An important feature of the book is the breadth of the area of application. In addition to treating mechanical systems of particles and rigid bodies, we include applications to electric circuits, to electromechanical transducers, and to continuous models of elastic structures, fluids, and electric transmission lines. The underlying unity of dynamic phenomena throughout this interdisciplinary range is stressed.

Another feature of the book is the consistent use of variational methods as an alternative technique for formulating dynamics problems. Hamilton's principle is introduced in the first chapter and acts as a major unifying element throughout the remainder of the text. Our experience in teaching this material during the past five years has indicated that the early introduction of the variational method as an alternative to the direct application of the fundamental principles has given students a better insight into the nature of the fundamental principles.

The first chapter contains an elementary introduction to static and dynamic variational principles developed in terms of simple particular cases. The second chapter presents a systematic treatment of dynamics of mechanical systems whose inertial elements are treated as particles. This chapter also includes a more thorough development of the variational approach. The dynamic properties of a rigid body are carefully examined in the third chapter, in preparation for Chapter 4, where a variety of rigid-body systems of engineering interest are analyzed.

In Chapter 5 the field of application shifts to electrical networks. This sets the stage for Chapter 6, where the mechanical and electrical threads are interwoven in the discussion of the dynamic behavior of electromechanical transducers. In Chapters 7 and 8 the development is extended to continuous models of mechanical and electrical systems. Elastic structures and electric transmission lines are treated in Chapter 7, and in Chapter 8 we consider dynamic systems involving fluids. In particular, we discuss acoustic waves and systems in which there is dynamic interaction between elastic structures and flowing fluids.

The major portion of the text is devoted to describing and illustrating principles and techniques in terms of particular examples of engineering importance. In the latter part of Chapter 2, however, we outline a generalized development of analytical dynamics. In Chapter 9 this is further extended to provide an introduction to the dynamical foundations of statistical mechanics. It is hoped that this material will constitute a useful preparation for the reader who plans to study statistical mechanics.

Modern engineering is increasingly concerned with difficult dynamics problems. To be effective, an engineer must develop an insight into, and an intuitive understanding of, dynamic behavior in general. These skills are not easily acquired. It is our conviction that the best preparation consists of thorough familiarization with the fundamental principles, together with experience in applying these principles to a carefully selected set of problems covering a wide range of applications. Through the acts of formulating, solving, and interpreting the solutions of dynamics problems, a student builds up a reservoir of experience which he can draw upon for the subsequent development of insight and intuition. The illustrative examples in the text and the problems for the reader at the end of each chapter have been chosen with these ends in view. Answers to most of the problems are given at the back of the book.

This text is the outcome of an educational experiment which began in the summer of 1961 and to which many people have contributed. On behalf of the authors I should like to acknowledge the helpful guidance of a special study group which included R. L. Bisplinghoff, S. I. Freedman, A. T. Murphy, P. L. Penfield, Jr., and E. Rabinowicz. Particularly valuable contributions were made by P. L. Penfield, Jr., who participated in teaching the course during its formative stage. A preliminary volume of notes was typed by Miss M. J. Clarke in 1962, and a major revision was typed by Mrs. de Clercq Zubli in 1966. I wish to express our appreciation to our typists and to our students, who have struggled with us during the development of this material. The generous support of the Ford Foundation at several stages in the process is gratefully acknowledged.

STEPHEN H. CRANDALL

CONTENTS

PREFACE   v

CHAPTER 1   INTRODUCTION TO VARIATIONAL
PRINCIPLES........................... 1

    1-1   Geometric variational principles for statics.... 2
    1-2   Force variational principle for statics......... 10
    1-3   Legendre transformations................... 13
    1-4   Variational principles for dynamics........... 22
        Problems................................. 35

CHAPTER 2   DYNAMICS OF MECHANICAL
SYSTEMS............................... 42

    2-1   Kinematics............................... 42
    2-2   Relative motion of rectangular frames........ 46
    2-3   Matrix representation of rotation............ 53
    2-4   Kinematical analysis utilizing intermediate
frames................................... 63
    2-5   Newtonian dynamics....................... 79
    2-6   Kinetic state functions.................... 87
    2-7   Work and potential energy................. 90
    2-8   Formulation of equations of motion.......... 101
    2-9   Techniques for assuring admissible variations.. 110
    2-10   Generalized variables...................... 115
    2-11   Hamilton's principle....................... 128
    2-12   Conservation laws......................... 135
    2-13   Momentum principles for a system of particles 141
        Problems................................. 146

CHAPTER 3   DYNAMICAL PROPERTIES OF A RIGID
BODY................................. 165

    3-1   Rigid bodies.............................. 165
    3-2   Angular momentum of a rigid body.......... 168

vii

3–3   The inertia tensor.......................... 171
3–4   Kinetic coenergy of a rigid body............. 175
3–5   Dynamic principles for a rigid body.......... 177
3–6   Parallel-axes theorem....................... 180
3–7   Rotational transformation of inertia-tensor
      components................................ 186
3–8   Principal directions........................ 190

      Problems.................................. 202

CHAPTER 4   DYNAMICS OF SYSTEMS CONTAINING
             RIGID BODIES........................ 208

4–1   Plane motion of rigid bodies................ 208
4–2   Balancing of rotors......................... 221
4–3   Euler's angles.............................. 224
4–4   Euler's equations........................... 226
4–5   Torque-free motions......................... 228
4–6   Gyroscopes.................................. 239

      Problems.................................. 246

CHAPTER 5   ELECTRICAL NETWORKS............... 257

5–1   Basic concepts.............................. 258
5–2   Passive circuit elements.................... 260
5–3   Electrical sources.......................... 266
5–4   Interconnection laws........................ 268
5–5   Electromechanical analogies................. 273
5–6   Hamilton's principle for electrical networks... 275
5–7   Generalized coordinates and Lagrange's
      equations................................. 280

      Problems.................................. 285

CHAPTER 6   ELECTROMECHANICAL TRANSDUCERS 291

6–1   Constitutive relations for transducers........ 291
6–2   Analysis of transducer systems.............. 300
6–3   Frequency response of transducers.......... 308
6–4   Variable mutual inductance................. 318

      Problems.................................. 326

CHAPTER 7   DYNAMICS OF CONTINUOUS SYSTEMS.   332

7–1   Continuum models........................... 332
7–2   Analysis of continuous systems.............. 334
7–3   Hamilton's principle for continuous systems... 336
7–4   The wave equation.......................... 351
7–5   Sinusoidal waves........................... 356

        Problems................................. 363

CHAPTER 8   DYNAMICS OF FLUID SYSTEMS....... 371

8–1   One-dimensional flow........................ 371
8–2   Equations of motion........................ 374
8–3   Acoustic approximation..................... 377
8–4   Eulerian description of fluid flow............ 385
8–5   Interaction of fluids and elastic solids........ 390

        Problems................................. 395

CHAPTER 9   PRELUDE TO STATISTICAL
                MECHANICS........................... 402

9–1   Hamilton's canonical equations.............. 402
9–2   A microscopic mechanical model............. 406
9–3   The distribution function................... 407
9–4   The phase fluid............................ 411

        Problems................................. 412

APPENDIX A   Variational calculus........................ 417

APPENDIX B   Matrix algebra........................... 425

APPENDIX C   Vector analysis.......................... 430

APPENDIX D   The complementary mechanical variational
               principle.................................. 433

APPENDIX E   Natural dependence of state-functions....... 441

ANSWERS TO PROBLEMS............................. 447

INDEX 457

DYNAMICS OF MECHANICAL AND ELECTROMECHANICAL SYSTEMS

CHAPTER ONE

Introduction to variational principles

The essence of dynamics is change. As a scientific discipline dynamics is the study of systems undergoing changes of state. The changes of state of a mechanical system are described by the motions of its parts. The changes of state of an electrical system are described by the time histories of its voltages and currents.

An engineer makes dynamic studies of a system for many reasons. He may simply wish to understand how the system works in a qualitative way, or he may require very detailed quantitative results. For example, if he is about to measure the performance of an existing system, he might make a rough study in order to decide what instruments to use and where they are to be placed. On the other hand, during the design of a proposed system, he may make quite thorough studies of the expected performance to ensure that the system, when built, will meet its specifications and to guide him in selecting parts of adequate strength and capacity.

A complete dynamic analysis of an engineering system can be conveniently subdivided into four steps:

1. Isolation and identification of an idealized model of the system under consideration
2. Formulation of equations of motion for the model by applying appropriate governing principles
3. Solution of the equations of motion
4. Interpretation of the solution

The examples and problems in this text have been chosen to give the student experience in carrying through these four steps for a rich vocabulary of dynamic models. It is this kind of experience which forms the foundation for the development of a qualitative feeling for dynamic behavior. We shall consider mechanical systems in which the inertial elements are treated successively as particles, as rigid bodies, and finally

1

as continuous deformable media. We shall consider electrical systems in which the circuit elements are first treated as distinct separate elements and then as continuously distributed entities. We shall also consider some mixed·systems in which there is dynamic coupling between·electrical and mechanical effects in transducers, and between elastic and fluid media in the flow of liquid through flexible tubes.

In this text considerable emphasis is laid on the second step in the list above, i.e., on the formulation of the equations of motion. We shall see that there are only a few basic requirements governing the dynamic changes of the systems we analyze. The *direct approach* to the equations of motion for a system is to translate these general requirements into a set of particular mathematical equations which must hold simultaneously during the change of state of the system. Algebraic reduction of the set then leads to one or more differential equations, called the equations of motion. The reader undoubtedly has had some experience with this approach. One of the purposes of this book is to broaden the appreciation of the basic dynamic requirements and the direct approach for obtaining the equations of motion.

A second purpose is to introduce the reader to variational methods which provide an alternative *indirect* approach for obtaining the equations of motion. The variational methods are valuable analytical tools in themselves, and in addition, they provide insights into the direct approach, which might be overlooked if the alternative approach were not available for comparison. Considered simply from a utilitarian point of view, we shall see that for some systems the direct approach leads more rapidly to the equations of motion, while for other systems the variational approach is quicker.

In this first chapter we present an elementary exposition of variational methods in terms of particular examples. We consider first some variational principles in mechanical statics, and then develop a special case of the dynamic variational principle which is used throughout the text. In subsequent chapters the principle is generalized and applied along with the direct method to study dynamics in mechanical, electrical, and electromechanical systems.

1-1   Geometric variational principles for statics

The basic requirements governing mechanical systems in equilibrium[1] can be divided into three categories:

1. Geometric constraints on the deformations
2. Equilibrium requirements on the forces

---

[1] See, for example, S. H. Crandall and N. C. Dahl (eds.), "An Introduction to the Mechanics of Solids," McGraw-Hill Book Company, New York, 1959.

3. Constitutive relations which fix the force-deformation behavior of
    the elements in the system

To illustrate, let us consider the system shown in Fig. 1–1.   Before
assembly there is an initial mismatch e in the length of the springs, as
indicated in Fig. 1–1a.   After assembly the system is subjected to a
loading force f, and assumes an equilibrium configuration, as shown in
Fig. 1–1b.   The problem is to analyze the forces and deformations in this
equilibrium configuration.   For the purpose of analysis the individual

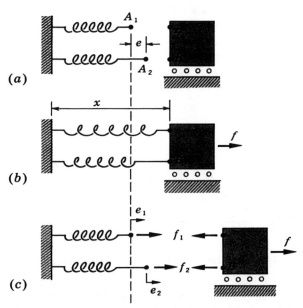

FIG. 1–1.   System of springs with initial mismatch e subjected to loading force f.

elongations e_1 and e_2 and the spring forces f_1 and f_2 are introduced in Fig.
1–1c.   Then the three basic requirements can be translated into equations
as follows:

$$\text{Geometric constraint:} \qquad e - e_1 + e_2 = 0 \qquad (1\text{--}1)$$

$$\text{Equilibrium requirement:} \qquad f - f_1 - f_2 = 0 \qquad (1\text{--}2)$$

$$\text{Constitutive relations:} \qquad f_1 = f_1(e_1) \qquad f_2 = f_2(e_2) \qquad (1\text{--}3)$$

The constitutive relations are taken to be known, not necessarily linear,
single-valued functions, as depicted in Fig. 1–2.

The four preceding equations represent a complete formulation for
determining the four quantities e_1, e_2, f_1, and f_2.   In the special case where

the constitutive relations take the linear form

$$f_1 = k_1 e_1 \qquad f_2 = k_2 e_2, \qquad (1\text{-}4)$$

these equations can be readily solved by substituting (1–4) into (1–2) and then eliminating e_2 between (1–1) and (1–2) to obtain

$$(k_1 + k_2)e_1 = f + k_2 \dot{e} \qquad (1\text{-}5)$$

as a single equation for e_1. The other unknowns may be found by substituting the value of e_1 given by (1–5) into (1–1) and (1–4).

The procedure that we just followed (i.e., the direct translation of the governing requirements into mathematical statements followed by algebraic elimination) is called the *direct* method. We next derive an alternative *indirect* method by starting from the formulation of Eqs. (1–1) to (1–3) and working backward to obtain a variational statement. This is actually the reverse of how a variational method is used in practice.

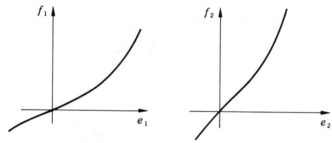

Fig. 1–2.   Constitutive relations for the springs in Fig. 1–1.

Once a variational principle is available, the analysis starts from the variational principle, and the end result is an equation or set of equations equivalent to those that would be obtained by the direct method.

CONSTRUCTION OF A VARIATIONAL PRINCIPLE

We begin by accepting the geometric constraint embodied in (1–1) and limiting ourselves to geometrically compatible displacement states. These are configurations in which the springs in Fig. 1–1 remain connected to the block B. Configurations which satisfy the geometric constraints are called *geometrically admissible states*. We also accept the constitutive relations (1–3), so that the spring forces corresponding to any configuration can be obtained. In general, however, the forces so obtained will *not* satisfy the equilibrium requirement (1–2). Equilibrium requirements are not necessarily satisfied by geometrically admissible states.

The idea underlying the variational method is that, from among all the geometrically admissible states, a *natural* state (i.e., a state which does satisfy the equilibrium requirements) can be detected by observing its behavior under infinitesimal geometrically admissible variations of state. These variations are hypothetical changes from an admissible state to neighboring admissible states which are made for the sake of comparison. In the older literature the adjective *virtual* was applied to such changes in state. We shall call them *admissible variations*.

The traditional symbol for the variation of a quantity x is δx. In Appendix A it is shown that the variational operator δ has many properties in common with a differential operator. In particular, if $f(x)$ is a function of x, then when x undergoes a variation δx, the variation of $f(x)$ is

$$\delta f = \frac{df}{dx} \, \delta x, \tag{1-6}$$

and if $F(x)$ is the integral,[1]

$$F(x) = \int_0^x f(\xi) \, d\xi, \tag{1-7}$$

then its variation is

$$\delta F = f(x) \, \delta x. \tag{1-8}$$

A variation in configuration in Fig. 1–1 will involve variations of e_1, e_2, and x. If the variation is to be geometrically admissible (i.e., if the springs are to remain connected to the block), we must have

$$\delta e_1 = \delta e_2 = \delta x. \tag{1-9}$$

We next construct a variational criterion for equilibrium. If the forces acting on the block B are not in equilibrium, the resultant unbalanced force, in the direction of increasing x, is $f - f_1 - f_2$. The increment of work done by this force under an admissible variation is

$$(f - f_1 - f_2) \, \delta x, \tag{1-10}$$

which is positive when the direction of the variation δx coincides with the direction of the unbalanced force acting on the block. We call (1–10) a *variational indicator* of equilibrium because it vanishes for arbitrary δx if,

---

[1] The integral (1–7) is a function of its upper limit x. The variable of integration ξ is sometimes called a *dummy variable* because the value of the integral does not depend on the name given to the integration variable. Any other symbol, such as η or ζ, can be used in place of ξ. At the risk of some ambiguity, the dummy variable can even be called x; that is, we can write

$$F(x) = \int_0^x f(x) \, dx \tag{1-7a}$$

if we agree to consider that the variable x under the integral sign represents a dummy variable that is distinct from the upper limit x. When there is no danger of confusion we shall often adopt the notation of (1–7a) in preference to that of (1–7) to save introducing superfluous symbols. As examples, see (1–12) and (1–31).

and only if, the equilibrium relation (1–2) is satisfied. In principle, the indicator (1–10) could be used to test whether a proposed geometrically admissible configuration was in equilibrium. The expression (1–10) would be evaluated for admissible variations in the neighborhood of the proposed configuration. If (1–10) did not vanish for arbitrary infinitesimal δx, the proposed configuration would not be in equilibrium.

Finally, it remains to transform the variational indicator (1–10) into a more useful form. First we insert the admissibility conditions (1–9) into (1–10) to obtain

$$\text{V.I.} = f \, \delta x - f_1 \, \delta e_1 - f_2 \, \delta e_2 \tag{1–11}$$

for the variational indicator (V.I.).

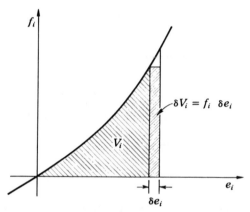

FIG. 1–3. Potential energy of a spring is defined as the integral of force with respect to extension.

Next we insert the constitutive relations (1–3) by introducing a potential-energy function. In Fig. 1–3 the constitutive relation for the ith spring is sketched. The potential energy V_i of the ith spring is defined as

$$V_i = \int_0^{e_i} f_i \, de_i. \tag{1–12}$$

When the extension e_i undergoes a variation δe_i, the corresponding variation in potential energy is, according to (1–8),

$$\delta V_i = f_i \, \delta e_i. \tag{1–13}$$

The total system potential energy V is defined as the sum of the potential energies of the individual elements, so that for the system of Fig. 1–1 we have

$$V = V_1 + V_2 \tag{1–14}$$

and
$$\delta V = f_1 \, \delta e_1 + f_2 \, \delta e_2. \tag{1–15}$$

When (1–15) is inserted in (1–10) the variational indicator takes the final form

$$\text{V.I.} = f\,\delta x - \delta V, \tag{1–16}$$

and the variational criterion for equilibrium can be stated as follows: *A geometrically admissible configuration is in equilibrium if, and only if, the variational indicator (1–16) vanishes for arbitrary geometrically admissible variations.*

The foregoing statement is a special case of a general geometric variational principle. Our demonstration has of course been in terms of a particular example. Before discussing any generalizations we shall illustrate how the variational principle would be applied to the special case of Fig. 1–1, where the constitutive relations take the linear form (1–4).

APPLICATION OF THE PRINCIPLE

The first step in the analysis would consist in identifying the geometric constraint (1–2) on admissible configurations and the constraint (1–9) on admissible variations. Then the potential energy would be constructed by using (1–14), (1–12), and (1–4).

$$V = \tfrac{1}{2}k_1 e_1^2 + \tfrac{1}{2}k_2 e_2^2. \tag{1–17}$$

The constraint (1–1) between e_1 and e_2 would be introduced as

$$V = \tfrac{1}{2}k_1 e_1^2 + \tfrac{1}{2}k_2(e - e_1)^2, \tag{1–18}$$

and the variation computed according to (1–6),

$$\delta V = [k_1 e_1 - k_2(e - e_1)]\,\delta e_1. \tag{1–19}$$

Finally, the variational indicator (1–16) would be put in the form

$$\text{V.I.} = f\,\delta x - \delta V = [f - (k_1 + k_2)e_1 + k_2 e]\,\delta e_1 \tag{1–20}$$

by using (1–9) to set $\delta x = \delta e_1$. The only way (1–20) could be zero for arbitrary admissible variations would be for the expression in square brackets to vanish, which implies

$$(k_1 + k_2)e_1 = f + k_2 e. \tag{1–21}$$

This equation is the result of applying the variational principle. Note that (1–21) is identical with (1–5), which was obtained by the direct method.

GENERALIZATION

The preceding demonstration has shown that, at least in a particular case, an alternative to the equilibrium requirements of statics is provided by a variational criterion involving work and energy increments in geometrically admissible variations. This alternative can be readily generalized to apply to a wide class of mechanical systems, although we shall not enter into the details here. The proof follows the same lines as in the preceding derivation. All that is necessary is a generalized notation capable of representing admissible configurations for a whole class of systems. Generalized coordinates having this capability will be introduced in Chap. 2.

Each term in the variational indicator (1–16) represents an increment of work under an admissible variation of configuration. The first term is the work of the external force f, and the second term accounts for the work done by both spring forces. Together they account for the work increment done by all forces acting on the single movable junction point, or node, in Fig. 1–1. In the extension to systems of greater complexity it can be shown that the variational indicator represents the work increment done by *all* forces acting on *all* movable nodes in a geometrically admissible variation of configuration. The general form is

$$\text{V.I.} = \sum_i \mathbf{f}_i \cdot \delta \mathbf{R}_i - \delta V, \tag{1–22}$$

where the system potential energy V may include the potential energies of several elements and where the summation extends over those node forces $\mathbf{f}_i$ whose work increments are not accounted for in the potential-energy decrement. In each term of the summation, $\mathbf{f}_i$ represents a vector force acting on a node and $\delta \mathbf{R}_i$ is the geometrically admissible vector displacement of the corresponding node. The generalized geometrical variational principle[1] which uses the indicator (1–22) is sometimes called the *principle of virtual work*, and may be stated as follows: *A geometrically admissible configuration of a mechanical system is in an equilibrium state if, and only if, the variational indicator (1–22) vanishes for arbitrary geometrically admissible variations of configuration.*

In Sec. 2–7 we shall see that the work increments of only a limited class of forces permit representation as decrements of potential energy. Forces of this class are said to be *conservative*. In the special case where the only forces which act in a system are conservative, and where the work of all forces is accounted for by a total potential energy V, the principle of virtual work reduces to the *principle of stationary potential energy: A geometrically admissible configuration of a conservative system is in an equilibrium state if, and only if, the total potential energy is stationary (i.e., $\delta V = 0$) for arbitrary geometrically admissible variations of configuration.*

The direct method and the variational method are alternative methods

---

[1] For an outline of the development of variational principles for statics, with references to the original sources, see G. Æ. Oravas and L. McLean, Historical Development of Energetic Principles in Elastomechanics, *Appl. Mech. Rev.*, I, **19**:647–658; II, **19**:919–933 (1966).

of analysis which lead to equivalent results. There may, however, be a considerable difference in their relative ease of application. To some extent this is a subjective matter, depending on the taste and experience of the analyst. As a general rule the variational method tends to become more attractive as the complexity of the system under analysis increases. One reason for this is that the indicator (1–22) involves only work and energy increments. In complex systems there are often many forces of constraint, such as reactions at frictionless guiding surfaces, which do no work and have no associated potential energy. In the direct method these forces must be included in the equilibrium requirements and later eliminated algebraically. In the variational method these forces do not enter at all.

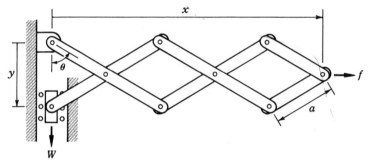

FIG. 1–4. Load W is held in equilibrium by lazy-tongs mechanism with applied force f.

Example 1–1. The system in Fig. 1–4 is a lazy-tongs mechanism assembled from bars and pins. The separation between pins is the same distance a on all bars. The problem is to find the magnitude of the force f required to balance the load W, under the assumptions that the bars are rigid and weightless and that there is no friction in the joints.

This is not a difficult problem by any method. The direct method does, however, require that free-body analyses be made of each of the six bars, with the interactions at each pin accounted for. Then a simple but tedious reduction of simultaneous equations is required before the final relation connecting f and W is obtained.

In the variational method, on the other hand, there is no need to consider the many internal forces since they do no net work in an admissible variation of configuration. The only forces which do work are W and f. There is no storage of potential energy in the system of bars. The variational indicator (1–22) for this system is simply the work expression

$$\text{V.I.} = \sum_i \mathbf{f}_i \cdot \delta\mathbf{R}_i = f\,\delta x + W\,\delta y, \qquad (1\text{–}23)$$

and it remains to draw out the consequences of the requirement that (1–23) must vanish for arbitrary admissible variations.

In an admissible variation of the configuration in Fig. 1–4 the increments δx and δy are not independent. In fact, geometric inspection of the figure provides us with the relations

$$x = 5a \sin \theta$$
$$y = 2a \cos \theta, \tag{1-24}$$

from which application of (1–6) yields

$$\delta x = 5a \cos \theta \; \delta\theta$$
$$\delta y = -2a \sin \theta \; \delta\theta. \tag{1-25}$$

We substitute (1–25) into the variational indicator (1–23) to obtain

$$\text{V.I.} = (5af \cos \theta - 2aW \sin \theta) \; \delta\theta. \tag{1-26}$$

Now (1–26) will vanish for arbitrary $\delta\theta$ if, and only if,

$$f = \tfrac{2}{5}W \tan \theta, \tag{1-27}$$

which is the desired relation between f and W.

1–2   Force variational principle for statics

In many mechanical systems there is a formal duality between the geometric-constraint conditions and the equilibrium requirements. It is therefore possible to enunciate variational principles which are the complements or duals of the principles described in the preceding section by interchanging the roles of the geometric and equilibrium requirements. As in the preceding section we shall demonstrate the validity of the complementary principles only in terms of the particular system of Fig. 1–1.

We begin by accepting the equilibrium, or force-balance, requirement embodied in (1–2) and limiting ourselves to self-equilibrating force states. A set of forces which satisfy the equilibrium requirements for a system is called an *admissible force state*. We also accept the constitutive relations (1–3), so that the elongations of the springs corresponding to given forces can be obtained. For most self-equilibrating force states the corresponding spring elongations will *not* satisfy the geometric constraint (1–1); i.e., the springs cannot be connected to the block without a gap or an overlap. Geometric requirements are not necessarily satisfied by an admissible force state.

Paralleling our previous development, we next construct a variational criterion for geometric compatibility which involves admissible variations in the forces in the neighborhood of an admissible force state. To do this we imagine that the second spring in Fig. 1-1a is connected to the block so that A_2 and B_2 coincide. Then any lack of geometric fit is concentrated in a gap (or overlap) between the points A_1 and B_1. In terms of the elongations e_1 and e_2 and the initial mismatch e, the length of the gap from A_1 to B_1 is $e - e_1 + e_2$. The tensile force acting across this gap is f_1, and the increment of *complementary work*,

$$(e - e_1 + e_2) \, \delta f_1, \qquad (1\text{-}28)$$

is positive for an increase in tension across an open gap (or a decrease in tension in the case of overlap). The expression (1-28) serves as a varia-

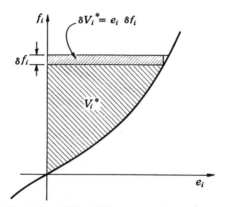

FIG. 1-5. Potential coenergy of a spring is defined as the integral of the extension with respect to the force

tional indicator for geometric compatibility since (1-28) will vanish for arbitrary δf if, and only if, the geometric requirement (1-1) is satisfied.

It remains to convert (1-28) into a more useful form. We first observe that for variations in force to be admissible we must have

$$\delta f_2 = -\delta f_1, \qquad (1\text{-}29)$$

which follows from direct observation of the balance of force on the block B or from application of the variational operator to (1-2). Insertion of (1-29) into (1-28) yields

$$\text{V.I.} = e \, \delta f_1 - e_1 \, \delta f_1 - e_2 \, \delta f_2 \qquad (1\text{-}30)$$

for the variational indicator.

We next introduce the constitutive relations (1-3) by introducing a potential-coenergy function. In Fig. 1-5 the *potential coenergy* V_i^* of the

ith spring is defined as

$$V_i^* = \int_0^{f_i} e_i \, df_i. \tag{1-31}$$

When f_i undergoes a variation δf_i, the corresponding variation in V_i^* is

$$\delta V_i^* = e_i \, \delta f_i. \tag{1-32}$$

The total potential coenergy V^* of the system of Fig. 1–1 is the sum of the coenergies of the two springs,

$$V^* = V_1^* + V_2^*, \tag{1-33}$$

and the variation of the total potential coenergy is

$$\delta V^* = e_1 \, \delta f_1 + e_2 \, \delta f_2. \tag{1-34}$$

Finally, insertion of (1–34) into (1–30) yields the following variational indicator for geometric compatibility:

$$\text{V.I.} = e \, \delta f_1 - \delta V^*. \tag{1-35}$$

An admissible force state is also geometrically compatible if, and only if, the indicator (1–35) vanishes for arbitrary admissible force variations.

The force variational principle just stated has been derived by working backward from the direct formulation of (1–1), (1–2), and (1–3). When the variational principle is used as an analytical tool, the analysis proceeds in the opposite direction. To illustrate, we shall indicate how the indicator (1–35) would be applied to the special case of Fig. 1–1, where the constitutive relations take the linear form (1–4).

APPLICATION OF THE PRINCIPLE

The analysis would begin with the identification of the force-admissibility condition (1–2) and the requirement (1–29) for admissible force variations. Then the potential coenergy for the system would be constructed using (1–33), (1–31), and the linear constitutive relations (1–4),

$$V^* = \frac{f_1^2}{2k_1} + \frac{f_2^2}{2k_2}. \tag{1-36}$$

The equilibrium requirement (1–2) relating f_1 and f_2 would then be inserted to yield

$$V^* = \frac{f_1^2}{2k_1} + \frac{(f - f_1)^2}{2k_2}. \tag{1-37}$$

Finally, the indicator (1–35) would be evaluated.

$$\text{V.I.} = e \, \delta f_1 - \delta V^* = \left(e - \frac{f_1}{k_1} + \frac{f - f_1}{k_2} \right) \delta f_1. \tag{1-38}$$

The condition for (1–38) to be zero for arbitrary δf_1 is that the contents of the parentheses must vanish:

$$e - \frac{f_1}{k_1} + \frac{f - f_1}{k_2} = 0, \tag{1–39}$$

which is a single equation for the unknown tension f_1. This is the final result of applying the force-variational principle. Note that (1–39) can be interpreted as a restatement of the geometric requirement (1–1) in terms of the force variable f_1.

GENERALIZATION

The demonstration of the force-variational principle given above was in terms of the particular system of Fig. 1–1. Although we shall not give the proof here, it is possible to generalize the principle to apply to a limited but important class of systems in mechanical equilibrium. This is the class of elastic structures for which the equilibrium requirements are independent of the deformation. If such an unloaded structure has initial gaps e_i, which are closed during assembly, the equilibrium configuration under subsequent loading can be determined from the following variational principle: *An admissible force state satisfies the geometric constraints if, and only if, the variational indicator*

$$\text{V.I.} = \sum_i e_i \, \delta f_i - \delta V^* \tag{1–40}$$

vanishes for arbitrary admissible force variations. In the indicator the system potential coenergy V^* is the sum of the potential coenergies of the individual elastic elements in the system, and the δf_i represent admissible increments in the tensions f_i acting to close the gaps e_i. In practical problems the gaps e_i may arise from mismatches in element sizes which require force fits when the system is assembled. They may also arise when analyzing thermal stresses, where it is customary to imagine that the individual elements are disassembled, permitted to expand differentially so as to produce gaps, and then reassembled.

In systems for which there are no such initial mismatches, the indicator (1–40) takes a particularly simple form. For such systems the variational principle reduces to the *principle of stationary potential coenergy: An admissible force state is geometrically compatible if, and only if, the total potential coenergy is stationary (i.e., $\delta V^* = 0$) for arbitrary admissible force variations.*

1–3   Legendre transformations

The complementary functions V and V^* encountered in the preceding developments represent particular cases of a functional duality which appears many times in this book. It is therefore appropriate to discuss the nature of this duality in somewhat greater detail.

Suppose that we have a device (such as a spring) which exhibits a constitutive relation between two variables f and x, as indicated by the curve OPR in Fig. 1–6. Let the current *state* of the device be indicated by the position of the operating point P. The constitutive relation can be considered as a function of x in the form $f = f(x)$, or inversely as a function of f in the form $x = x(f)$. Instead of dealing directly with the constitutive relation, it is sometimes more convenient to employ one of the *state functions* V or V^* indicated in Fig. 1–6. The function $V(x)$ is the integral of $f(x)$ with respect to x out to the operating point. The

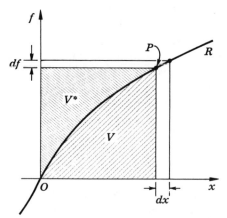

FIG. 1–6. Complementary state functions defined by a single constitutive relation.

function $V^*(f)$ is the integral of $x(f)$ with respect to f up to the operating point. The derivative of V is

$$\frac{dV}{dx} = f, \tag{1–41}$$

and the derivative of V^* is

$$\frac{dV^*}{df} = x. \tag{1–42}$$

If the value of x at P is known, the corresponding value of f may be obtained directly from the constitutive relation $f = f(x)$, or indirectly by evaluating dV/dx. Similarly, if the value of f at P is known, the corresponding value of x may be obtained directly from the constitutive relation $x = x(f)$, or indirectly by evaluating dV^*/df. Either state function thus provides a complete description of the constitutive relation.

The complementary nature of $V(x)$ and $V^*(f)$ is displayed in (1–41) and (1–42). Both equations describe the same constitutive relation, but they do so in terms of different independent variables. If one of the state functions, say, $V(x)$, is known, it is possible to evaluate the complementary function $V^*(f)$ directly by using the relation

$$V^* = xf - V(x), \tag{1–43}$$

which follows from Fig. 1–6 since the total area of the rectangle $V + V^*$ is just the product xf evaluated at the operating point. It is also possible to establish the properties of V^* without reference to the shaded area in Fig. 1–6 if we take (1–43) to be the basic definition of V^*. It is implied that the right side of (1–43) is to be evaluated only at those points in the xf plane which satisfy the constitutive relation, i.e., at points where (1–41) is valid. The total differential of (1–43) is thus

$$dV^* = x\,df + f\,dx - dV$$
$$= x\,df + 0 \tag{1–44}$$

if (1–41) is used. This indicates that V^* is a function only of f at the operating point, with the property that its derivative dV^*/df gives the current value of x. This conclusion is, of course, identical with (1–42), which was obtained from the interpretation of $V^*(f)$ as a shaded area in Fig. 1–6.

The definition of V^* by (1–43) is a simple example of a Legendre transformation. It provides a means of changing from the independent variable x in $V(x)$ to the independent variable $f = \partial V/\partial x$ in $V^*(f)$ without loss of information concerning the underlying constitutive behavior. In general, a *Legendre transform* V^* of a function $V(x, y, \ldots)$ is a function which provides an equivalent representation of the fundamental functional relationship involved and has the property that the set of independent variables upon which V^* depends contains one or more of the derived variables $\partial V/\partial x, \partial V/\partial y, \ldots$, in place of the corresponding variable or variables x, $y, \ldots$ which appear in V. For example, suppose that we have a device which exhibits constitutive relations of the form

$$f = f(x,y)$$
$$g = g(x,y), \tag{1–45}$$

i.e., a device which is characterized by two pairs of variables f, g and x, y. Suppose, in addition, that there exists a state function $V(x,y)$ which has the property that the first derivatives of V with respect to x and y at the operating point are f and g, respectively,

$$\frac{\partial V}{\partial x} = f \qquad \frac{\partial V}{\partial y} = g. \tag{1–46}$$

[We may note that such a state function exists only under the special circumstance in which the functions of (1–45) satisfy the *integrability requirement* $\partial f/\partial y = \partial g/\partial x$, which is necessary in order that $\partial^2 V/\partial y\, \partial x = \partial^2 V/\partial x\, \partial y$.] Under these circumstances the state function $V(x,y)$ provides a compact representation of the constitutive behavior (1–45).

An alternative representation of the same constitutive relations is provided by the complementary function V_{10}^* defined by a Legendre transformation similar to (1–43),

$$V_{10}^* = xf - V(x,y). \tag{1-47}$$

In this transformation the variable x in V is transformed to the variable f in V_{10}^*, and is called the *active* variable of the transformation. The variable y which appears in both V and V_{10}^* is called a *passive* variable under the transformation. The properties of V_{10}^* follow on computing its total differential subject to the constitutive behavior implied by (1–46),

$$dV_{10}^* = x\, df + f\, dx - \frac{\partial V}{\partial x}\, dx - \frac{\partial V}{\partial y}\, dy$$

$$= x\, df - g\, dy. \tag{1-48}$$

Thus $V_{10}^* = V_{10}^*(f,y)$, and the first derivatives of V_{10}^* with respect to f and y at the operating point are the current values of x and $-g$, respectively.

By a simple extension of (1–47) we can establish a Legendre transformation in which both x and y are active variables. Consider the function

$$V_{11}^* = xf + yg - V(x,y). \tag{1-49}$$

The total differential of V_{11}^* subject to (1–46) is

$$dV_{11}^* = x\, df + f\, dx + y\, dg + g\, dy - \frac{\partial V}{\partial x}\, dx - \frac{\partial V}{\partial y}\, dy$$

$$= x\, df + y\, dg, \tag{1-50}$$

which implies that $V_{11}^* = V_{11}^*(f,g)$ and that its first derivatives with respect to f and g at the operating point are the current values of x and y, respectively.

The preceding transformations break down if the underlying constitutive equations cannot be inverted. Thus the transformation (1–43) fails if the constitutive equation $f = f(x)$ cannot be inverted to obtain a unique relation of the form $x = x(f)$. This occurs, for example, with the constitutive relation $f = c$, where c is a constant. Given x, we find f uniquely, but if $f = c$ is given, x is undetermined. For this case $V = cx$ is a state function,[1] but V^* does not exist. In terms of the function $V(x)$, the condition for local invertibility, and hence for the existence of the transformation (1–43), is that the second derivative of V should be dif-

---

[1] The potential energy of a mass in a uniform gravitational field is of this form.

ferent from zero:

$$\frac{d^2V}{dx^2} = \frac{df}{dx} \neq 0. \tag{1–51}$$

The corresponding condition for the existence of the transformation (1–49) is[1] that the following determinant should be different from zero:

$$\begin{vmatrix} \dfrac{\partial^2 V}{\partial x^2} & \dfrac{\partial^2 V}{\partial x\,\partial y} \\[2mm] \dfrac{\partial^2 V}{\partial x\,\partial y} & \dfrac{\partial^2 V}{\partial y^2} \end{vmatrix} \neq 0. \tag{1–52}$$

THERMODYNAMIC EXAMPLES

The Legendre transformation is employed extensively in thermodynamics.[2] For example, the constitutive relations for a gas in equilibrium may be considered to consist of statements as to how the temperature T and pressure p depend on the entropy S and volume V. This information is conveniently contained in the *internal-energy* function $U(S,V)$, together with the formulas

$$\frac{\partial U}{\partial S} = T \qquad \frac{\partial U}{\partial V} = -p. \tag{1–53}$$

An alternative description of the same constitutive behavior is provided by the *enthalpy*,

$$H(S,p) = pV + U(S,V). \tag{1–54}$$

Note that (1–54) is a Legendre transformation involving a change of variable from V to p and that (except for an arbitrary minus sign) there is a complete parallelism between (1–46), (1–47) and (1–53), (1–54). The total differential of the enthalpy is, in analogy with (1–48),

$$dH = T\,dS + V\,dp. \tag{1–55}$$

Another alternative description of the same information is provided by the *Gibbs' free energy*,

$$\Phi(T,p) = pV - TS + U(S,V). \tag{1–56}$$

Note that (1–56) is a Legendre transformation of the same type as (1–49), in which the independent variables are changed from S and V to T and p. In analogy with (1–50) we have

$$d\Phi = -S\,dT + V\,dp. \tag{1–57}$$

[1] See, for example, F. B. Hildebrand, "Advanced Calculus for Applications," Prentice-Hall, Inc., Englewood Cliffs, N.J., 1962, p. 343.

[2] See H. B. Callen, "Thermodynamics," John Wiley & Sons, Inc., New York, 1960, pp. 90–102.

Thus the same fundamental constitutive behavior can be equally well described by any one of the functions $U(S,V)$, $H(S,p)$, and $\Phi(T,p)$, which are related to one another by Legendre transformations.

APPLICATION TO KINETIC STATE FUNCTIONS

We shall see that the Legendre transformation plays an important role in variational principles for dynamics. To provide a simple introduction, let us consider a particle traveling with *momentum p* along the axis of x under the influence of a force f, as shown in Fig. 1-7. We shall compute the work done by f, using the fact that force equals the time rate of change of momentum,

$$f = \frac{dp}{dt}. \tag{1-58}$$

This requirement is a basic principle in newtonian mechanics, which also has validity[1] in the theory of special relativity. If the *velocity* of the

FIG. 1-7.   A force f acts on a particle traveling with momentum p.

particle is $v = dx/dt$, the increment of displacement during the motion can be written $dx = v\,dt$. The increment of work done by f on the particle is thus

$$f\,dx = \frac{dp}{dt}\,dx = \frac{dp}{dt}\,v\,dt = v\,dp, \tag{1-59}$$

and the total work done in increasing the momentum from zero to the value p is the kinetic state function

$$T(p) = \int_0^p v\,dp, \tag{1-60}$$

which is called the *kinetic energy*. Note that T is a function of the instantaneous momentum with derivative equal to the instantaneous velocity,

$$\frac{dT}{dp} = v. \tag{1-61}$$

Up to now an explicit relation between velocity and momentum has not been introduced. This relation is similar to a constitutive relation for a deformable element, but it depends on the particular dynamic theory

[1] See, for example, "Physics: A New Introductory Course," M.I.T. Science Teaching Center, 1964, p. 21–20, or P. G. Bergmann, "Introduction to the Theory of Relativity," Prentice-Hall, Inc., Englewood Cliffs, N.J., 1942, p. 103.

adopted, rather than on the special properties of an individual device. If we adopt newtonian dynamics, we take the velocity-momentum relation

$$p = mv, \tag{1-62}$$

where m is the mass of the particle. The kinetic energy of newtonian

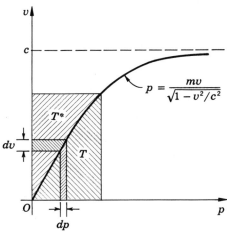

Fig. 1-8. Illustration of the complementary kinetic state functions for the velocity-momentum relation of special relativity.

dynamics is found by using (1-62) to eliminate v in (1-60) and then integrating to obtain

$$T(p) = \frac{p^2}{2m}. \tag{1-63}$$

On the other hand, if we adopt the special theory of relativity, the velocity-momentum relation is[1]

$$p = \frac{mv}{\sqrt{1 - v^2/c^2}}, \tag{1-64}$$

where m now is the *rest mass*, and c is the velocity of light. This relation is sketched in Fig. 1-8. The inverse of (1-64) is

$$v = \frac{p/m}{\sqrt{1 + p^2/m^2c^2}}. \tag{1-65}$$

[1] See "Physics: A New Introductory Course," M.I.T. Science Teaching Center, 1964, p. 21–21, or R. P. Feynman, R. B. Leighton, and M. Sands, "The Feynman Lectures on Physics," vol. 1, Addison Wesley Publishing Co., Reading, Mass., 1963, p. 16–6.

The kinetic energy in special relativity is found by inserting (1–65) in (1–60) and then integrating to obtain[1]

$$T(p) = mc^2 \left(\sqrt{1 + \frac{p^2}{m^2c^2}} - 1 \right). \tag{1-66}$$

Let us return to the general kinetic state function $T(p)$ of (1–60). Irrespective of the velocity-momentum relation, the complementary state function is

$$T^*(v) = \int_0^v p \, dv. \tag{1-67}$$

The properties of T^* can be inferred directly from a diagram similar to Fig. 1–8 or from the Legendre transformation

$$T^* = vp - T(p), \tag{1-68}$$

which is analogous to (1–43). The complementary kinetic state function T^* is a function of the instantaneous velocity with derivative equal to the instantaneous momentum,

$$\frac{dT^*}{dv} = p. \tag{1-69}$$

We shall call $T^*(v)$ the *kinetic coenergy.*

As soon as the velocity-momentum relation is fixed, it is possible to evaluate the kinetic coenergy (1–67). Thus, in newtonian dynamics, with the linear relation (1–62), we find

$$T^* = \tfrac{1}{2}mv^2 \tag{1-70}$$

for the kinetic coenergy of a particle. In special relativity with the nonlinear relation (1–64), the corresponding result is[1]

$$T^* = mc^2 \left(1 - \sqrt{1 - \frac{v^2}{c^2}} \right). \tag{1-71}$$

Several observations can be made concerning the newtonian kinetic state functions (1–63) and (1–70) and the relativistic kinetic state functions (1–66) and (1–71). It is well known that the two dynamical theories become indistinguishable when the particle velocities are very small in comparison with the speed of light. It is to be expected, then, that the relativistic state functions should approach their newtonian counterparts when $v/c \to 0$. This is in fact the case, as can easily be shown.[2] The complementary relativistic functions T and T^* are never strictly

---

[1] See Prob. 1–15.

[2] See Prob. 1–16.

equal in magnitude for any operating point except at the origin of Fig. 1–8. This is because of the *nonlinear* velocity-momentum relation (1–64). On the other hand, the complementary newtonian state functions T and T^* have equal magnitudes for every operating point. This is because of the *linearity* of (1–62).

In relativistic dynamics the kinetic state functions T and T^* are of some importance in connection with variational principles,[1] but the kinetic state function most commonly encountered is actually neither of these but is, instead, the total energy[2]

$$E = T + mc^2 = mc^2 \sqrt{1 + \frac{p^2}{m^2 c^2}}, \qquad (1\text{–}72)$$

which is the sum of the kinetic energy T of (1–66) and the so-called rest energy mc^2. It is also common to introduce the relativistic mass m_{rel}, which is related to the rest mass m by the equations

$$m_{\text{rel}} = m \sqrt{1 + \frac{p^2}{m^2 c^2}} = \frac{m}{\sqrt{1 - v^2/c^2}}. \qquad (1\text{–}73)$$

Then the total energy of (1–72) becomes, simply,

$$E = m_{\text{rel}} c^2. \qquad (1\text{–}74)$$

In newtonian mechanics the kinetic energy $T = p^2/2m$ and the kinetic coenergy $T^* = mv^2/2$ play very important roles. The reader should be alerted to the fact that in maintaining a distinction here between energy and coenergy the authors are going beyond established practice in engineering mechanics.[3] It can be argued that, since within the confines of newtonian dynamics the magnitudes of T and T^* are always equal, it is pointless to make such a distinction. It is our opinion, however, that the structure of the theory as a whole is clarified when the formal distinction between T and T^* is maintained. Many of the results of new-

[1] See W. Yourgrau and S. Mandelstam, "Variational Principles in Dynamics and Quantum Theory," Pitman Publishing Corporation, New York, 1960, pp. 39–40.

[2] See "Physics: A New Introductory Course," M.I.T. Science Teaching Center, 1964, p. 17–12.

[3] The distinction has been treated in research papers, e.g., R. A. Toupin, A Variational Principle for the Mesh-type Analysis of a Mechanical System, *Trans. ASME*, **74**, *J. Appl. Mech.*, **19**:151–152 (1952), and S. H. Crandall, Complementary Extremum Principles for Dynamics, *Ninth Intern. Congr. Appl. Mech.*, vol. 5, pp. 80–87, Brussels, 1957. The distinction has also been maintained in recent books on electromechanical energy conversion, e.g., D. C. White and H. H. Woodson, "Electromechanical Energy Conversion," John Wiley & Sons, Inc., New York, 1959, and J. Meisel, "Principles of Electromechanical Energy Conversion," McGraw-Hill Book Company, New York, 1966.

tonian dynamics can be immediately extended to relativistic dynamics by giving T and T^* their relativistic values (1–66) and (1–71), provided the distinction between T and T^* has been retained[1] in the newtonian analysis.   In addition, by emphasizing the formal complementarity of these state functions in mechanics, we provide a foundation for the subsequent extension to electrical systems and to mixed mechanical and electrical systems.   There is also a basis for contending that kinetic energy T has a *natural* dependence on momentum and that kinetic coenergy T^* has a *natural* dependence on velocity.   The argument for this contention is presented in Appendix E.

1–4   Variational principles for dynamics

In the static variational principles described in Secs. 1–1 and 1–2, the variational criteria involve comparisons of neighboring admissible *states*.

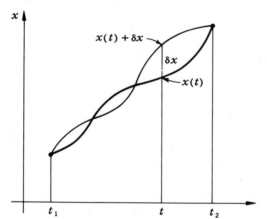

FIG. 1–9.   The variation δx between two admissible trajectories varies with time.

In dynamics there are corresponding variational principles, but the criteria involve the comparison of admissible *motions*.   Instead of static configurations, we must compare complete time histories.   For example, in the case of a single variable $x(t)$, as sketched in Fig. 1–9, we must compare entire trajectories of admissible motions during finite intervals of time.   The variation between a specified trajectory $x(t)$ and a neighboring trajectory is indicated by the notation δx.   For a particular neighboring trajectory the variation δx is a function of time.   The velocity on the

[1] See, for example, P. Penfield, Jr., Hamilton's Principle for Fluids, *Phys. Fluids*, 9:1184–1194 (1966), and P. Penfield, Jr. and H. A. Haus, Hamilton's Principle for Electromagnetic Fluids, *Phys. Fluids*, 9:1195–1204 (1966).

specified trajectory is dx/dt, while the velocity on the neighboring trajectory is

$$\frac{d}{dt}(x + \delta x) = \frac{dx}{dt} + \frac{d}{dt}(\delta x). \qquad (1\text{-}75)$$

The variation in velocity is thus

$$\delta \frac{dx}{dt} = \frac{d}{dt}(\delta x), \qquad (1\text{-}76)$$

which can be interpreted as a statement that the operations of variation and time differentiation are *commutative*. An alternative development of the commutativity property appears in Appendix A.

We next proceed to a derivation of a variational principle for dynamics in terms of a particular system. In Fig. 1-10a a mass constrained to

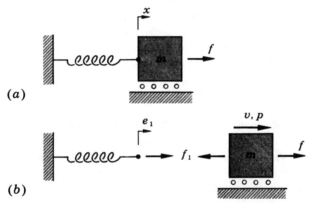

(a)

(b)

FIG. 1-10. Spring-mass system subjected to force f.

move parallel to the axis of x is restrained by a spring and acted on by a prescribed force f. It is desired to formulate the equations of motion for this system. We shall first describe the direct method and then work backward to obtain a variational principle.

The basic requirements governing mechanical motion are similar to the requirements for mechanical equilibrium but somewhat more extensive:

1. Geometric requirements on the motions
2. Dynamic requirements on the forces
3. Constitutive relations for the deformable elements and velocity-momentum relations for the masses

The dynamic requirements for forces are that unbalanced forces on masses must equal the time rates of change of the corresponding momenta of the masses. For the system of Fig. 1-10 these general requirements can be translated into specific statements in terms of the symbols introduced in Fig. 1-10b.

Geometric requirements: $x = e_1$ $\dfrac{dx}{dt} = v$ (1-77)

Dynamic-force requirement: $f - f_1 = \dfrac{dp}{dt}$ (1-78)

Constitutive relation: $f_1 = f_1(e_1)$ (1-79)

Velocity-momentum relation: $v = v(p)$ (1-80)

The constitutive relation (1-79) is taken to be a known function like those shown in Fig. 1-2, and the velocity-momentum relation (1-80) is taken to be either the newtonian relation (1-62) or the relativistic relation (1-65). There are five independent equations in the set (1-77) to (1-80) for the five unknown variables $e_1, f_1, p, v,$ and x. In principle it is possible to eliminate four of the variables and obtain a single differential equation of motion for the system. For the case of a *linear* constitutive relation of the form $f_1 = ke_1$ and a *linear* velocity-momentum relation $p = mv$, elimination of $e_1, f_1, p,$ and v leads readily to

$$m\frac{d^2x}{dt^2} + kx = f \qquad (1-81)$$

as a single equation of motion. This illustrates the direct method for formulating dynamic equations of motion.

An alternative indirect method can be derived by starting from the direct statements in the list (1-77) to (1-80) and transforming them into a variational criterion. The procedure that we follow is similar to the one described in Sec. 1-1 for deriving a geometrical variational principle for statics. Some important additions are, however, required to extend the static case to the dynamic case.

CONSTRUCTION OF A DYNAMIC VARIATIONAL PRINCIPLE

We begin by accepting the geometric constraints embodied in (1-77) and restrict ourselves to geometrically compatible motions. These are motions in which the connection between spring and mass is always maintained and for which the time histories of $e_1, x,$ and v always satisfy (1-77). We also accept the constitutive relation (1-79) and the velocity-momentum relation (1-80) so that the time histories of the spring force f_1 and the momentum p associated with a geometrically compatible motion can be obtained. For most such motions the force and momentum will *not* satisfy the dynamic-force requirement (1-78) at every instant. We shall employ the phrase *geometrically admissible motion* to denote a motion which is always geometrically compatible but which does not necessarily satisfy the dynamic-force requirements at all times.

To compare neighboring admissible motions we introduce the concept of an *admissible variation*. A variation in the motion will involve variations δe_1, δx, and δv in the histories of e_1, x, and v. If the variation is to be geometrically admissible, we must have at all times

$$\delta e_1 = \delta x$$

$$\delta v = \delta \frac{dx}{dt} = \frac{d}{dt}(\delta x). \tag{1-82}$$

The first of (1-82) implies that the spring remains connected to the mass, and the second follows from (1-77) and (1-76).

Next we construct a variational criterion for the satisfaction of the dynamic-force requirement (1-78). If (1-78) is not satisfied, the resultant unbalanced force acting on the mass in the direction of increasing x is $f - f_1 - dp/dt$. The increment of work done by this force under an admissible variation,

$$\left(f - f_1 - \frac{dp}{dt}\right)\delta x, \tag{1-83}$$

is positive when the direction of δx coincides with the direction of the unbalanced force acting on the mass. Note that (1-83) must vanish identically for arbitrary time histories of δx if (1-78) is always satisfied. The converse is also true. If (1-83) remains zero for arbitrary histories of δx, it is necessary for the dynamic-force requirement (1-78) to be satisfied at all times. The expression (1-83) is therefore a suitable variational indicator for the satisfaction of the dynamic-force requirement. It now remains to transform (1-83) into a more useful form.

We begin by inserting the geometric admissibility condition $\delta e_1 = \delta x$ into (1-83) to get

$$f\,\delta x - f_1\,\delta e_1 - \frac{dp}{dt}\,\delta x. \tag{1-84}$$

The potential energy V_1 of the spring in Fig. 1-10 is defined in (1-12). The middle term of (1-84) is then recognized as δV_1 according to (1-13). Since in this system V_1 is also the total potential energy V, we write the variational indicator (1-84) as

$$f\,\delta x - \delta V - \frac{dp}{dt}\,\delta x. \tag{1-85}$$

Up to now the development has been similar to that for a static-variational principle. Note that the first two terms of (1-85) are identical in form with the static-variational indicator (1-16). We are confronted with the essentially dynamic aspect of the development for the first time when we consider the third term of (1-85). It is tempting to expect that this term can be directly transformed into the variation of a *kinetic* state function in much the same way as the term involving the spring force was transformed into the variation of a *potential* energy. There is an apparent similarity between the term $(dp/dt)\,dx$ in (1–59) which led directly to the definition of kinetic energy T and the term $(dp/dt)\,\delta x$ in (1–85); however, the similarity is illusory since the displacements dx and δx have completely different interpretations. The displacement dx is *along* a particular trajectory as it unfolds in time, whereas the displacement δx measures the separation *between* different trajectories at a given instant.

Now, although there is no direct relation between the term $(dp/dt)\,\delta x$ and the kinetic energy T, there does exist a connection between this term and the kinetic coenergy T^*. This connection arises out of the following simple but perhaps unexpected identity:

$$\frac{d}{dt}\,(p\,\delta x) = \frac{dp}{dt}\,\delta x + p\,\frac{d}{dt}\,(\delta x)$$

$$= \frac{dp}{dt}\,\delta x + p\,\delta v$$

$$= \frac{dp}{dt}\,\delta x + \delta T^*, \tag{1–86}$$

where the substitutions on the right follow from (1–82) and the definition (1–67) of kinetic coenergy. When $(dp/dt)\,\delta x$ from (1–86) is inserted in (1–85), the variational indicator becomes

$$f\,\delta x - \delta V + \delta T^* - \frac{d}{dt}\,(p\,\delta x). \tag{1–87}$$

The identity (1–86) has permitted the introduction of the kinetic state function T^* into the indicator (1–87), but it has also introduced the time-derivative term, which is not convenient to evaluate. In an attempt to eliminate this term we next integrate (1–87) over a time interval extending from t_1 to t_2 to obtain a new variational indicator,

$$\text{V.I.} = \int_{t_1}^{t_2} (\delta T^* - \delta V + f\,\delta x)\,dt -- [p\,\delta x]_{t_1}^{t_2}. \tag{1–88}$$

It can be shown[1] that (1–88) vanishes for arbitrary admissible variations if, and only if, (1–87) vanishes for arbitrary admissible variations at every instant within the interval from t_1 to t_2. Thus the vanishing of the integrated expression (1–88) for arbitrary admissible variations can be used as a criterion for satisfaction of the dynamic-force requirements at all time t in the closed interval $t_1 \leq t \leq t_2$.

The times t_1 and t_2 which appear in (1–88) have no special significance; they simply mark the end points of a time interval long enough to com-

[1] See Appendix A.

pletely include the period of dynamic behavior under study.   The form
of the indicator (1–88) can, however, be further simplified if we agree to
the following somewhat artificial restriction.   Let us restrict our considerations to variation time histories $\delta x(t)$ which are arbitrary within the
interval $t_1 < t < t_2$ *but for which* $\delta x = 0$ *at* $t = t_1$ *and* $t = t_2$.   The vanishing of δx at the end points of the interval implies that we are restricting
ourselves to a family of admissible motions which begin from a specified
configuration at $t = t_1$ and terminate together at some other specified
configuration at $t = t_2$.   Imparting this special significance to t_1 and t_2
does not, however, alter the usefulness of (1–88) as an indicator for satisfaction of the dynamic-force requirements at all times t within the open
interval $t_1 < t < t_2$.   Thus, if we agree to consider that $\delta x = 0$ at t_1 and
t_2, the expression between limits on the right of (1–88) drops out, and we
can state the following *variational principle: A geometrically admissible motion of the system between prescribed configurations at t_1 and t_2 satisfies
the dynamic-force requirements if, and only if, the variational indicator*

$$\text{V.I.} = \int_{t_1}^{t_2} (\delta T^* - \delta V + f\, \delta x)\, dt \qquad (1\text{–}89)$$

*vanishes for arbitrary geometrically admissible variations of the motion
within the interval $t_1 < t < t_2$.*
     The structure of the variational indicator (1–89) should be noted.
Each term in the integrand can be interpreted as an increment of work
done on the mass under an admissible variation.   The term $f\, \delta x$ is the
work of the external force f acting *on* the mass in Fig. 1–10.   The term
$-\delta V$ is the decrease in the potential energy of the spring, and represents the work done by the spring force acting *on* the mass.   The interpretation of the term δT^* is less direct because of the identity (1–86).
A suggestive significance can, however, be given to this term if we introduce the concept of *inertia force*.   The inertia force of a mass particle is
defined to be a vector equal in magnitude and oppositely directed to the
particle's time rate of change of momentum.   It then can be shown[1]
that a time integral of the work done by the inertia force acting on the
mass equals the corresponding time integral of δT^* whenever the geometric variations vanish at the end points of the time interval.   Thus,
with this extended interpretation, the term δT^* in (1–89) can be considered to account for the work done by the inertia force acting *on* the
mass.
     Example 1–2.   To provide a brief preview of how variational prin-
ciples are applied in dynamics we shall illustrate how the preceding principle can be used to formulate the equation of motion for the special case

[1] See Prob. 1–18.

of Fig. 1–10 in which the spring has the linear constitutive relation $f_1 = ke_1$ and for which the newtonian velocity-momentum relation $p = mv$ is employed.

The first step consists in identifying the geometric constraints (1–77) on admissible motions and the constraints (1–82) on admissible variations. By using these constraints to express all geometric quantities in terms of the displacement x, we automatically ensure geometric admissibility. Next the potential energy (1–12),

$$V = \tfrac{1}{2}ke_1^2 = \tfrac{1}{2}kx^2, \tag{1-90}$$

and kinetic coenergy (1–70),

$$T^* = \tfrac{1}{2}mv^2 = \tfrac{1}{2}m\left(\frac{dx}{dt}\right)^2, \tag{1-91}$$

are constructed, and the variational integral (1–89) established,

$$\text{V.I.} = \int_{t_1}^{t_2}\left[m\left(\frac{dx}{dt}\right)\delta\left(\frac{dx}{dt}\right) - kx\,\delta x + f\,\delta x\right]dt. \tag{1-92}$$

The remainder of the analysis consists in drawing out the consequences of the requirement that (1–92) be zero for arbitrary histories of δx which vanish at $t = t_1$ and $t = t_2$. The manipulations involved are part of the technique of the *calculus of variations* (see Appendix A). The most important step is an integration by parts of the first term in (1–92). This manipulation may be viewed as reversing the effect of the identity (1–86) used in establishing the variational principle. By using the commutativity relation (1–76), the first term of the integral (1–92) can be written

$$\int_{t_1}^{t_2} m\,\frac{dx}{dt}\,\frac{d}{dt}\,(\delta x)\,dt \tag{1-93}$$

and then integrated by parts to yield

$$\int_{t_1}^{t_2} m\,\frac{dx}{dt}\,\frac{d}{dt}\,(\delta x)\,dt = \left[m\,\frac{dx}{dt}\,\delta x\right]_{t_1}^{t_2} - \int_{t_1}^{t_2} m\,\frac{d^2x}{dt^2}\,\delta x\,dt. \tag{1-94}$$

The integrated term $m(dx/dt)\,\delta x$ vanishes at the limits because of the agreement[1] that $\delta x = 0$ at $t = t_1$ and $t = t_2$.

---

[1] Note that if the agreement were not made and if the variational indicator (1–88) were employed instead of (1–89), the integrated term in (1–94) would cancel with the last term on the right of (1–88).

When the remaining integral on the right of (1-94) is substituted for the first term of (1-92), the variational indicator becomes

$$\text{V.I.} = \int_{t_1}^{t_2} \left(-m \frac{d^2x}{dt^2} - kx + f \right) \delta x \, dt. \qquad (1\text{-}95)$$

Now if (1-95) is to vanish for arbitrary histories of δx within the interval from t_1 to t_2, it is necessary for the expression in parentheses to vanish identically throughout the interval:

$$-m \frac{d^2x}{dt^2} - kx + f = 0. \qquad (1\text{-}96)$$

The variational principle thus requires that the geometrically admissible motion $x(t)$ must satisfy (1-96) if it is to meet the dynamic-force requirements. Note that (1-96) is equivalent to the equation of motion (1-81) obtained by algebraic elimination.

GENERALIZATION

The variational principle stated on page 27, although derived with reference to a particular system, is a special case of a principle of very broad applicability. We shall describe the form of the generalized principle without entering into detailed justifications at this juncture.

For a mechanical system of arbitrary complexity the variational indicator is a time integral over an interval from t_1 to t_2 of the increments of work done by *all* forces (including inertia forces) acting on *all* masses in a geometrically admissible variation. The general form of the indicator is

$$\text{V.I.} = \int_{t_1}^{t_2} \left(\delta T^* - \delta V + \sum_i \mathbf{f}_i \cdot \delta \mathbf{R}_i \right) dt, \qquad (1\text{-}97)$$

where now T^* is the *sum* of the kinetic coenergies of all the individual mass particles in the system, and V is the *sum* of the potential energies of the individual energy-storage elements. The remaining summation in (1-97) extends over any forces $\mathbf{f}_i$ whose work increments are not accounted for in δT^* and $-\delta V$. In each term of the summation, $\mathbf{f}_i$ represents a vector force acting *on* a mass particle, and $\delta \mathbf{R}_i$ is the geometrically admissible vector displacement of that particle.

The indicator (1-97) is considered for infinitesimal geometrically admissible variations in the neighborhood of geometrically admissible motions. The agreement is also made to consider only variational histories which vanish at the end points t_1 and t_2 of the time interval. Then the variational principle on page 27 for the particular system of Fig. 1-10 can be extended to apply to a system of arbitrary complexity by simply replacing the particular variational indicator (1–89) by the generalized indicator (1–97).

This variational principle can be restated in a more concise form which lends itself to broader generalizations by introducing some additional abbreviations in notation and terminology. The state functions T^* and V occur so frequently in the combination $T^* - V$ that it has become common practice to use a single symbol $\mathcal{L}$, called the *lagrangian*, for the difference between the kinetic coenergy and the potential energy:

$$\mathcal{L} = T^* - V. \tag{1–98}$$

In the following statement we shall use the abbreviation *admissible* to mean geometrically admissible. We shall also use the terminology *natural motion* to mean a geometrically admissible motion *which also satisfies the dynamic-force requirements*. With these abbreviations the preceding variational principle can be stated in a generalized form, which we shall refer to as *Hamilton's principle*:[1]

An admissible motion of a dynamic system between specified configurations at t_1 and t_2 is a natural motion if, and only if, the variational indicator

$$\text{V.I.} = \int_{t_1}^{t_2} \left(\delta\mathcal{L} + \sum_i \mathbf{f}_i \cdot \delta\mathbf{R}_i \right) dt \tag{1–99}$$

vanishes for arbitrary admissible variations.

We shall see that, with the definitions given, this statement of Hamilton's principle applies to the mechanical systems treated in the first four chapters of this book. With slight extensions of the definitions of the lagrangian and the work expression and with a broadening of the concepts of admissible motions and natural motions, we shall see that the same principle applies to electrical networks (in Chap. 5) and to electromechanical systems (in Chap. 6). In Chaps. 7 and 8 we shall see that

---

[1] Sir William Rowan Hamilton (1805–1865) introduced the principle in articles, On a General Method in Dynamics, *Phil. Trans. Roy. Soc. London*, **A34** 247–308 (1834), **A35**:95–144 (1835). Strictly speaking, the principle he dealt with is what we now call the principle of stationary action. For simplicity of nomenclature we have attached Hamilton's name to the slightly extended form of the principle which permits inclusion of forces f_i, whose work increments are not accounted for by the lagrangian. For a stimulating treatment of dynamical-variational principles, including philosophical aspects and a brief outline of the historical development, see C. Lanczos, "The Variational Principles of Mechanics," University of Toronto Press, Toronto, Canada, 1949.

this same principle can be interpreted to apply to continuous mechanical and electrical systems.

Hamilton's principle provides an alternative to the direct method for formulating the equations of motion for dynamic systems. One of the reasons for our development of the indirect variational method in this book is the additional insights which this gives into the direct methods. A more strictly utilitarian reason is that for some systems the indirect procedure proves to be easier for the analyst to apply than the direct method. As was stated with respect to the static-variational principles, the relative advantage of the indirect method tends to increase with the complexity of the system under analysis. The variational method provides equations of motion without involving the forces of workless constraints, whereas in the direct method these forces must be included and subsequently eliminated. Variational methods are also of interest in their own right as tools for studying optimal-control systems[1] and optimization processes in general. Several important analytical approximation techniques in engineering science[2] are based on variational methods.

In Hamilton's principle for mechanical systems the geometric constraints are embodied in the admissibility conditions, and the dynamic-force requirements are embodied in the variational criterion. For a limited class of mechanical systems it is possible to enunciate a complementary dynamic variational principle in which the admissibility conditions are based on dynamic-force requirements and the variational criterion represents geometric-compatibility requirements. A brief description of the principle and a discussion of its limitations are contained in Appendix D.

PRINCIPLE OF STATIONARY ACTION

The generalized Hamilton's principle takes an especially simple form when applied to systems in which the work of all forces is accounted for by the lagrangian. In these cases the work expression in (1–99) is not required, and the variational indicator can be written

$$\text{V.I.} = \delta \int_{t_1}^{t_2} \mathcal{L} \, dt \qquad (1\text{--}100)$$

by interchanging the operations of integration and variation. Let us call

---

[1] See, for example, C. W. Merriam, III, "Optimization Theory and the Design of Feedback Control Systems," McGraw-Hill Book Company, New York, 1964.

[2] See, for example, S. H. Crandall, "Engineering Analysis," McGraw-Hill Book Company, New York, 1956, secs. 1–4, 2–5, 3–6, 4–4, 4–5, 5–4, 5–5, and 6–4.

the time integral of the lagrangian along an admissible motion the *action*[1] of that motion,

$$A = \int_{t_1}^{t_2} \mathcal{L} \, dt. \qquad (1\text{--}101)$$

The action A is a scalar quantity having the dimensions of the product of energy and time (or alternatively, of the product of momentum and displacement). The magnitude of the action depends on the terminal configurations at t_1 and t_2 and on the trajectory of the motion between these terminal configurations. Given any admissible trajectory, it is possible, at least in principle, to evaluate the lagrangian (1–98) at each point along the motion and then to determine the action (1–101) by integration. The action A is thus a scalar label which characterizes an admissible trajectory between the terminal configurations.

In terms of the action, the variational indicator (1–100) is simply δA and the criterion based on its vanishing is the *principle of stationary action:*

For a dynamic system in which the work of all forces is accounted for in the lagrangian, an admissible motion between specified configurations at t_1 and t_2 is a natural motion if, and only if, the action A is stationary for arbitrary admissible variations.

It is interesting to note that the principle of stationary action bears the same relationship to Hamilton's principle as the principle of stationary potential energy bears to the principle of virtual work for statics (see page 8). In this text the main application of dynamic-variational principles is to the formulation of equations of motion. For this purpose there is no special advantage of the stationary-action principle over the generalized Hamilton's principle. In both cases the calculus of variations is applied to a variational integral to obtain differential equations of motion, and the presence or absence of work terms $\mathbf{f}_i \cdot \delta \mathbf{R}_i$ does not greatly alter the procedure.

The principle of stationary action does, however, have a remarkable conceptual interpretation which is worth noting, even though we make no direct practical use of the interpretation. The criterion for a natural motion provided by the principle of stationary action is quite different in spirit from that provided by the equations of motion. The differential equations of motion dictate the course of a natural motion on a point-by-point basis. In order to obtain the complete trajectory it is necessary to

---

[1] In advanced theoretical mechanics three different kinds of action are recognized: maupertuis action, lagrangian action, and hamiltonian action. Although similar, they differ because of different admissibility conditions involved in their definitions. For each type of action there is a corresponding variational principle. See J. L. Synge, Classical Dynamics, pp. 105, 110, and 137, in S. Flügge (ed.), "Handbuch der Physik," vol. III/1, Springer Verlag OHG, Berlin, 1960. In this book we discuss only the lagrangian action (1–101).

integrate the equations of motion. On the other hand, the principle of stationary action deals, at least conceptually, with complete trajectories from the start. We can imagine an infinity of admissible trajectories emanating from one configuration at $t = t_1$ and converging on another configuration at $t = t_2$. With each trajectory is associated a particular value of action A. Now, in principle, we could pick out a natural trajectory by simply observing how the value of A varied from trajectory to trajectory. As a practical matter this process cannot be implemented exactly[1] (except by returning to the differential equations and integrating) because of the infinity of independent admissible trajectories which would

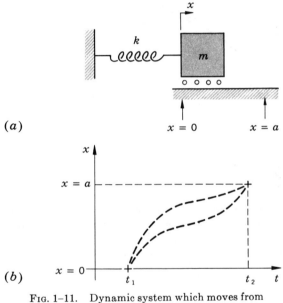

FIG. 1–11. Dynamic system which moves from $x = 0$ at $t = t_1$ to $x = a$ at $t = t_2$.

have to be compared. The behavior demanded by the principle of stationary action can, however, be readily illustrated by a numerical example.

Example 1–3. Consider the spring-mass system shown in Fig. 1–11a. This system is identical with that of Fig. 1–10, except that the external force f has been removed. Geometrically admissible trajectories are represented by single-valued functions $x(t)$ provided the spring elongation

---

[1] There are systematic approximate procedures for implementing this process. A well-known procedure is the *Rayleigh-Ritz* method (see, for example, Crandall, *op. cit.*, p. 232). More recently the technique of *dynamic programming* has been suggested for the same purpose (see J. T. Tou, "Modern Control Theory," McGraw-Hill Book Company, New York, 1964, chap. 7, and S. E. Dreyfus, "Dynamic Programming and the Calculus of Variations," Academic Press Inc., New York, 1965).

is taken equal to x and the mass velocity is taken equal to dx/dt. Let us consider trajectories which emanate from $x = 0$ at $t = t_1$ and which converge on $x = a$ at $t = t_2$, as indicated in Fig. 1–11b. The action A associated with each trajectory is obtained by inserting the potential energy V of (1–90) and the kinetic coenergy T^* of (1–91) into the

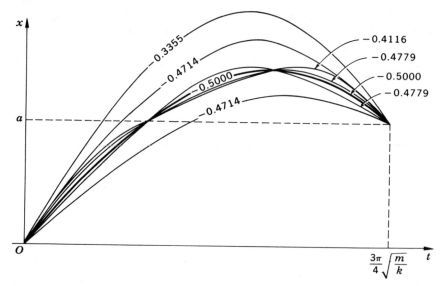

FIG. 1–12. Action is stationary in the neighborhood of the natural trajectory. Number attached to a trajectory, when multiplied by $a^2 \sqrt{km}$, is the action A.

lagrangian (1–98) and integrating according to (1–101).

$$A = \int_{t_1}^{t_2} \left[\tfrac{1}{2} m \left(\frac{dx}{dt} \right)^2 - \tfrac{1}{2} k x^2 \right] dt. \qquad (1\text{–}102)$$

For any given trajectory, $x(t)$, it is a routine calculation to obtain the corresponding value of the action A from (1–102).

In Fig. 1–12 several trajectories have been drawn for the particular case in which $t_1 = 0$ and $t_2 = (3\pi/4) \sqrt{m/k}$. Among these trajectories the natural motion[1] $x = \sqrt{2}\, a \sin \sqrt{k/m}\, t$ is represented by a heavy line. The numerical value of the action along each trajectory, as calculated from (1–102), is indicated directly on the figure. Note that, at least in terms of the neighboring trajectories shown in the figure, the action appears to be stationary in the neighborhood of the natural motion. The action $A = -0.5000 a^2 \sqrt{km}$ along the natural motion is actually a *minimum* in this particular case. Generally speaking, the stationary action (in the principle of stationary action) is actually a minimum, or

[1] See Prob. 1–20.

least, action if the time interval between t_1 and t_2 is short enough. For longer time intervals[1] the stationarity may be of a saddle-point nature.

PROBLEMS

1–1. The system shown consists of a weightless lever on a frictionless pivot stabilized by three linear springs, each with spring constant k. Prior to assembly there are the gaps e_1 and e_2. Find the equilibrium angle θ which results when the springs are connected and the gaps closed by using

(a) The direct method
(b) **The force variational principle**
(c) The principle of stationary potential coenergy

Assume that the angle θ is sufficiently small to permit the approximations $\sin \theta \approx \theta$ and $\cos \theta \approx 1$.

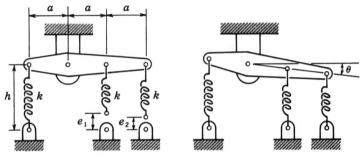

PROB. 1–1

1–2. The sections of shaft act as linear torsional springs with torsional-spring constant k. Assuming that there is no torque at the supports A and B prior to the application of the torques τ_1 and τ_2, find the equilibrium torque at the support A after τ_1 and τ_2 are applied by using

(a) The direct method
(b) The geometric variational principle
(c) The principle of stationary potential coenergy

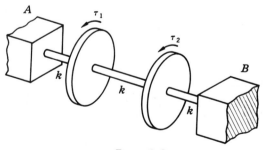

PROB. 1–2

[1] See Prob. 1–22.

1-3. A nonlinear spring has the constitutive relation shown. Find the potential energy and potential coenergy for this element.

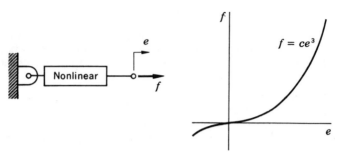

PROB. 1-3

1-4. The nonlinear element of Prob. 1–3 is connected (without a force fit) in parallel with two linear springs. Obtain an equation for the displacement x due to the force f by using the principle of virtual work.

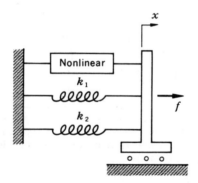

PROB. 1-4

1-5. The nonlinear element of Prob. 1–3 is connected in series with two linear springs (without a force fit) between the unyielding supports A and B. Obtain an equation for the reaction force at A due to the applied force f by using the principle of stationary potential coenergy.

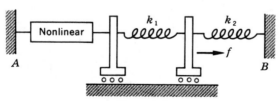

PROB. 1-5

1–6. The upper wheel of the differential chain hoist has sprocket radii r_1 and r_2. Estimate the force P required to balance the weight W by neglecting friction and the weight of the chain and pulley.

PROB. 1–6

1–7. A threaded rod of diameter d engages a fixed nut at B and is attached to a tension spring with spring constant k at A. A string connected to the rod is wrapped around the rod several times, and then a weight W is hung from the string. If the unstretched length of the spring is L_0, what is the equilibrium length L when the weight W is attached? Neglect friction and assume that the spring connection offers no resistance to twisting.

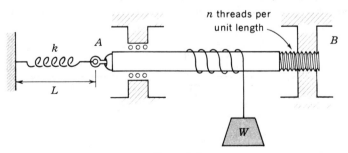

PROB. 1–7

1-8. Find the range of values for W which will hold the 100-lb block in equilibrium on the inclined plane if the coefficient of static friction is 0.5.   Neglect friction in the pulley.

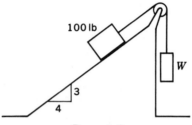

100 lb

W

3

4

PROB. 1-8

1-9. Use the principle of virtual work to find the angle θ for which the weight W can be supported by the rigid massless rod against the frictionless wall and the frictionless pivot.   Note that in an admissible variation of configuration the rod must remain in contact with both the wall and the pivot.

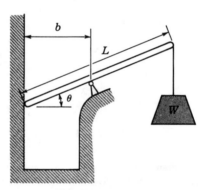

b

L

θ

W

PROB. 1-9

1-10. The principle of virtual work can often be used to find equilibrium force distributions in systems which do not permit variations of configuration in the neighborhood of the equilibrium position by considering related systems which do allow such variations.   For example, in (a), the rigid horizontal rod cannot be moved downward

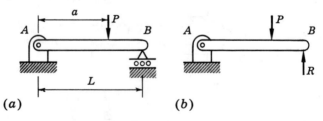

a

P

A B

L

(a)

P

A B

R

(b)

PROB. 1-10

at B because of the unyielding support, and if it is moved upward, it loses contact with the support. The equilibrium value of the support force R can, however, be obtained by applying the principle of virtual work to the related system (*b*) in which the reaction R moves with the bar in an admissible small rotation about A. Evaluate R by this technique.

1–11. Use the technique described in Prob. 1–10 to find the equilibrium value of the horizontal reaction force H due to the load P. Neglect friction and the weight of the bars.

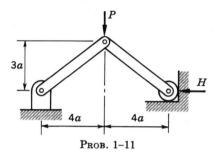

PROB. 1–11

1–12. Consider a general nonlinear elastic system in equilibrium under the loads f and g.

(*a*) Let the potential energy V be a function of the displacements u and v as well as other independent geometric variables. Use the principle of virtual work to show that

$$\frac{\partial V}{\partial u} = f \qquad \frac{\partial V}{\partial v} = g.$$

(*b*) In the same system let the potential coenergy V^* be a function of the forces f and g as well as other independent force variables. Use the force-variational principle to show that

$$\frac{\partial V^*}{\partial f} = u \qquad \frac{\partial V^*}{\partial g} = v.$$

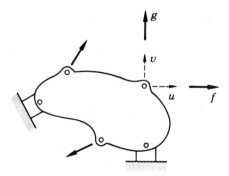

PROB. 1–12

1–13. Develop the properties of the Legendre transformation

$$V^*_{01} = yg - V(x,y)$$

of the state function V described in connection with Eq. (1–46). What are the active and passive variables of the transformation? Evaluate the first partial derivatives of V^*_{01}.

1–14. The *Helmholtz free energy* F is a state function for a gas which is related to the internal energy $U(S,V)$ by a Legendre transformation

$$F = U - TS.$$

What are the active and passive variables of this transformation? Evaluate the first partial derivatives of F.

1–15. Verify (1–66) and (1–71) by carrying out the indicated integrations or by showing that the derivatives of (1–66) and (1–71) have the required properties.

1–16. Show that when $v/c \to 0$, then (1–64) $\to$ (1–62), (1–66) $\to$ (1–63), and (1–71) $\to$ (1–70). The binomial expansion

$$(1 + x)^n = 1 + nx + \frac{n(n-1)}{2!} x^2 + \cdots,$$

valid for $|x| < 1$, is useful in this connection.

1–17. It is desired to obtain the differential equations which describe the motion of the two masses in a free vibration which takes place after the masses have been disturbed from the equilibrium configuration $x_1 = 0$, $x_2 = 0$. Obtain T^* and V for this system, and determine the requirements for geometric admissibility. Apply Hamilton's principle to obtain the equations of motion. Note that arbitrary admissible variations involve *independent* variations of x_1 and x_2.

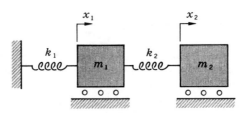

PROB. 1–17

1–18. The *inertia force* f_{in} acting on the particle in Fig. 1–7 is defined to be $-(dp/dt)$. Show that, in this case,

$$\int_{t_1}^{t_2} f_{in}\, \delta x \, dt = \int_{t_1}^{t_2} \delta T^* \, dt,$$

if the agreement is made to assume that the variation δx vanishes at t_1 and t_2.

1–19. Obtain the equation of motion for the system of Fig. 1–10 when the velocity-momentum relation (1–64) of the special theory of relativity is considered to apply. Use Hamilton's principle in a procedure paralleling Example 1–2.

1–20. Apply the calculus of variations to (1–102) to obtain the equation of motion implied by the requirement that $\delta A = 0$ for arbitrary variations. The procedure parallels that for Example 1–2. Verify that $x = \sqrt{2}\, a \sin \sqrt{k/m}\, t$ satisfies the equation of motion and is therefore the natural motion for Example 1–3.

1-21. Consider for newtonian dynamics a mass particle m which moves parallel to the axis of x with no forces acting on it. Evaluate the action for admissible motions which emanate from $x = 0$ at $t_1 = 0$ and converge on $x = a$ at $t_2 = a/v_0$. Show that the uniform-velocity trajectory with $dx/dt = v_0$ has less action than any other admissible trajectory.

1-22. Consider for newtonian dynamics the system of Fig. 1–11a. Set up the action integral for trajectories which emanate from $x = a$ at $t = 0$ and return to $x = a$ at $t = 2\pi \sqrt{m/k}$. Use the results of Prob. 1–20 to verify that $x = a \cos \sqrt{k/m}\, t$ is a natural motion. What is the action along the natural trajectory? Show that $x = a$ is an admissible trajectory which has *smaller* action. The natural motion here has stationary action, but not minimum action.

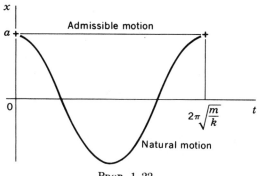

PROB. 1–22

CHAPTER TWO

Dynamics of mechanical systems

In Chap. 1 a brief introduction to Hamilton's principle is developed in terms of particular mechanical examples. In this chapter we study in a more systematic manner the dynamics of mechanical systems composed of mass particles, springs, dashpots, and force fields. The equations of motion for such systems are formulated directly by using Newton's laws and indirectly by using Hamilton's principle. After treating several particular systems, we introduce generalized coordinates and give a generalized derivation of Hamilton's principle and Lagrange's equations. In the process, most of the techniques involved in applying Hamilton's principle are developed and a foundation is laid for the unified treatment of electromechanical systems in Chap. 6 and of systems of continuous media in Chaps. 7 and 8.

2–1 Kinematics

Kinematics can be regarded as a study of the *geometry of motion*. It centers on the behavior in time of *displacements, velocities, and accelerations*. Kinematic considerations are basic ingredients in the dynamic analysis of any mechanical system. In Secs. 2–1 to 2–4 we discuss the fundamental relationships of kinematics for points and rigid frames which move with respect to a reference frame.

We begin by considering a rigid rectangular reference frame $OXYZ$ and a point P which moves with respect to the reference frame as indicated in Fig. 2–1. Let $\mathbf{R}$ denote the *position vector* OP. The vector $\mathbf{R}$ has the dimension of *length*. Common units are feet and meters (1 ft = 0.3048 m). In general, the vector $\mathbf{R}$ will vary with time t; i.e., $\mathbf{R} = \mathbf{R}(t)$. We shall be interested in changes in $\mathbf{R}$ *measured with respect to the reference frame $OXYZ$*. It is important to note that we leave open the possibility that the reference frame $OXYZ$ may in fact be moving with respect to

42

some other reference frame. We are considering only the relative motion of P with respect to the reference frame $OXYZ$ as viewed from $OXYZ$. We also make the classical (nonrelativistic) assumption that there is a universal time t which is independent of the spatial reference frame employed. The common unit of time is the *second*.

In Fig. 2-1 the path, or trajectory, of the point P with respect to the reference frame is indicated by the dotted curve PP_1P_2. The position vector $\mathbf{R}$ is shown at two times which differ by the time increment Δt. The corresponding *displacement vector* is $\Delta \mathbf{R} = \overrightarrow{PP_1}$. The *velocity* $\mathbf{v}$ of the point P with respect to the reference frame is the time rate of change of position,

$$\mathbf{v} = \lim_{\Delta t \to 0} \frac{\Delta \mathbf{R}}{\Delta t} = \frac{d\mathbf{R}}{dt}. \tag{2-1}$$

The magnitude of $\mathbf{v}$ is the instantaneous *speed* along the path at P, and

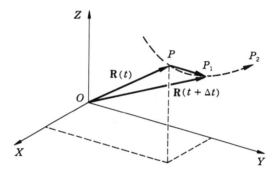

Fig. 2-1. The displacement of the point P with respect to the frame $OXYZ$ during the interval Δt is $\Delta \mathbf{R} = \overrightarrow{PP_1}$.

the direction of $\mathbf{v}$ is the limiting direction of the displacement $\Delta \mathbf{R}$, i.e., *tangent* to the path at P. The vector $\mathbf{v}$ has the dimension of *length per unit time*. Common units are feet per second and meters per second.

In general, the velocity vector $\mathbf{v}$ will also vary with time; i.e., $\mathbf{v} = \mathbf{v}(t)$. The time rate of change of $\mathbf{v}$ is the *acceleration* $\mathbf{a}$ of the point P with respect to the reference frame,

$$\mathbf{a} = \frac{d\mathbf{v}}{dt} = \frac{d^2\mathbf{R}}{dt^2}. \tag{2-2}$$

The vector $\mathbf{a}$ has the dimension of *length per unit time per unit time*. Common units are feet per second per second (ft/sec²) and meters per second per second (m/sec²). Accelerations are often quoted in multiples of g, a standardized sea-level acceleration at latitude 45°. In the British system g is taken as 32.1740 ft/sec², and in the metric system g is taken as 9.80665 m/sec².

We have called displacements, velocities, and accelerations *vectors*. This implies that they can be represented by directed line segments, and it also implies that they can be compounded or decomposed according to the usual laws of vector algebra.[1] For example, the total displacement of a point resulting from two individual displacements is represented by a vector which is the vector sum of the vectors representing the individual displacements.   We shall shortly meet entities which can be represented by directed line segments but which are not compounded according to the laws of vector addition.

Example 2–1.   A circular disk of radius *b rolls* without slip in the *XY* plane along the axis of *X*.   The path of a point *P* on the periphery of the

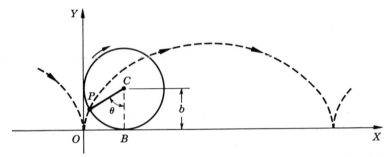

Fɪɢ. 2–2.   Example 2–1.   Trajectory of point *P* on the rim of the rolling disk is a cycloid.

disk is indicated in Fig. 2–2.   It is desired to express the velocity and acceleration of *P* in terms of the time history of the angle θ.

In Fig. 2–2 the no-slip requirement implies that the circular arc *PB* on the disk has the same length as the distance *OB* on the roadbed.   The coordinates of the point *P* in the *XY* frame can thus be written

$$X = b\theta - b \sin \theta$$
$$Y = b - b \cos \theta. \tag{2-3}$$

These are the parametric equations of a *cycloid*.   The position vector **R** extending from the origin to the point *P* is

$$\mathbf{R} = b\mathbf{u}_X(\theta - \sin \theta) + b\mathbf{u}_Y(1 - \cos \theta), \tag{2-4}$$

where $\mathbf{u}_X$ and $\mathbf{u}_Y$ are *unit* vectors parallel to the *X* and *Y* axes.   We take

[1] See, for example, F. B. Hildebrand, "Advanced Calculus for Applications," Prentice-Hall, Inc., Englewood Cliffs, N.J., 1962, p. 263.

the time history $\theta(t)$ to be known and introduce the notation

$$\dot{\theta} = \frac{d\theta}{dt}$$

$$\ddot{\theta} = \frac{d^2\theta}{dt^2} \tag{2-5}$$

to represent the time derivatives of $\theta(t)$. The velocity **v** of the point P is then obtained by differentiating (2-4) according to (2-1).

$$\mathbf{v} = b\dot{\theta}[\mathbf{u}_X(1 - \cos \theta) + \mathbf{u}_Y \sin \theta]. \tag{2-6}$$

Note that the unit vectors $\mathbf{u}_X$ and $\mathbf{u}_Y$ do not change in magnitude or direction and are therefore treated as constants in the differentiation. The velocities at representative points along the trajectory are shown in

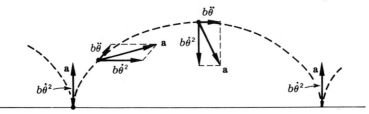

Fig. 2-3.   Instantaneous velocity vector **v** is tangent to the trajectory.

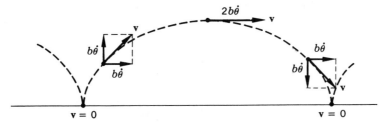

Fig. 2-4.   Instantaneous acceleration vector **a** depends in general
on both θ and $\dot{\theta}$.

Fig. 2-3.   Observe that where the trajectory comes in contact with the roadbed ($\theta = 0$, $\pm 2\pi$, etc.), the velocity of P is instantaneously zero independently of the speed with which the trajectory is traversed.

The acceleration **a** of the point P is obtained according to (2-2) by a further differentiation of (2-6).

$$\mathbf{a} = b\ddot{\theta}[\mathbf{u}_X(1 - \cos \theta) + \mathbf{u}_Y \sin \theta] + b\dot{\theta}^2[\mathbf{u}_X \sin \theta + \mathbf{u}_Y \cos \theta]. \tag{2-7}$$

Accelerations at representative points along the trajectory are indicated in Fig. 2-4. Note that where the trajectory comes in contact with the roadbed the acceleration of P is directed away from the roadbed and has the magnitude $b\dot{\theta}^2$ independently of the value of $\ddot{\theta}$.

2-2   Relative motion of rectangular frames

In many applications the velocity or acceleration of a point P with respect to a reference frame $OXYZ$ is required but the motion of P is most easily described in terms of an intermediate frame $oxyz$, which is itself moving

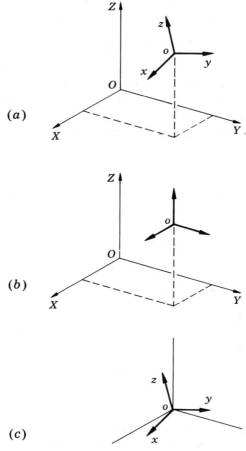

FIG. 2-5. A general displacement (a) of the frame $oxyz$ with respect to $OXYZ$ may be regarded as the superposition of a translation (b) and a rotation (c).

with respect to $OXYZ$. For example, the motion of the tip of an airplane propeller with respect to the ground is conveniently described with respect to an intermediate frame which is fixed in the airplane and which moves with respect to the ground. In order to make use of such intermediate frames we turn now to a study of the relative motion of rigid frames.

In Fig. 2–5a the rigid rectangular frame *oxyz* is shown displaced from the reference frame *OXYZ*. As before, all displacements and their changes are measured with respect to *OXYZ*; i.e., we are considering the

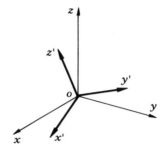

FIG. 2–6. Rectangular frame *ox'y'z'* is angularly displaced with respect to the frame *oxyz*.

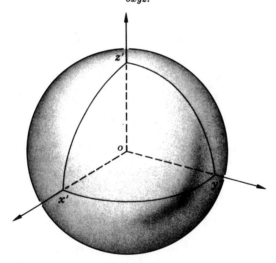

FIG. 2–7. An octant of a sphere attached to the frame *ox'y'z'*.

relative motion of *oxyz* with respect to *OXYZ* as viewed from *OXYZ*. A displacement of the frame *oxyz* in which the axes remain parallel to their original orientations is called a *translation*. Such a displacement is completely described by the displacement of a single point (e.g., the origin) in the *oxyz* frame. A displacement of the frame *oxyz* in which one point (e.g., the origin) does not move but in which the orientations of the axes are altered is called a *rotation*.   An arbitrary displacement of the frame *oxyz* can generally be decomposed into a translation plus a rotation.   If the frame *oxyz* in Fig. 2–5 is considered to have been displaced from an initial position coincident with *OXYZ* to the final position shown in Fig. 2–5*a*, then this total displacement can be considered to be the sum of the translation indicated in Fig. 2–5*b* and the rotation indicated in Fig. 2–5*c*.

The precise description of rotation or angular displacement in three dimensions is one of the more intricate parts of kinematics.   To provide a simple introduction, we give here a brief qualitative discussion of the main features of rotation before entering into a quantitative analysis in

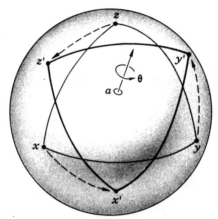

Fig. 2–8.   Rotation of the frame of Fig. 2–7 through the angle about the axis *oa* carries the octant *xyz* into the octant *x′y′z′*.

Sec. 2–3.   We consider the situation shown in Fig. 2–6, where the origins of the frames *oxyz* and *ox′y′z′* coincide.   To visualize the angular displacements of the frame *ox′y′z′* (with respect to *oxyz*) it is convenient to imagine a spherical surface centered on the origin and a spherical octant rigidly attached to the frame *ox′y′z′*, as shown in Fig. 2–7.   Then, as the frame *ox′y′z′* rotates about the origin, the spherical octant slides about on the surface of the sphere.

ANGULAR DISPLACEMENT AND ANGULAR VELOCITY

An angular displacement resulting from a single finite rotation of the frame *ox′y′z′* through the angle θ about an axis *oa* is sketched in Fig. 2–8. The initial and final positions of the octant of Fig. 2–7 are indicated. The rotation itself can be designated by a directed line segment $\boldsymbol{\theta}$ (along the axis *oa*) whose length represents the magnitude of the angle θ and whose sense is fixed by the right-hand rule (when the axis *oa* is grasped by the right hand so that the fingers curl around the axis in the sense of the angle θ, the right thumb points in the sense of the directed line segment **θ**). Angular displacements are *dimensionless*. The fundamental unit is the *radian*.

Any angular displacement of the frame *oxyz* which carries the octant *xyz* into an octant *x'y'z'* on the same sphere can always be achieved by a single finite rotation of the type illustrated in Fig. 2–8. If the initial and final positions of the octant are known, the location of the axis *oa* can be

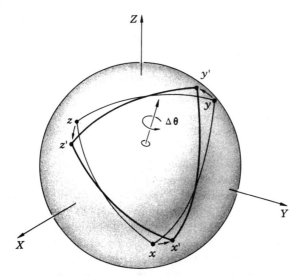

Fig. 2–9. Origin of moving frame *oxyz* coincides with origin of reference frame *OXYZ*. Angular velocity ω of frame *oxyz* with respect to *OXYZ* is defined as limit of quotient Δθ/Δ*t* as Δ*t* → 0.

determined from the fact that the three great-circle bisectors of the great-circle arcs *xx'*, *yy'*, and *zz'* all intersect at point *a*. Once the axis of rotation has been determined, the magnitude and sense of the rotation angle fix the directed line segment **θ**. There is a possible ambiguity here because a rotation of angle θ in one sense yields the same final displacement as a rotation of $2\pi - \theta$ in the opposite sense, but this ambiguity can be removed by arbitrarily limiting the angle θ to the range $0 \leq \theta \leq \pi$.

If the angular displacement of a frame *oxyz* with respect to a reference frame *OXYZ* is changing with respect to time, as indicated in Fig. 2–9, the *angular velocity* ω of *oxyz* is defined as the limiting value of the quotient of the angular-displacement increment Δθ divided by the time increment Δt during which the displacement occurs,

$$\omega = \lim_{\Delta t \to 0} \frac{\Delta \theta}{\Delta t}. \qquad (2\text{--}8)$$

In Fig. 2–9 the octant xyz represents the position of the frame $oxyz$ at time t and the octant $x'y'z'$ represents the position at the same frame at time $t + \Delta t$. The corresponding angular displacement is $\Delta\theta$. The angular velocity ω can be represented by a directed line segment. The direction and sense of ω are determined by the limiting direction and sense of $\Delta\theta$. The line of action of ω is called the *instantaneous axis* of rotation of the frame $oxyz$ with respect to the reference frame. The length $|\omega| = \omega$ of the directed line segment ω is determined by the magnitude of the rate of rotation. Common units for angular velocity are *radians per second* and *revolutions per minute*.

The discussion of angular displacements and velocities of a frame has so far paralleled the treatment of linear displacements and velocities of a point. In both cases it is possible to represent displacements and velocities by directed line segments. There is, however, a major difference between the nature of the directed line segments which represent *angular* displacements and the directed line segments which represent *linear* displacements. This difference appears when we consider combining a number of separate displacements. If a point P makes several displacements in succession, the resulting total displacement can be represented by a directed line segment which is the *vector sum* of the directed line segments which represent the individual displacements. When, however, a frame $oxyz$ makes several angular displacements about different axes in succession, the situation is essentially different. Although each individual angular displacement can be represented by a directed line segment, the directed line segment which represents the total angular displacement is *not* the vector sum of the individual directed line segments. Angular displacements can be represented by directed line segments, but these directed line segments do not in general[1] combine according to the laws of *vector* algebra. In Sec. 2–3 it is shown that angular displacements can be combined in a systematic manner by using *matrix* algebra. Here we merely emphasize the nonvector nature of angular displacements by demonstrating that they do not satisfy the commutative requirement of vector addition, i.e., $\mathbf{A} + \mathbf{B} = \mathbf{B} + \mathbf{A}$. When two angular displacements are performed in sequence, the resulting total angular displacement depends on the order in which the individual displacements are executed. A striking example of this is shown in Fig. 2–10, where two 90° rotations are compounded in different orders. In Fig. 2–10a the rotation about

[1] The algebra of angular displacements does degenerate to vector algebra when all rotations are about the *same* axis.

ox is followed by the rotation about oy', while in Fig. 2–10b the rotation about oy is followed by the rotation about ox'. The resultant angular displacement is entirely different in the two cases.

Now, although angular displacements do not in general combine according to the laws of vector addition, it is shown in Sec. 2–3 that angular velocities *do*. Thus, if the angular velocity of a moving rectangular frame $oxyz$ with respect to a reference frame $OXYZ$ is ω_1, and if the angular velocity of a second moving frame $ox'y'z'$ with respect to $oxyz$ is ω_2, then the angular velocity of $ox'y'z'$ with respect to the reference

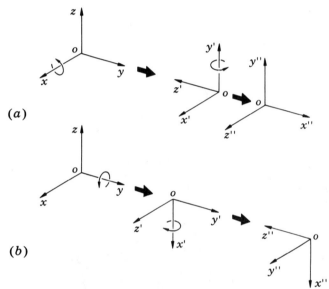

FIG. 2–10. Demonstrating that the order of finite rotations influences the resultant angular displacement.

frame $OXYZ$ can be represented by a vector which is the vector sum of ω_1 and ω_2.

We next give an elementary discussion of the change in orientation of a vector which is carried by a rotating frame. An alternative treatment is given in Sec. 2–3.

RATE OF CHANGE OF A VECTOR FIXED IN A ROTATING FRAME

In Fig. 2–11 the rectangular frame $oxyz$ has angular velocity ω with respect to the reference frame $OXYZ$. An arbitrary vector $\mathbf{A}$ is fixed in $oxyz$ and is carried around by the motion of the moving frame. From the reference frame the vector $\mathbf{A}$ appears to be changing its orientation. We shall show that the rate of change of $\mathbf{A}$ observed in the reference frame is

$$\frac{d\mathbf{A}}{dt} = \boldsymbol{\omega} \times \mathbf{A} \qquad (2\text{-}9)$$

at the instant when the angular velocity of $oxyz$ is $\boldsymbol{\omega}$. The demonstration of (2–9) is somewhat more direct when $\mathbf{A}$ is not an arbitrary vector but is a vector extending from the origin (such as $\mathbf{a}_1$ and $\mathbf{a}_2$ in Fig. 2–11). How-

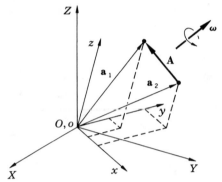

FIG. 2–11. Vector $\mathbf{A}$ is fixed in the frame $oxyz$, which has angular velocity $\boldsymbol{\omega}$ with respect to the reference frame $OXYZ$.

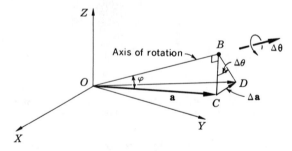

FIG. 2–12. Vector $\mathbf{a}$ is given increment $\Delta\mathbf{a}$ by rotation $\Delta\boldsymbol{\theta}$ during time interval Δt.

ever, if we can show that

$$\frac{d\mathbf{a}_1}{dt} = \boldsymbol{\omega} \times \mathbf{a}_1 \qquad \frac{d\mathbf{a}_2}{dt} = \boldsymbol{\omega} \times \mathbf{a}_2, \qquad (2\text{-}10)$$

then since $\mathbf{A} = \mathbf{a}_1 - \mathbf{a}_2$, we have only to subtract the two equations of (2–10) to obtain (2–9).

We therefore consider in Fig. 2–12 a vector $\mathbf{a}$ which originates at O and is carried around by the rotating frame. During a short time interval Δt

there is an angular displacement $\Delta\theta$ of the rotating frame (with $\Delta\theta \approx \omega\,\Delta t$) which carries the vector **a** from OC to OD. The magnitude of Δ**a** is, approximately,

$$|\Delta\mathbf{a}| = \widehat{CD} \approx \widehat{BC}\,\Delta\theta = \widehat{OC}\sin\varphi\,\Delta\theta = |\mathbf{a}|\,|\Delta\theta|\sin\varphi. \qquad (2\text{-}11)$$

The direction of Δ**a** is very nearly perpendicular to the plane containing **a** and the axis of rotation. The derivative $d\mathbf{a}/dt$ is obtained by a limiting process as $\Delta t \to 0$. The magnitude has the limit

$$\left|\frac{d\mathbf{a}}{dt}\right| = |\mathbf{a}|\,|\omega|\sin\varphi, \qquad (2\text{-}12)$$

and the limiting direction is perpendicular to the plane containing **a** and ω. Next we note that the vector cross product $\omega \times \mathbf{a}$ has the same direction and sense and, moreover, has precisely the magnitude (2-12). Hence we conclude that

$$\frac{d\mathbf{a}}{dt} = \omega \times \mathbf{a} \qquad (2\text{-}13)$$

for any vector **a** which originates from the origin of the rotating frame and is fixed in that frame. Finally, working backward through (2-10), we verify the truth of (2-9) for an arbitrary vector fixed in the rotating frame.

The relation (2-9) is of fundamental importance in kinematics. Several applications are illustrated in Sec. 2-4. In Sec. 2-3 angular displacements are represented by matrices, and an alternative derivation of (2-9) is given. If the reader wishes to postpone the study of angular displacement, he can proceed directly to Sec. 2-4 and the kinematical applications of (2-9).

2-3   Matrix representation of rotation

In Fig. 2-13 we consider the rigid right-handed rectangular frame $ox'y'z'$ which has been displaced from the right-handed rectangular reference frame $oxyz$. Let $\mathbf{u}_x$, $\mathbf{u}_y$, and $\mathbf{u}_z$ be a triad of *unit* vectors along the axes of the reference frame, and let $\mathbf{u}_{x'}$, $\mathbf{u}_{y'}$, and $\mathbf{u}_{z'}$ be a triad of *unit* vectors along the axes of the displaced frame. To relate the vectors of one triad to those of the other we note that the component of $\mathbf{u}_{x'}$ along $\mathbf{u}_y$, for example, is just the cosine of the angle between the axes ox' and oy. We shall use the notation $c_{x'y}$ to represent such *direction cosines*. In terms of the vector dot product we have

$$c_{x'y} = \mathbf{u}_{x'} \cdot \mathbf{u}_y = \mathbf{u}_y \cdot \mathbf{u}_{x'}. \qquad (2\text{-}14)$$

Resolution of the vectors $\mathbf{u}_{x'}$, $\mathbf{u}_{y'}$, and $\mathbf{u}_{z'}$ into their components along the reference axes ox, oy, and oz can then be indicated as follows:

$$\mathbf{u}_{x'} = c_{x'x}\mathbf{u}_x + c_{x'y}\mathbf{u}_y + c_{x'z}\mathbf{u}_z$$
$$\mathbf{u}_{y'} = c_{y'x}\mathbf{u}_x + c_{y'y}\mathbf{u}_y + c_{y'z}\mathbf{u}_z \qquad (2\text{-}15)$$
$$\mathbf{u}_{z'} = c_{z'x}\mathbf{u}_x + c_{z'y}\mathbf{u}_y + c_{z'z}\mathbf{u}_z.$$

If the nine direction cosines which appear in (2-15) are specified, the primed triad of unit vectors is uniquely located with respect to the unprimed triad.   Conversely, the vectors $\mathbf{u}_x$, $\mathbf{u}_y$, $\mathbf{u}_z$ can be resolved into their components along the axes ox', oy', and oz'.

$$\mathbf{u}_x = c_{x'x}\mathbf{u}_{x'} + c_{y'x}\mathbf{u}_{y'} + c_{z'x}\mathbf{u}_{z'}$$
$$\mathbf{u}_y = c_{x'y}\mathbf{u}_{x'} + c_{y'y}\mathbf{u}_{y'} + c_{z'y}\mathbf{u}_{z'} \qquad (2\text{-}16)$$
$$\mathbf{u}_z = c_{x'z}\mathbf{u}_{x'} + c_{y'z}\mathbf{u}_{y'} + c_{z'z}\mathbf{u}_{z'}.$$

The angular displacement of Fig. 2-13 is thus completely described by the nine direction cosines which appear in (2-15) and in (2-16).   We take

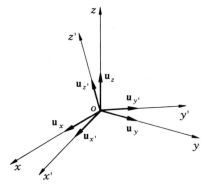

FIG. 2-13.   The triad of orthogonal unit vectors $(\mathbf{u}_{x'}, \mathbf{u}_{y'}, \mathbf{u}_{z'})$ is angularly displaced with respect to the reference triad $(\mathbf{u}_x, \mathbf{u}_y, \mathbf{u}_z)$.

this set of nine direction cosines as our basic quantitative representation of angular displacement.   At first sight a set of nine quantities appears to be an awkward way to describe an angular displacement which can be represented by a single directed line segment as in Fig. 2-8.   The reason why nine direction cosines contain the same information as a single directed line segment (which can be fixed by three independent coordinates) is that there is a sixfold redundancy among the direction cosines. Despite this redundancy, or perhaps because of it, the algebra of finite rotations takes its simplest form in terms of sets of direction cosines. This simplicity is particularly evident if matrix notation[1] is employed.

Let us place the direction cosines as they are arranged in (2-15) into the square matrix

$$[C] = \begin{bmatrix} c_{x'x} & c_{x'y} & c_{x'z} \\ c_{y'x} & c_{y'y} & c_{y'z} \\ c_{z'x} & c_{z'y} & c_{z'z} \end{bmatrix}. \qquad (2\text{-}17)$$

This matrix is called the *rotational-transformation* matrix, or simply, the *rotation* matrix, associated with the angular displacement shown in Fig. 2-13. We shall shortly investigate the special properties of this matrix. If we also arrange the unit vector triads into the column matrices

$$\{\mathbf{U}'\} = \begin{Bmatrix} \mathbf{u}_{x'} \\ \mathbf{u}_{y'} \\ \mathbf{u}_{z'} \end{Bmatrix} \qquad \{\mathbf{U}\} = \begin{Bmatrix} \mathbf{u}_{x} \\ \mathbf{u}_{y} \\ \mathbf{u}_{z} \end{Bmatrix}, \qquad (2\text{-}18)$$

then the three equations of (2-15) can be represented in matrix shorthand as

$$\{\mathbf{U}'\} = [C]\{\mathbf{U}\}, \qquad (2\text{-}19)$$

which states that the orientation of the displaced (primed) triad with respect to the reference (unprimed) triad is obtained by matrix multiplication of the rotation matrix into the reference triad.

Equation (2-16) can also be represented in matrix notation. Here we notice that the arrangement of direction cosines in (2-16) is just the *transpose* of the arrangement in (2-15) and (2-17). Thus the matrix form of (2-16) can be written

$$\{\mathbf{U}\} = [C]^t\{\mathbf{U}'\}. \qquad (2\text{-}20)$$

Now we also note that (2-19) and (2-20) are *inverse* relations, so that

$$[C]^{-1} = [C]^t, \qquad (2\text{-}21)$$

or

$$[C]^t[C] = [C][C]^t = [1] \equiv \begin{bmatrix} 1 & 0 & 0 \\ 0 & 1 & 0 \\ 0 & 0 & 1 \end{bmatrix}. \qquad (2\text{-}22)$$

The relations (2-21) and (2-22) describe most of the special properties of the rotation matrix. If the equations of (2-22) are written out in full, there are nine equations, three of which are repeated, constraining the elements of $[C]$; i.e., there are six independent relations[2] among the nine direction cosines.

For a square matrix $[C]$ to be a rotation matrix, it is necessary that its

[1] See Appendix B.

[2] See Prob. 2-9.

transpose be identical with its inverse.   This is not, however, a sufficient condition.   It happens that the matrix of the direction cosines between a *right-handed* orthogonal triad and a *left-handed* orthogonal triad with arbitrary orientation also satisfies (2–21) and (2–22).   Such matrices are called *improper orthogonal transformations.*   A rotation matrix and an improper orthogonal-transformation matrix differ analytically in the values of their determinants.   The determinant of a rotation matrix is $+1$, while the determinant of an improper orthogonal transformation is -1.   Thus the necessary and sufficient conditions for $[C]$ to be a rotation matrix are that $[C]$ satisfy (2–21) and that the determinant of $[C]$ be unity.

The rotation matrix (2–17) has many uses.   In this section we shall use (2–19) to study the motion of one triad with respect to another.   Another

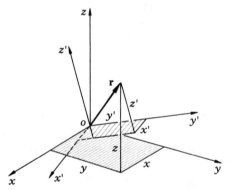

Fig. 2–14.   The vector **r** has components x, y, and z referred to the frame *oxyz* and components x', y', and z' referred to the frame
ox'y'z'.

application arises in Chap. 3, where we shall be faced with the problem sketched in Fig. 2–14.   The same vector **r** is simultaneously observed in the frames *oxyz* and *ox'y'z'*, which are rotated with respect to one another. If the components of **r** in the *oxyz* frame are x, y, and z, respectively, the problem is to determine the corresponding components x', y', and z' in the *ox'y'z'* frame.   By introducing the unit vectors of Fig. 2–13, we have

$$\mathbf{r} = x\mathbf{u}_x + y\mathbf{u}_y + z\mathbf{u}_z = x'\mathbf{u}_{x'} + y'\mathbf{u}_{y'} + z'\mathbf{u}_{z'}. \qquad (2\text{--}23)$$

The solution of this for x', y', and z' can be obtained by using (2–16) to eliminate the unprimed unit vectors in favor of the primed unit vectors, collecting terms, and then setting the separate coefficients of $\mathbf{u}_{x'}$, $\mathbf{u}_{y'}$, and $\mathbf{u}_{z'}$ equal to zero.   This calculation can be carried out in detail, or matrix algebra can be used to obtain the same result[1] in a much abbreviated

[1] See Prob. 2–10.

manner.   The desired relation between the primed and unprimed components is

$$\begin{Bmatrix} x' \\ y' \\ z' \end{Bmatrix} = [C] \begin{Bmatrix} x \\ y \\ z \end{Bmatrix}. \tag{2-24}$$

Given the components of a vector in the unprimed frame, the components
of the same vector in the primed frame are obtained by matrix multiplication of the rotation matrix into the column of unprimed components.
Note that (2–24) has the same form as (2–19) but that it has a considerably different interpretation.   The elements of the column matrices in
(2–19) are unit *vectors*, whereas the elements of the columns in (2–24) are
the *scalar* components of the vector **r**.   The relation (2–24) provides us
with a useful alternative interpretation of the rotational-transformation
matrix.   The rotation matrix operates on the components of a vector to
provide the corresponding components of the same vector in the rotated
frame.

THE ALGEBRA OF FINITE ROTATIONS

We next investigate how compound angular displacements can be
represented.   Equation (2–19) relates the primed triad to the unprimed
triad through the rotation matrix [C].   Suppose that an additional right-
handed orthogonal triad {U''} with unit vectors $u_{x''}$, $u_{y''}$, and $u_{z''}$ has
the same origin as the triads in Fig. 2–13 and that the orientation of the
double-primed triad with respect to the single-primed triad is described by
a matrix of direction cosines [C'], so that we have

$$\{U''\} = [C']\{U'\}. \tag{2-25}$$

The relation between the double-primed triad and the unprimed triad is
obtained by eliminating {U'} between (2–25) and (2–19).

$$\{U''\} = [C'][C]\{U\}. \tag{2-26}$$

This implies that the matrix product

$$[C'][C] = [C''] \tag{2-27}$$

gives the direction cosines linking the unprimed and double-primed triads

$$\{U''\} = [C'']\{U\}. \tag{2-28}$$

From an analytical standpoint it is easy to verify[1] that the product of two
arbitrary rotation matrices is always another rotation matrix.   The
result just obtained can thus be interpreted in the following way: Let
the primed and double-primed triads represent successive positions of a

[1] See Prob. 2–11.

frame which is undergoing angular displacement with respect to the (unprimed) reference frame. The first displacement (in Fig. 2–13) is characterized by the rotation matrix $[C]$. The second displacement is characterized by the rotation matrix $[C']$. The total angular displacement which results is also a rotation, and the rotation matrix $[C'']$ which characterizes it is just the matrix product (2–27) of the individual rotations. When angular displacements are compounded, their rotation matrices are *multiplied*. The difference between the behavior of linear displacements of points and the behavior of angular displacements of rectangular frames may be described by saying that linear displacements are compounded according to the laws of *vector addition*, whereas angular displacements are compounded according to the laws of *matrix multiplication*. The peculiarities of the algebra of finite rotations are just the peculiarities of matrix multiplication. For example, the importance of the order in which finite rotations are performed (which was illustrated in Fig. 2–10) is mirrored by the fact that matrix multiplication is not *commutative*.

There is a unique relation between the directed-line-segment representation $\boldsymbol{\theta}$ of an angular displacement and the direction-cosine representation $[C]$ of the same angular displacement. This relation is not simple for finite rotation angles. For completeness we shall quote the general result without proof. Let the rotation $\boldsymbol{\theta}$ be an angular displacement of magnitude θ about an axis whose direction cosines with respect to the *oxyz* frame are l, m, and n, as indicated in Fig. 2–15a. Under the rotation the unprimed frame of Fig. 2–15b is carried into the primed frame. The rotation matrix $[C]$ which gives the direction cosines of the angles between the primed and unprimed axes can then be represented[1] as follows:

$$[C] = \cos\theta \begin{bmatrix} 1 & 0 & 0 \\ 0 & 1 & 0 \\ 0 & 0 & 1 \end{bmatrix} + (1 - \cos\theta) \begin{bmatrix} l^2 & lm & ln \\ lm & m^2 & mn \\ ln & mn & n^2 \end{bmatrix}$$
$$+ \sin\theta \begin{bmatrix} 0 & n & -m \\ -n & 0 & l \\ m & -l & 0 \end{bmatrix}. \quad (2\text{–}29)$$

Given θ and the axis of rotation, the matrix $[C]$ can be constructed directly from (2–29). Conversely, if the matrix $[C]$ is given, it is possible to use the representation (2–29) to deduce[2] the value of θ and the location of the axis of rotation.

---

[1] See, for example, J. L. Synge, Classical Dynamics, pp. 15–18, in S. Flügge (ed.), "Encyclopedia of Physics," vol. III/1, Principles of Classical Mechanics and Field Theory, Springer-Verlag OHG, 1960.

[2] See Prob. 2–12.

SMALL ROTATIONS

It is a remarkable fact that angular velocity is in many respects easier to deal with than angular displacement. The underlying reason for this is that angular displacement enters the definition of angular velocity only in the limit as the magnitude of the angular displacement approaches zero. We shall see that the treatment of angular displacements simplifies considerably when the rotations are small.

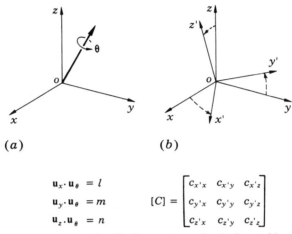

$$\mathbf{u}_x \cdot \mathbf{u}_\theta = l$$
$$\mathbf{u}_y \cdot \mathbf{u}_\theta = m \qquad [C] = \begin{bmatrix} c_{x'x} & c_{x'y} & c_{x'z} \\ c_{y'x} & c_{y'y} & c_{y'z} \\ c_{z'x} & c_{z'y} & c_{z'z} \end{bmatrix}$$
$$\mathbf{u}_z \cdot \mathbf{u}_\theta = n$$

FIG. 2-15. Angular displacement can be designated by (a) directed line segment θ along the axis of rotation or by (b) rotation matrix $[C]$ of direction cosines of the angles between primed and unprimed axes.

In preparation for allowing a rotation $\Delta\theta$ to approach zero, we shall introduce the *order* notation

$$O[(\Delta\theta)^n] \tag{2-30}$$

to represent an expression which approaches zero at least as fast as a fixed multiple of $(\Delta\theta)^n$ when $\Delta\theta$ approaches zero. The trigonometric functions of such an angle may be written

$$\sin \Delta\theta = \Delta\theta + O[(\Delta\theta)^3]$$
$$\cos \Delta\theta = 1 + O[(\Delta\theta)^2]. \tag{2-31}$$

Let us now consider a small angular displacement which can be represented by the directed line segment $\Delta\theta$ shown in Fig. 2-16. The magnitude of the rotation is $\Delta\theta$, and the direction cosines of the axis of rotation are

$$l = \frac{\Delta\theta_x}{\Delta\theta} \qquad m = \frac{\Delta\theta_y}{\Delta\theta} \qquad n = \frac{\Delta\theta_z}{\Delta\theta}. \tag{2-32}$$

Let us designate the corresponding rotation matrix as $[C(\Delta\boldsymbol{\theta})]$. If we insert (2–31) in the general formula (2–29), the rotation matrix simplifies to

$$[C(\Delta\boldsymbol{\theta})] = \begin{bmatrix} 1 & 0 & 0 \\ 0 & 1 & 0 \\ 0 & 0 & 1 \end{bmatrix} + \begin{bmatrix} 0 & \Delta\theta_z & -\Delta\theta_y \\ -\Delta\theta_z & 0 & \Delta\theta_x \\ \Delta\theta_y & -\Delta\theta_x & 0 \end{bmatrix} + O[(\Delta\theta)^2], \quad (2\text{–}33)$$

where all the higher-order terms have been placed in the second-order remainder term. Note that the first-order contribution of the small rotation is a *skew-symmetric* matrix constructed from the components of the directed line segment $\Delta\boldsymbol{\theta}$. As an alternative to deriving (2–33)

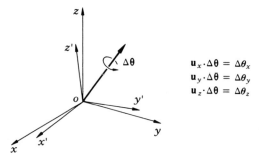

$$\mathbf{u}_x \cdot \Delta\boldsymbol{\theta} = \Delta\theta_x$$
$$\mathbf{u}_y \cdot \Delta\boldsymbol{\theta} = \Delta\theta_y$$
$$\mathbf{u}_z \cdot \Delta\boldsymbol{\theta} = \Delta\theta_z$$

FIG. 2–16. Small angular displacement. The components of the directed line segment $\Delta\boldsymbol{\theta}$ are $\Delta\theta_x$, $\Delta\theta_y$, and $\Delta\theta_z$.

from the general formula (2–29) it is possible[1] to build up (2–33) by compounding three small rotations about mutually perpendicular axes.

The representation (2–33) can be used to show that, to first order, small rotations can be compounded according to the laws of vector algebra. It is a straightforward exercise[2] in matrix multiplication to verify that

$$[C(\Delta\boldsymbol{\theta}_1)][C(\Delta\boldsymbol{\theta}_2)] = [C(\Delta\boldsymbol{\theta}_2)][C(\Delta\boldsymbol{\theta}_1)] + O[(\Delta\theta_1)^2 + (\Delta\theta_2)^2], \quad (2\text{–}34)$$

which means that, to first order, the total rotation resulting from applying one small rotation after another is independent of the order in which the rotations are performed. Furthermore, the resulting rotation matrix is, to first order, the same[2] as that corresponding to a single rotation designated by a directed line segment which is the *vector sum* $\Delta\boldsymbol{\theta}_1 + \Delta\boldsymbol{\theta}_2$,

$$[C(\Delta\boldsymbol{\theta}_1 + \Delta\boldsymbol{\theta}_2)] = [C(\Delta\boldsymbol{\theta}_2)][C(\Delta\boldsymbol{\theta}_1)] + O[(\Delta\theta_1)^2 + (\Delta\theta_2)^2]. \quad (2\text{–}35)$$

Thus we can say that while finite rotations do not obey the rules of vector addition, small rotations *almost* do. As the magnitude of a rota-

[1] See Prob. 2–16.

[2] See Prob. 2–17.

tion decreases, the relative importance of the second-order remainder terms decreases and the more nearly do the first-order terms represent the entire rotation. To completely remove the effects of the higher-order terms it is necessary to pass to the limit of vanishing angle.

ANGULAR VELOCITY

Consider a rectangular frame *oxyz* whose orientation with respect to a reference frame *OXYZ* changes with time. At time t let the triad $\mathbf{U}(t)$ of unit vectors for the frame *oxyz* be $\mathbf{u}_x(t)$, $\mathbf{u}_y(t)$, and $\mathbf{u}_z(t)$, as shown in Fig. 2–17. At time $t + \Delta t$ this triad becomes $\mathbf{U}(t + \Delta t)$, with unit vectors $\mathbf{u}_x(t + \Delta t)$, $\mathbf{u}_y(t + \Delta t)$, and $\mathbf{u}_z(t + \Delta t)$. The angular displacement

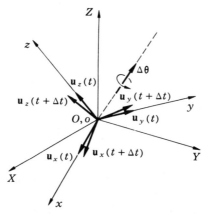

FIG. 2–17. The triad $\mathbf{U}(t)$ undergoes rotation $\Delta\boldsymbol{\theta}$ during the interval Δt and becomes the triad $\mathbf{U}(t + \Delta t)$.

of the triad during the interval Δt (as observed from the reference frame) is indicated by the rotation vector $\Delta\boldsymbol{\theta}$. The *angular velocity* $\boldsymbol{\omega}$ of the frame *oxyz* with respect to the reference frame is the limiting value of the quotient of the displacement $\Delta\boldsymbol{\theta}$ divided by the time interval Δt.

$$\boldsymbol{\omega} = \lim_{\Delta t \to 0} \frac{\Delta\boldsymbol{\theta}}{\Delta t}. \tag{2-36}$$

The properties of the vector $\boldsymbol{\omega}$ follow from the properties of the angular rotation $\Delta\boldsymbol{\theta}$ in the limit. The direction of the vector $\boldsymbol{\omega}$ is along the *instantaneous axis* of rotation of the frame *oxyz* with respect to *OXYZ*. The right-hand rule gives the sense, and the length $|\boldsymbol{\omega}| = \omega$ gives the magnitude of the *rate of rotation* about the instantaneous axis. The vector $\boldsymbol{\omega}$ can be compounded or decomposed according to the laws of vector algebra. In particular, the angular velocity can be decomposed into components along the axes of the frame $oxyz$,

$$\boldsymbol{\omega} = \omega_x \mathbf{u}_x + \omega_y \mathbf{u}_y + \omega_z \mathbf{u}_z, \tag{2-37}$$

where the components are

$$\omega_x = \lim_{\Delta t \to 0} \frac{\Delta \theta_x}{\Delta t}$$

$$\omega_y = \lim_{\Delta t \to 0} \frac{\Delta \theta_y}{\Delta t} \tag{2-38}$$

$$\omega_z = \lim_{\Delta t \to 0} \frac{\Delta \theta_z}{\Delta t}.$$

As the frame $oxyz$ rotates, the triad of unit vectors $\mathbf{U}(t)$ is carried through a succession of different orientations. We shall find the rate of change of $\mathbf{U}(t)$ by a limiting process. The angular displacement $\Delta\boldsymbol{\theta}$ which carries $\mathbf{U}(t)$ into $\mathbf{U}(t + \Delta t)$ in Fig. 2–17 can be represented by the rotation matrix $[C(\Delta\boldsymbol{\theta})]$ of (2–33). Thus, according to (2–19), we have

$$\{\mathbf{U}(t + \Delta t)\} = [C(\Delta\boldsymbol{\theta})]\{\mathbf{U}(t)\}. \tag{2-39}$$

In preparation for taking the limit, we form the quotient

$$\frac{\{\Delta \mathbf{U}\}}{\Delta t} = \frac{\{\mathbf{U}(t + \Delta t)\} - \{\mathbf{U}(t)\}}{\Delta t} = \frac{[C(\Delta\boldsymbol{\theta})] - [1]}{\Delta t}\{\mathbf{U}(t)\}$$

$$\begin{Bmatrix} \dfrac{\Delta \mathbf{u}_x}{\Delta t} \\[2mm] \dfrac{\Delta \mathbf{u}_y}{\Delta t} \\[2mm] \dfrac{\Delta \mathbf{u}_z}{\Delta t} \end{Bmatrix} = \begin{bmatrix} 0 & \dfrac{\Delta \theta_z}{\Delta t} & -\dfrac{\Delta \theta_y}{\Delta t} \\[2mm] -\dfrac{\Delta \theta_z}{\Delta t} & 0 & \dfrac{\Delta \theta_x}{\Delta t} \\[2mm] \dfrac{\Delta \theta_y}{\Delta t} & -\dfrac{\Delta \theta_x}{\Delta t} & 0 \end{bmatrix} \begin{Bmatrix} \mathbf{u}_x \\[2mm] \mathbf{u}_y \\[2mm] \mathbf{u}_z \end{Bmatrix} + \frac{O[(\Delta\theta)^2]}{\Delta t}. \tag{2-40}$$

On passing to the limit as Δt and $\Delta \theta$ approach zero, the remainder term vanishes and we find

$$\begin{Bmatrix} \dfrac{d\mathbf{u}_x}{dt} \\[2mm] \dfrac{d\mathbf{u}_y}{dt} \\[2mm] \dfrac{d\mathbf{u}_z}{dt} \end{Bmatrix} = \begin{bmatrix} 0 & \omega_z & -\omega_y \\[2mm] -\omega_z & 0 & \omega_x \\[2mm] \omega_y & -\omega_x & 0 \end{bmatrix} \begin{Bmatrix} \mathbf{u}_x \\[2mm] \mathbf{u}_y \\[2mm] \mathbf{u}_z \end{Bmatrix}, \tag{2-41}$$

or more concisely,

$$\left\{ \frac{d\mathbf{U}}{dt} \right\} = [\omega]\{\mathbf{U}\}, \tag{2-42}$$

where $[\omega]$ denotes the skew-symmetric square matrix of the components of $\boldsymbol{\omega}$. An alternative interpretation of (2–41) is obtained by carrying out the indicated multiplication on the right-hand side and identifying the resulting terms with vector cross products.   In this way we obtain

$$\frac{d\mathbf{u}_x}{dt} = \omega \times \mathbf{u}_x \qquad \frac{d\mathbf{u}_y}{dt} = \omega \times \mathbf{u}_y \qquad \frac{d\mathbf{u}_z}{dt} = \omega \times \mathbf{u}_z. \qquad (2\text{-}43)$$

The rates of change of the unit vectors fixed in the $oxyz$ frame, due to the angular velocity of the frame, are thus described in matrix notation by (2-42) or in vector notation by (2-43).   It may be noted that each of the equations in (2-43) represents a special case of the general formula (2-9).   The general result can be deduced from (2-43) since an arbitrary vector fixed in the $oxyz$ frame can be expressed as

$$\mathbf{A} = A_x\mathbf{u}_x + A_y\mathbf{u}_y + A_z\mathbf{u}_z, \qquad (2\text{-}44)$$

where the coefficients A_x, A_y, and A_z do not change with time.   With the help of (2-43) the time derivative of (2-44) is

$$\begin{aligned}
\frac{d\mathbf{A}}{dt} &= A_x\omega \times \mathbf{u}_x + A_y\omega \times \mathbf{u}_y + A_z\omega \times \mathbf{u}_z \\
&= \omega \times (A_x\mathbf{u}_x + A_y\mathbf{u}_y + A_z\mathbf{u}_z) \\
&= \omega \times \mathbf{A}, \qquad (2\text{-}45)
\end{aligned}$$

which provides an alternative derivation of (2-9).

2-4   Kinematical analysis utilizing intermediate frames

We now employ the results of Sec. 2-2 concerning rectangular frames in relative motion to study the motion of a point with the aid of an intermediate reference frame.   In Fig. 2-18 the basic reference frame is $OXYZ$.   The intermediate frame $oxyz$, with the triad $\mathbf{U}$ of unit vectors $\mathbf{u}_x$, $\mathbf{u}_y$, and $\mathbf{u}_z$ affixed, translates and rotates with respect to the basic reference frame.   The point P is located with respect to the basic reference frame by the position vector $\mathbf{R}$ and with respect to the intermediate frame by the relative position vector $\mathbf{r}$.   Our problem is to determine the velocity and acceleration of P with respect to the basic reference frame $OXYZ$.

Let us first examine the motion of P with respect to the intermediate frame $oxyz$.   The relative position vector may be written in component form,

$$\mathbf{r} = x\mathbf{u}_x + y\mathbf{u}_y + z\mathbf{u}_z. \qquad (2\text{-}46)$$

We shall use the special notation $(\partial/\partial t)_{\text{rel}}$ to indicate time derivatives observed from the intermediate frame.   The velocity and acceleration of

P relative to $oxyz$ are then

$$\mathbf{v}_{\text{rel}} = \left(\frac{\partial \mathbf{r}}{\partial t}\right)_{\text{rel}} = \dot{x}\mathbf{u}_x + \dot{y}\mathbf{u}_y + \dot{z}\mathbf{u}_z$$

$$\mathbf{a}_{\text{rel}} = \left(\frac{\partial \mathbf{v}_{\text{rel}}}{\partial t}\right)_{\text{rel}} = \left(\frac{\partial^2 \mathbf{r}}{\partial t}\right)_{\text{rel}} = \ddot{x}\mathbf{u}_x + \ddot{y}\mathbf{u}_y + \ddot{z}\mathbf{u}_z,$$

(2–47)

according to (2–1) and (2–2). The superior dots are used to indicate time differentiation; i.e., $\dot{x} = dx/dt$ and $\ddot{x} = d^2x/dt^2$.

In order to obtain the velocity and acceleration of P relative to $OXYZ$ it is necessary to take into account the motion of the intermediate frame with respect to the basic reference frame. Let $\mathbf{R}_o$ be the position vector

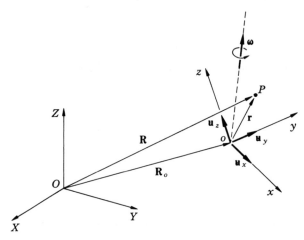

FIG. 2–18.   Intermediate frame $oxyz$ translates and rotates
with respect to basic reference frame $OXYZ$.

of the origin of the intermediate frame with respect to the origin of the basic reference frame, and let $\boldsymbol{\omega}$ denote the angular-velocity vector of the intermediate frame with respect to the basic reference frame. We shall now compare the time rates of change of a vector as viewed from the two frames. Let $\mathbf{B}$ be an *arbitrary* vector whose instantaneous component representation referred to $oxyz$ is

$$\mathbf{B} = B_x\mathbf{u}_x + B_y\mathbf{u}_y + B_z\mathbf{u}_z. \tag{2–48}$$

In general, the scalar components B_x, B_y, and B_z will vary with time and the unit vectors will vary in orientation when viewed from $OXYZ$. Using the notation d/dt to indicate time derivatives observed from the basic reference frame, the derivative of (2–48) is

$$\frac{d\mathbf{B}}{dt} = \dot{B}_x\mathbf{u}_x + \dot{B}_y\mathbf{u}_y + \dot{B}_z\mathbf{u}_z + B_x\frac{d\mathbf{u}_x}{dt} + B_y\frac{d\mathbf{u}_y}{dt} + B_z\frac{d\mathbf{u}_z}{dt}. \tag{2–49}$$

The time derivatives of the scalar components indicated by the superior dots are independent of the frame of observation.

The right side of (2–49) splits into two parts. The first three terms give the rate of change of **B** as viewed from the intermediate frame, and the last three give the contribution due to rotation of the intermediate frame. The first three terms can be denoted by $(\partial \mathbf{B}/\partial t)_{\text{rel}}$ in analogy with the first of (2–47). The last three terms can be simplified by applying (2–9) to evaluate the derivatives of the unit vectors.

$$B_x \frac{d\mathbf{u}_x}{dt} + B_y \frac{d\mathbf{u}_y}{dt} + B_z \frac{d\mathbf{u}_z}{dt} = B_x \boldsymbol{\omega} \times \mathbf{u}_x + B_y \boldsymbol{\omega} \times \mathbf{u}_y + B_z \boldsymbol{\omega} \times \mathbf{u}_z$$

$$= \boldsymbol{\omega} \times (B_x \mathbf{u}_x + B_y \mathbf{u}_y + B_z \mathbf{u}_z)$$

$$= \boldsymbol{\omega} \times \mathbf{B}. \tag{2-50}$$

Consequently, (2–49) can be written

$$\frac{d\mathbf{B}}{dt} = \left(\frac{\partial \mathbf{B}}{\partial t}\right)_{\text{rel}} + \boldsymbol{\omega} \times \mathbf{B}. \tag{2-51}$$

This result connects the time rates of change of an arbitrary vector as viewed from different reference frames. The time derivative observed from the basic reference frame is the sum of the time derivative observed from the intermediate frame plus the term $\boldsymbol{\omega} \times \mathbf{B}$, which accounts for the angular velocity of the intermediate frame with respect to the basic reference frame. When the vector **B** is frozen into the intermediate frame so that $(\partial \mathbf{B}/\partial t)_{\text{rel}} = 0$, we note that (2–51) becomes equivalent to (2–9). We also note that if **B** is parallel to $\boldsymbol{\omega}$ (i.e., parallel to the axis of rotation), the cross product vanishes and the rates of change of **B** as viewed from the two frames are identical. In particular, when the rate of change of $\boldsymbol{\omega}$ itself is viewed from the two frames, we have

$$\frac{d\boldsymbol{\omega}}{dt} = \left(\frac{\partial \boldsymbol{\omega}}{\partial t}\right)_{\text{rel}}. \tag{2-52}$$

Since the vector **B** in (2–51) is completely arbitrary, we may rewrite (2–51) in the *operational form*

$$\frac{d}{dt} = \left(\frac{\partial}{\partial t}\right)_{\text{rel}} + \boldsymbol{\omega} \times, \tag{2-53}$$

with the understanding that the operation on the right when applied to any vector gives the same result as the operation on the left.

VELOCITY AND ACCELERATION FORMULAS FOR FRAMES IN RELATIVE MOTION

Returning now to Fig. 2–18, the position vector **R** for the point P with respect to the origin of the basic reference frame is at all times the vector sum of $\mathbf{R}_o$ and $\mathbf{r}$. The *velocity* of P with respect to the basic reference frame is, then,

$$\mathbf{v} = \frac{d\mathbf{R}}{dt} = \frac{d\mathbf{R}_o}{dt} + \frac{d\mathbf{r}}{dt}. \qquad (2\text{--}54)$$

Applying the operational relation (2–53) to the second term on the right, we find

$$\mathbf{v} = \frac{d\mathbf{R}_o}{dt} + \mathbf{v}_{rel} + \boldsymbol{\omega} \times \mathbf{r}, \qquad (2\text{--}55)$$

where $\mathbf{v}_{rel}$ is defined by the first of (2–47). We can interpret the first term on the right of (2–55) as the contribution due to the *translation* of the intermediate frame. The second term is the velocity of P with respect to the intermediate frame, and the final term is the contribution due to the *rotation* of the intermediate frame. Note that the $\mathbf{v}$ and $\mathbf{v}_{rel}$ are generally different[1] except when the intermediate frame is fixed (i.e., not translating and not rotating) with respect to the basic reference frame.

An interesting and useful interpretation of (2–55) is provided by introducing the concept of the *vehicle point* P_{veh}. This is the point *fixed* in the *intermediate frame* which at the instant under consideration coincides with the actual point P; i.e., the vehicle point has the same relative position vector $\mathbf{r}$ as the actual point, but since it is fixed in the intermediate frame, it has no relative velocity or relative acceleration. The velocity of the vehicle point is

$$\mathbf{v}_{veh} = \frac{d\mathbf{R}_o}{dt} + \boldsymbol{\omega} \times \mathbf{r}, \qquad (2\text{--}56)$$

which is obtained from (2–55) by deleting the relative velocity. Using (2–56), we may rewrite (2–55) as follows:

$$\mathbf{v} = \mathbf{v}_{veh} + \mathbf{v}_{rel}. \qquad (2\text{--}57)$$

As an example of the application of (2–57), consider the exhaust gas passing through the nozzle of a rocket in flight. With the basic reference frame fixed to the earth and the intermediate frame fixed in the rocket, the velocity of the gas with respect to the earth is the vector sum of the velocity of the nozzle with respect to the earth plus the relative velocity of the gas with respect to the rocket.

The *acceleration* of P with respect to the basic reference frame is

$$\mathbf{a} = \frac{d\mathbf{v}}{dt} = \frac{d^2\mathbf{R}_o}{dt^2} + \frac{d}{dt}(\mathbf{v}_{rel} + \boldsymbol{\omega} \times \mathbf{r}), \qquad (2\text{--}58)$$

if we substitute for $\mathbf{v}$ from (2–55). Using the operational identity (2–53)

---

[1] An exception occurs when the intermediate frame is rotating and the velocity of P is directed along the axis of rotation.

to expand the second term on the right,

$$\mathbf{a} = \frac{d^2\mathbf{R}_o}{dt^2} + \left[\left(\frac{\partial}{\partial t}\right)_{rel} + \boldsymbol{\omega} \times\right] \mathbf{v}_{rel} + \dot{\boldsymbol{\omega}} \times \mathbf{r} + \boldsymbol{\omega} \times \left[\left(\frac{\partial}{\partial t}\right)_{rel} + \boldsymbol{\omega} \times\right] \mathbf{r}$$

$$= \frac{d^2\mathbf{R}_o}{dt^2} + \mathbf{a}_{rel} + 2\boldsymbol{\omega} \times \mathbf{v}_{rel} + \dot{\boldsymbol{\omega}} \times \mathbf{r} + \boldsymbol{\omega} \times (\boldsymbol{\omega} \times \mathbf{r}), \qquad (2\text{-}59)$$

where we have introduced the definition (2-47) for $\mathbf{a}_{rel}$ and have used $\dot{\boldsymbol{\omega}}$ to stand for the derivative given in (2-52). The resultant acceleration of P with respect to the frame $OXYZ$ can thus be decomposed into five contributions: The first term of (2-59) is the acceleration of the origin of the intermediate frame with respect to $OXYZ$. The second term is the relative acceleration of P with respect to $oxyz$. The third term $2\boldsymbol{\omega} \times \mathbf{v}_{rel}$ is called the *Coriolis* acceleration. The fourth term gives the contribution of the *angular acceleration* $\dot{\boldsymbol{\omega}}$, and the last term is called the *centripetal acceleration*. We may note that the acceleration of P with respect to the basic reference frame is not generally the same as the acceleration with respect to the intermediate frame except when there is no angular motion between the frames ($\boldsymbol{\omega} = 0$) and the translation is limited to *uniform* relative velocity ($d^2\mathbf{R}_o/dt^2 = 0$).

A useful alternative representation of (2-59) is obtained by introducing the acceleration of the *vehicle point* P_{veh}, i.e., the point fixed in $oxyz$ which instantaneously coincides with the actual point P. Deleting $\mathbf{v}_{rel}$ and $\mathbf{a}_{rel}$ from (2-59), we find the acceleration of the vehicle point,

$$\mathbf{a}_{veh} = \frac{d^2\mathbf{R}_o}{dt^2} + \frac{d\boldsymbol{\omega}}{dt} \times \mathbf{r} + \boldsymbol{\omega} \times (\boldsymbol{\omega} \times \mathbf{r}). \qquad (2\text{-}60)$$

Then, introducing the acceleration of the vehicle point into (2-59), we obtain

$$\mathbf{a} = \mathbf{a}_{veh} + \mathbf{a}_{rel} + 2\boldsymbol{\omega} \times \mathbf{v}_{rel} \qquad (2\text{-}61)$$

for the acceleration of the actual point P. This representation provides an opportunity to give an interpretation of the Coriolis term $2\boldsymbol{\omega} \times \mathbf{v}_{rel}$. Consider the special case in which the actual point P moves with *uniform* velocity relative to the intermediate frame so that $\mathbf{a}_{rel} = 0$. In this case the difference between the acceleration of the actual point and the vehicle point is just the Coriolis acceleration. For example, consider an automobile traveling at uniform speed along a straight flat road, as viewed from a basic reference frame fixed in the solar system. If an intermediate frame is assumed to be fixed in the earth, the vehicle point P_{veh} may be taken as the point on the road over which the actual point P, the auto, is passing at a particular instant. Because the earth rotates with respect to the solar system, the acceleration of the point on the road is not the same as the acceleration of the auto (when both are viewed from the solar system), although the relative velocity of the auto with respect to the road is uniform. In the Northern Hemisphere the component of the Coriolis acceleration parallel to the earth's surface is directed toward the driver's left. The magnitude of this component varies from zero at the equator to a maximum at the North Pole. For automobiles this magnitude is insignificant (at 60 mph the Coriolis acceleration is less than 0.04 percent of the acceleration of gravity) in comparison with the magnitudes of the accelerations involved in the normal operations of starting, stopping, and turning. The accelerations of large air masses moving across the globe are, however, of the same order of magnitude as the Coriolis acceleration, and the Coriolis term in (2–61) plays an important role in the establishment of weather patterns.

Further insight into the significance of the various terms in the velocity

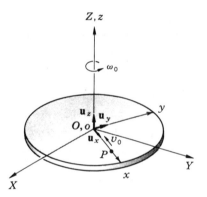

FIG. 2–19.   Example 2–2.   Particle on
a turntable.

expression (2–55) and the acceleration expression (2–59) can be obtained from studying the following examples. Most engineering problems in kinematical analysis can be solved by repeated application of these two formulas.

Example 2–2. A flat turntable rotates with uniform angular speed ω_0 with respect to the reference frame $OXYZ$, as shown in Fig. 2–19. A particle P moves radially inward with uniform speed v_0 with respect to the turntable. It is desired to obtain the velocity and acceleration of P with respect to $OXYZ$. We shall obtain quantitative results for the case where $\omega_0 = 1$ rad/sec and $v_0 = 1$ ft/sec at the instant when $\overline{OP} = 1$ ft.

As an intermediate frame we take $oxyz$, which is fixed in the turntable, with the z axis coincident with the axis of rotation OZ. To obtain the velocity of P we apply (2–55), where, in this case,

$$\mathbf{R}_o \equiv 0 \qquad \boldsymbol{\omega} = \omega_0 \mathbf{u}_z \qquad \mathbf{r} = x\mathbf{u}_x \qquad \mathbf{v}_{\text{rel}} = -v_0 \mathbf{u}_x. \qquad (2\text{–}62)$$

On substituting these values into (2–55), we find

$$\mathbf{v} = \frac{d\mathbf{R}_o}{dt} + \mathbf{v}_{\text{rel}} + \boldsymbol{\omega} \times \mathbf{r}$$

$$= 0 - v_0 \mathbf{u}_x + \omega_0 x \mathbf{u}_y, \qquad (2\text{–}63)$$

in general, and for the particular numerical values assumed,

$$\mathbf{v} = -\mathbf{u}_x + \mathbf{u}_y \qquad \text{ft/sec,} \qquad (2\text{–}64)$$

as sketched in Fig. 2–20. The tangential component may be interpreted as the velocity of the vehicle point, i.e., the point P_{veh} of the turntable over which the particle P is passing at this instant. The radial component is the relative velocity. The total velocity is their vector sum.

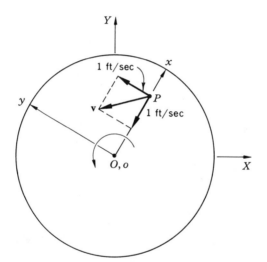

Fig. 2–20. Top view of turntable showing velocity components.

The acceleration of P is obtained by application of (2–59). Here we note that, in addition to (2–62), we have

$$\frac{d\boldsymbol{\omega}}{dt} = 0 \qquad \text{and} \qquad \mathbf{a}_{\text{rel}} = 0. \qquad (2\text{–}65)$$

Substitution in (2–59) yields

$$\mathbf{a} = \frac{d^2\mathbf{R}_o}{dt^2} + \mathbf{a}_{\text{rel}} + 2\boldsymbol{\omega} \times \mathbf{v}_{\text{rel}} + \frac{d\boldsymbol{\omega}}{dt} \times \mathbf{r} + \boldsymbol{\omega} \times (\boldsymbol{\omega} \times \mathbf{r})$$

$$= 0 + 0 - 2\omega_0 v_0 \mathbf{u}_y + 0 - \omega_0^2 \mathbf{u}_x. \qquad (2\text{–}66)$$

For the particular numerical values assumed, this becomes

$$\mathbf{a} = -2\mathbf{u}_y - \mathbf{u}_x \qquad \text{ft/sec}^2, \qquad (2\text{-}67)$$

as sketched in Fig. 2–21. The radial component is the *centripetal* acceleration. It may be interpreted as the acceleration of the vehicle point P_{veh}. This point is fixed in the turntable and travels in a circle at a uniform rate. Its acceleration is directed toward the center of the circle. The tangential component is the *Coriolis* acceleration. Note that for the rotation sense shown, the direction of the Coriolis acceleration can always be found by facing in the direction of the $\mathbf{v}_{\text{rel}}$ and extending the left hand. For fixed values of ω_0 and v_0, the magnitude of the Coriolis acceleration is

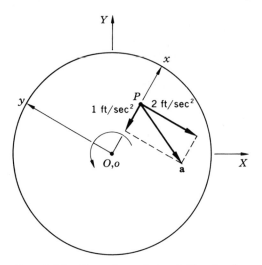

FIG. 2–21. Top view of turntable showing
acceleration components.

independent of where P is on the turntable or in which direction it is headed.

Example 2–3.   In the preceding example the motion was confined to a single plane. To illustrate a truly three-dimensional case we consider the problem of obtaining the velocity and acceleration of a point P with respect to the $OXYZ$ frame when the position of the point is described by the *spherical coordinates* r, θ, and φ, shown in Fig. 2–22.

We begin by selecting the intermediate frame $oxyz$ shown, where ox is along the radius OP, oy lies in the plane ZOP, and oz is normal to the plane ZOP. Instead of expressing our results in terms of the triad $\mathbf{u}_x$, $\mathbf{u}_y$, and $\mathbf{u}_z$, we shall use the parallel triad $\mathbf{u}_r$, $\mathbf{u}_\theta$, and $\mathbf{u}_\varphi$ emanating from P. Within the intermediate frame the relative position vector $\mathbf{r}$ for the point P is

$$\mathbf{r} = r\mathbf{u}_r, \tag{2-68}$$

and hence we can write

$$\mathbf{v}_{rel} = \dot{r}\mathbf{u}_r \quad \text{and} \quad \mathbf{a}_{rel} = \ddot{r}\mathbf{u}_r. \tag{2-69}$$

Since the origin of the intermediate frame coincides with the origin of the basic reference frame, the vector $\mathbf{R}_o$ in (2–55) and (2–59) is always zero.

To obtain the angular velocity $\boldsymbol{\omega}$ of the intermediate frame, we utilize the fact that angular velocities are compounded according to the laws of vector algebra. A small rotation of the $oxyz$ frame may be considered to

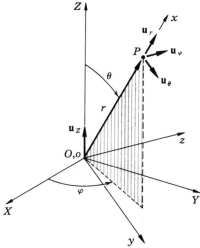

Fig. 2–22. Example 2–3. Point P is located by the spherical coordinates r, θ, and φ.

consist of a rotation of the plane ZOP about OZ through an angle $\Delta\varphi$, followed by a rotation of xoy about oz through an angle $\Delta\theta$. Passing to the limit, we have

$$\boldsymbol{\omega} = \dot{\varphi}\mathbf{u}_Z + \dot{\theta}\mathbf{u}_\varphi. \tag{2-70}$$

This is a mixed expression since $\mathbf{u}_Z$ is a unit vector in the basic reference frame, while $\mathbf{u}_\varphi$ is parallel to the unit vector $\mathbf{u}_z$ in the intermediate frame. To express $\boldsymbol{\omega}$ in terms of the triad $\mathbf{u}_r$, $\mathbf{u}_\theta$, and $\mathbf{u}_\varphi$, we have only to note from Fig. 2–22 that

$$\mathbf{u}_Z = \cos\theta\,\mathbf{u}_r - \sin\theta\,\mathbf{u}_\theta \tag{2-71}$$

and substitute in (2–70) to obtain

$$\boldsymbol{\omega} = \dot{\varphi}\cos\theta\,\mathbf{u}_r - \dot{\varphi}\sin\theta\,\mathbf{u}_\theta + \dot{\theta}\mathbf{u}_\varphi. \tag{2-72}$$

The angular acceleration follows on using (2–52).

$$\dot{\boldsymbol{\omega}} = \left(\frac{\partial \boldsymbol{\omega}}{\partial t}\right)_{\text{rel}} = (\ddot{\varphi}\cos\theta - \dot{\varphi}\dot{\theta}\sin\theta)\mathbf{u}_r - (\ddot{\varphi}\sin\theta + \dot{\varphi}\dot{\theta}\cos\theta)\mathbf{u}_\theta + \ddot{\theta}\mathbf{u}_\varphi.$$
$$(2\text{–}73)$$

The velocity of P can now be obtained by substituting in (2–55).

$$\mathbf{v} = \frac{d\mathbf{R}_o}{dt} + \mathbf{v}_{\text{rel}} + \boldsymbol{\omega} \times \mathbf{r}$$
$$= 0 + \dot{r}\mathbf{u}_r + r\dot{\theta}\mathbf{u}_\theta + r\dot{\varphi}\sin\theta\,\mathbf{u}_\varphi. \qquad (2\text{–}74)$$

This expression for $\mathbf{v}$ is directly useful in dynamics problems when spherical coordinates are employed.

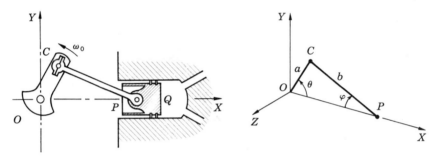

FIG. 2–23.   Example 2–4.   Crank, connecting rod, and piston.

The acceleration of P is obtained in a similar fashion by substituting in (2–59). The term $\mathbf{a}_{\text{rel}}$ is given by (2–69); the Coriolis term is

$$2\boldsymbol{\omega} \times \mathbf{v}_{\text{rel}} = 2\dot{r}\dot{\theta}\mathbf{u}_\theta + 2\dot{r}\dot{\varphi}\sin\theta\,\mathbf{u}_\varphi; \qquad (2\text{–}75)$$

the angular-acceleration term is

$$\dot{\boldsymbol{\omega}} \times \mathbf{r} = r\ddot{\theta}\mathbf{u}_\theta + (r\ddot{\varphi}\sin\theta + r\dot{\varphi}\dot{\theta}\cos\theta)\mathbf{u}_\varphi; \qquad (2\text{–}76)$$

and the centripetal-acceleration term is

$$\boldsymbol{\omega} \times (\boldsymbol{\omega} \times \mathbf{r}) = -(r\dot{\theta}^2 + r\dot{\varphi}^2\sin^2\theta)\mathbf{u}_r - r\dot{\varphi}^2\sin\theta\cos\theta\,\mathbf{u}_\theta + r\dot{\varphi}\dot{\theta}\cos\theta\,\mathbf{u}_\varphi.$$
$$(2\text{–}77)$$

Gathering these all together according to (2–59), we find

$$\mathbf{a} = (\ddot{r} - r\dot{\theta}^2 - r\dot{\varphi}^2\sin^2\theta)\mathbf{u}_r + (r\ddot{\theta} + 2\dot{r}\dot{\theta} - r\dot{\varphi}^2\sin\theta\cos\theta)\mathbf{u}_\theta$$
$$+ (r\ddot{\varphi}\sin\theta + 2\dot{r}\dot{\varphi}\sin\theta + 2r\dot{\varphi}\dot{\theta}\cos\theta)\mathbf{u}_\varphi \qquad (2\text{–}78)$$

for the acceleration of P.

Example 2–4.   A mechanism consisting of a crank OC, connecting rod CP, and piston PQ is sketched in Fig. 2–23. It is desired to obtain the velocity and acceleration of the point P with respect to the frame $OXYZ$

when the crank rotates with uniform angular speed $\dot{\theta} = \omega_0$. We shall use the formulas (2-55) and (2-59) to obtain, first, the velocity and acceleration of the crankpin C and then, with a new choice of intermediate frame, the velocity and acceleration of P. This procedure illustrates how the basic formulas (2-55) and (2-59) can be systematically applied to trace velocities and accelerations throughout mechanisms of arbitrary complexity.

To obtain the velocity and acceleration of the crankpin we select an intermediate frame $oxyz$ fixed to the crank OC as sketched in Fig. 2-24,

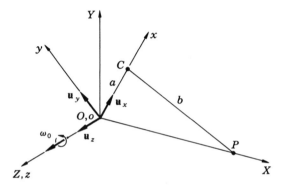

FIG. 2-24. Intermediate frame $oxyz$ for obtaining velocity and acceleration of crankpin C.

where, in preparation for applying (2-55) and (2-59), we recognize

$$\mathbf{R}_o = 0 \qquad \mathbf{r} = a\mathbf{u}_x \qquad \boldsymbol{\omega} = \omega_0\mathbf{u}_z. \qquad (2\text{-}79)$$

Since $\mathbf{v}_{rel}$, $\mathbf{a}_{rel}$, and $\dot{\boldsymbol{\omega}}$ are zero, the velocity and acceleration of C are given by (2-55) and (2-59) as

$$\mathbf{v}_C = \boldsymbol{\omega} \times \mathbf{r} = a\omega_0\mathbf{u}_y$$
$$\mathbf{a}_C = \boldsymbol{\omega} \times (\boldsymbol{\omega} \times \mathbf{r}) = -a\omega_0^2\mathbf{u}_x, \qquad (2\text{-}80)$$

as indicated in Fig. 2-25.

To proceed from the crankpin to the piston, we repeat the preceding step, but now we select an intermediate frame $Cx'y'z'$ fixed to the connecting rod CP as shown in Fig. 2-26. The velocity and acceleration of the origin C of this frame have already been determined in (2-80). The magnitudes of the angular velocity ω_1 and the angular acceleration $\dot{\omega}_1$ of this frame are initially unknown. We shall, however, be able to determine them in the process of finding the velocity and acceleration of P by utilizing the fact that P is constrained to remain on the axis OX.

To apply (2-55) and (2-59) to Fig. 2-26, we make the following identifications:

$$\mathbf{R}_o = a\mathbf{u}_x \quad \text{and} \quad \mathbf{r} = b\mathbf{u}_{z'}. \tag{2-81}$$

Then, since $\mathbf{v}_{\text{rel}}$ is zero, the velocity of P according to (2–55) is

$$\mathbf{v}_P = \mathbf{v}_C + \boldsymbol{\omega}_1 \times \mathbf{r} = a\omega_0\mathbf{u}_y + b\omega_1\mathbf{u}_{y'}. \tag{2-82}$$

In writing (2–82) we have used the fact that the orientation of ω_1 is as indicated in Fig. 2–26, although the scalar magnitude ω_1 is unknown. Similarly, $\mathbf{v}_P$ is not completely unknown since its orientation must be

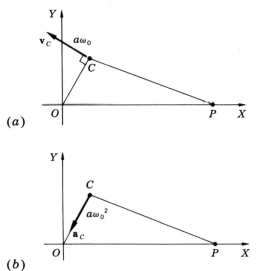

(a)

(b)

FIG. 2–25.   (a) Crankpin velocity and (b) acceleration.

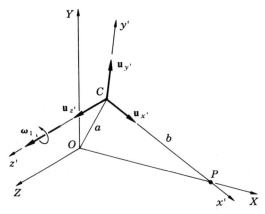

FIG. 2–26.   Intermediate frame $Cx'y'z'$ for obtaining velocity and acceleration of P.

along the line OX; i.e., we can set $\mathbf{v}_P = v_P\mathbf{u}_X$ with only the scalar magnitude v_P unknown. It is possible to solve for both scalar magnitudes ω_1 and v_P from the single vector equation (2–82) either analytically or graphically. A graphical solution is indicated in Fig. 2–27. The vector $\mathbf{v}_C$ is erected from point P, and a dotted line parallel to $\mathbf{u}_{y'}$ is drawn through its tip. The intersection of the dotted line with OX determines $\mathbf{v}_P$. The lower vector triangle in Fig. 2–27 is simply a graphical representation of

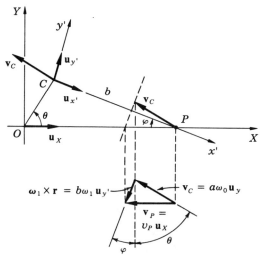

FIG. 2–27. Graphical construction for velocity of
P based on Eq. (2–82).

(2–82). An analytical solution can be obtained by writing the components of (2–82) parallel to OX and to OY, respectively,

$$v_P = -a\omega_0 \sin\theta + b\omega_1 \sin\varphi$$
$$0 = a\omega_0 \cos\theta + b\omega_1 \cos\varphi,$$

(2–83)

and solving for ω_1 and v_P. After eliminating φ from the results (by using the geometrical relation $a \sin\theta = b \sin\varphi$), we find

$$\omega_1 = -\omega_0 \frac{\cos\theta}{\sqrt{b^2/a^2 - \sin^2\theta}}$$
$$v_P = -a\omega_0 \sin\theta \left(1 + \frac{\cos\theta}{\sqrt{b^2/a^2 - \sin^2\theta}}\right).$$

(2–84)

The manner in which the piston velocity v_P varies with crank angle θ is shown in Fig. 2–28 for several values of the ratio b/a. In order to permit complete revolutions of the crank, the connecting-rod radius b must be greater than the crank throw a. Note that for very long connecting rods

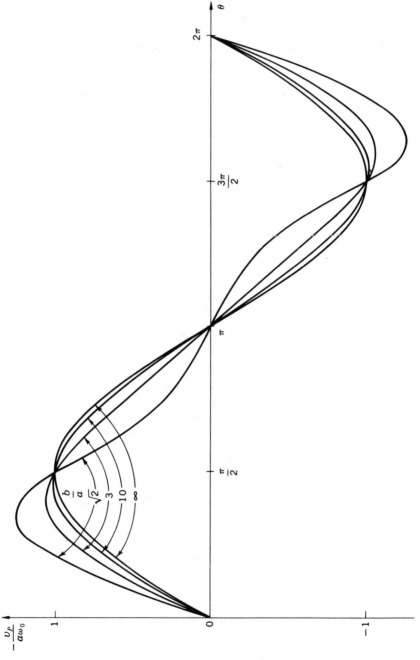

Fɪɢ. 2–28.   Piston velocity v_P as a function of crank angle θ.

the velocity of the piston is almost sinusoidal, but that for short connecting rods the piston motion deviates considerably from simple harmonic motion.

Returning to the acceleration of P in Fig. 2-26 and noting that $\mathbf{a}_{rel}$ is zero, we apply (2-59) to find

$$\mathbf{a}_P = \mathbf{a}_C + \dot{\boldsymbol{\omega}}_1 \times \mathbf{r} + \boldsymbol{\omega}_1 \times (\boldsymbol{\omega}_1 \times \mathbf{r})$$

$$a_P \mathbf{u}_X = -a\omega_0^2 \mathbf{u}_x + b\dot{\omega}_1 \mathbf{u}_{y'} - b\omega_1^2 \mathbf{u}_{x'}. \tag{2-85}$$

Here the magnitude $\dot{\omega}_1$ of the angular acceleration of the rod and the magnitude a_P of the piston acceleration are unknown, but as in the case

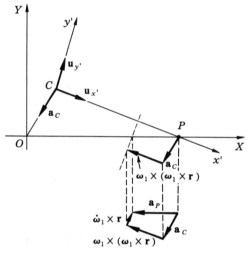

FIG. 2-29. Graphical construction for acceleration of P based on Eq. (2-85).

of the velocities, two scalar unknowns can be found from the single vector equation. A graphical solution is indicated in Fig. 2-29. The acceleration $\mathbf{a}_C$ of (2-80) is erected at P, and the centripetal acceleration $\boldsymbol{\omega}_1 \times (\boldsymbol{\omega}_1 \times \mathbf{r})$, computed from the previously obtained $\boldsymbol{\omega}_1$, is added tail to head. Then a dotted line parallel to $\mathbf{u}_{y'}$ is drawn through the sum of $\mathbf{a}_C$ and $\boldsymbol{\omega}_1 \times (\boldsymbol{\omega}_1 \times \mathbf{r})$. The intersection of the dotted line and OX simultaneously fixes the magnitude of $\mathbf{a}_P$ and the angular-acceleration term $\dot{\boldsymbol{\omega}}_1 \times \mathbf{r}$. The lower vector polygon in Fig. 2-29 is a graphical representation of Eq. (2-85). Analytical expressions for the magnitudes of $\mathbf{a}_P$ and ω_1 can also be obtained, but we omit the details here[1] and simply show the resulting variation of a_P with crank angle for various values of the ratio b/a in Fig. 2-30.

[1] See Prob. 2-8.

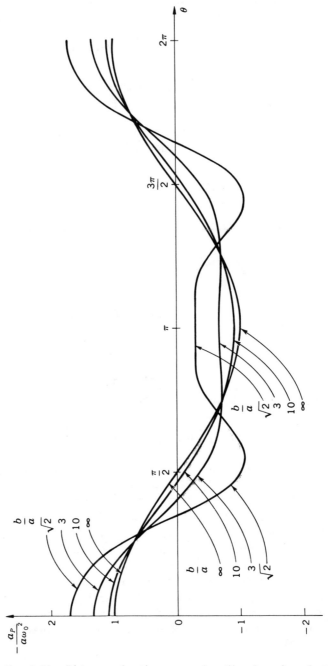

FIG. 2–30.   Piston acceleration a_P as a function of crank angle θ.

2–5   Newtonian dynamics

In this section we review the concepts of momentum and force and their interrelationship within the framework of classical newtonian particle dynamics.   We begin with the concept of a *particle* as a point endowed with *mass m* and *charge q*.   In classical mechanics a particle preserves its identity, and its mass does not vary with time or with motion.   The *linear momentum* **p** of a particle in newtonian dynamics is the product of the mass of the particle and the velocity of the particle,

$$\mathbf{p} = m\mathbf{v}. \tag{2–86}$$

Since the velocity of the particle depends on the reference frame in use, the particle's momentum also depends on the reference frame.

PROPERTIES OF FORCE

Interactions between particles are conveniently described in terms of *forces*.   For example, an attraction between particles m_1 and m_2 is shown

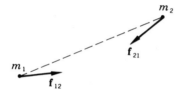

FIG. 2–31.   Interaction forces.
Action of m_2 upon m_1 is represented by $\mathbf{f}_{12}$.   Action of m_1 on m_2 is represented by $\mathbf{f}_{21}$.

in Fig. 2–31.   The force of m_2 acting on m_1 is indicated by the vector $\mathbf{f}_{12}$, and the force of m_1 acting on m_2 is indicated by the vector $\mathbf{f}_{21}$.   Force is an extremely useful concept[1] in engineering, although its existence must be inferred indirectly from observable effects such as the deformation of a spring or change of motion of a mass.   In many applications involving a force interaction such as that depicted in Fig. 2–31, we confine our analysis to one of the particles, say, m_1.   In this case the effect of the external particle m_2 is represented by the vector $\mathbf{f}_{12}$ acting on m_1.   This implies that the observable effects produced by the interaction can be associated with a unique magnitude and vector direction.   When a given particle m interacts with more than one external particle, it is a funda-

[1] For a discussion of the development of the concept of force from the philosophical point of view, see M. Jammer, "Concepts of Force: A Study in the Foundations of Dynamics," Harvard University Press, Cambridge, Mass., 1957, reprinted by Harper & Row, Publishers, New York, 1962.

mental postulate based on observation that the effect of the several inter-action forces acting simultaneously on m is the same as that of a single force which is the *vector sum* of the individual forces acting on m; i.e., forces are compounded according to the laws of *vector addition*.

The fundamental types of force presently recognized in physics are gravitational forces, electromagnetic forces, and nuclear forces. Complete descriptions of these interactions require relativistic and quantum-mechanical frameworks. For applications to macroscopic systems with speeds which are negligible compared with the speed of light, the simpler descriptions of classical mechanics are entirely adequate. In classical mechanics force interactions are divided into forces *which act at a distance* and *contact* forces. The macroscopic forces which act at a distance are those due to gravity and electromagnetism. Of these the gravitational and electrostatic interactions are the simplest. For two interacting particles the situation is as shown in Fig. 2–32. The interaction forces

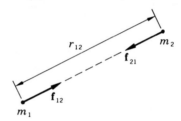

Fig. 2–32. Gravitational and electrostatic interactions involve forces which satisfy $\mathbf{f}_{21} = -\mathbf{f}_{12}$ and which are central.

$\mathbf{f}_{12}$ and $\mathbf{f}_{21}$ satisfy Newton's third law (action is equal and opposite to reaction); i.e., $\mathbf{f}_{21} = -\mathbf{f}_{12}$ at all times. This implies that there is no time lag associated with action at a distance. Furthermore, for gravitational and electrostatic interactions the forces are *central*, which means that both $\mathbf{f}_{12}$ and $\mathbf{f}_{21}$ act along the straight line joining m_1 and m_2. In addition, the magnitudes of the forces satisfy an *inverse square law*, which means that the magnitude f_{12} varies inversely with the square of the distance r_{12}.

FORCE FIELDS

When more than two particles interact, the net force acting on a particular particle, such as m in Fig. 2–33, could be described by giving the locations of all the external interacting particles and by application of the individual force laws. An alternative description is possible if the effect of m on changing the motions of the external particles can be neglected. Then it is possible to describe in advance the *force field* $\mathbf{f} = \mathbf{f}(\mathbf{R},t)$, which is simply a catalog of the values of net force which m will experience when it occupies any position **R** at any time t within a certain range.   As far as the analysis of m alone is concerned, it is much simpler to have the force-field description.   All the details of the motions of the external particles and of the individual force laws are absorbed in the construction of the force field.

For *gravitational* and *electrostatic* interactions it is sufficient to employ a single field which depends on time and on the position **R** of the particle m which experiences the force.   For *electromagnetic* interactions a *moving* particle m experiences a force which depends on time and on both its position **R** and its velocity **v**.   Here the only convenient macroscopic description is in terms of a field description.   Two fields are required:

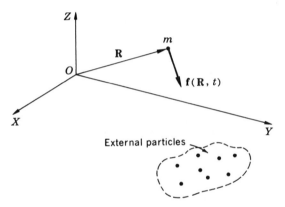

Fig. 2–33.   A *force field* **f**($\mathbf{R}$,t) is a specification of the force **f** that will be experienced by particle m when it occupies position **R** at time t.

the electric field **E**($\mathbf{R}$,t) and the magnetic induction **B**($\mathbf{R}$,t).   The force **f** experienced by the particle m with charge q at location **R** and traveling with velocity **v** is then given by the Lorentz force law,

$$\mathbf{f} = q(\mathbf{E} + \mathbf{v} \times \mathbf{B}). \qquad (2\text{–}87)$$

The field values **E** and **B** are obtained from electromagnetic field theory and are presumed to depend only on the environment external to m.

CONTACT FORCES

Force interactions which can be localized at the points of contact of interacting bodies are called *contact* forces.   They include the pressure and friction forces of bodies in contact, and also the shear and normal forces within a particular body.   These are macroscopically observable manifestations of enormous numbers of individual microscopic interactions (principally electromagnetic). It is quite possible that some day we shall be able to predict the nature of contact forces entirely from considerations of the microscopic interactions, but this still remains a gap in physical theory. In engineering we deal with a great many different types of contact forces, e.g., friction, elastic, plastic, viscous, and viscoelastic forces. In each of these cases phenomenological behavior has been established on the basis of experiment and macroscopic theories. In most cases contact interactions can be represented by equal and opposite resultant-force vectors at the point of contact, as indicated in Fig. 2–34. The magnitude and direction of the resultant-force vectors can often be deduced from the macroscopic behavior of the systems in contact.

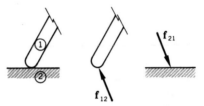

FIG. 2–34. Contact interaction involves an enormous number of microscopic interactions but can be represented on a macroscopic level by resultant forces $\mathbf{f}_{12}$ and $\mathbf{f}_{21}$ with $\mathbf{f}_{12} = -\mathbf{f}_{21}.$

NEWTON'S SECOND LAW

Consider a single mass particle m subjected to a resultant force $\mathbf{f}$. It is assumed that both m and $\mathbf{f}$ are independent of the reference frame used to observe the motion of the particle. The velocity and acceleration of the particle do, however, depend on the reference frame selected. The momentum $\mathbf{p}$ of (2–86) also depends on the choice of reference frame. Newton's second law of motion is equivalent to the assertion that there exists at least one reference frame such that in this frame *the force vector is always equal to the time rate of change of the momentum vector;* i.e.,

$$\mathbf{f} = \frac{d\mathbf{p}}{dt}. \tag{2–88}$$

An alternative statement obtained by introducing (2–86) and (2–2) is

$$\mathbf{f} = m\mathbf{a}, \tag{2–89}$$

or *the force vector is always equal to the mass times the acceleration vector.*

INERTIAL REFERENCE FRAMES

In connection with (2–59) we noted that accelerations observed from two frames in *uniform* relative velocity are identical. This means that if there does exist one reference frame in which (2–88) and (2–89) are always valid, then these relations are also valid in any other frame which is in uniform translation with respect to this one frame. Any such frame in which the relations (2–88) and (2–89) are valid is called an *inertial frame* because in these frames, when the force **f** vanishes, the particle's momentum (and velocity) remain unchanged.

A true inertial frame is an idealized concept. In practice, the choice of a reference frame for dynamics is governed by the scale of the system and by the accuracy required. In celestial mechanics remarkably precise predictions are obtained within the solar system under the assumptions that a reference frame with origin at the center of the solar system (very nearly at the center of the sun) and which does not rotate with respect to the so-called fixed stars is actually an inertial frame. Such a frame is probably not, however, a true inertial frame because of the acceleration of the solar system within our galaxy. A reference frame fixed in the earth is even less of an inertial frame because of the acceleration of the earth's center with respect to the solar system and because of the daily rotation of the earth. Nevertheless, for many engineering problems involving relatively short distances and short times and for which only modest precision is required, it is quite adequate to assume that a frame fixed in the earth is actually an inertial frame. A reference frame fixed in a vehicle such as a ship at sea is even further from a true inertial frame. Nevertheless, for the limited purpose of interpreting local high-level shock and vibration measurements, it is often adequate to treat a frame fixed in the vehicle as an inertial reference frame. In the remainder of this book we shall assume, unless specific mention is made to the contrary, that the reference frames employed may be taken as inertial frames with adequate accuracy.

UNITS

There is a wide variety of sets of units in which Newton's law (2–89) can be expressed. Several possibilities are indicated in Table 2–1. The standard of mass is a platinum *kilogram* (kg) kept by the International Bureau of Weights and Measures. The *pound mass* (lbm) is defined to be 0.453592 kg. The force units newton and poundal are defined implicitly by Newton's law. The *newton* is the force required to give one kilogram of mass an acceleration of one meter per second per second. The *poundal* is the force required to give one pound mass an acceleration of one foot per second per second. The *pound force* (lbf) is defined to be the gravitational force exerted on a pound mass at a location where the acceleration of gravity is 32.1740 ft/sec². One pound force is equivalent to 4.4482 newtons. The *kilogram force* (kgf) is defined to be the gravitational force exerted on a kilogram mass at a location where the acceleration of gravity is 9.80665 m/sec². The *slug* is a mass unit defined by Newton's law in terms of the pound force. It is the mass which, when acted on by one pound of force, accelerates with an acceleration of one foot per second per second. A slug has the dimensions of lbf-sec²/ft and is equivalent to 32.1740 lbm. The corresponding metric unit (undesignated) has the dimensions of kgf-sec²/m and is equivalent to 9.80665 kg. The accelera-

Table 2–1
Systems of units for Newton's second law

Force unit	poundal	newton	pound force	kilogram force	pound force	kilogram force
Mass unit	pound mass	kilogram mass	slug	$\dfrac{\text{kgf-sec}^2}{\text{m}}$	pound mass	kilogram mass
Acceleration unit	ft/sec²	m/sec²	ft/sec²	m/sec²	g	g

tion unit g is the standardized sea-level acceleration, equivalent to 32.1740 ft/sec², or 9.80665 m/sec². It has the dimensions of lbf/lbm or kgf/kgm.

ANGULAR MOMENTUM FOR A PARTICLE

A useful corollary to the force-momentum relation is obtained if we take moments of both sides of (2–88) with respect to a point B. We shall see that a simple result is obtained for a special class of points B. At the outset, however, let us consider that point B in Fig. 2–35 is an arbitrary point with position vector $\mathbf{R}_B$ and velocity $\mathbf{v}_B = d\mathbf{R}_B/dt$. The particle m in Fig. 2–35 has momentum $\mathbf{p}$ with respect to the inertial frame $OXYZ$ and is acted on by the resultant force $\mathbf{f}$.

Now taking moments of (2–88) with respect to B, we cross multiply both sides by the vector $\mathbf{r}$.

$$\mathbf{r} \times \mathbf{f} = \mathbf{r} \times \frac{d\mathbf{p}}{dt}. \tag{2-90}$$

To develop the right-hand side we utilize an identity which follows from the differentiation law for a product.

$$\mathbf{r} \times \mathbf{f} = \frac{d}{dt}(\mathbf{r} \times \mathbf{p}) - \frac{d\mathbf{r}}{dt} \times \mathbf{p}$$

$$= \frac{d}{dt}(\mathbf{r} \times \mathbf{p}) - (\mathbf{v} - \mathbf{v}_B) \times \mathbf{p}. \tag{2-91}$$

The term $\mathbf{v} \times \mathbf{p}$ on the right is zero because the particle's velocity and momentum are always parallel. The term $\mathbf{v}_B \times \mathbf{p}$ is not always zero, but it will be zero if

1. B is *fixed* in the inertial frame; i.e., $\mathbf{v}_B = 0$.
2. B moves so that its velocity is always *parallel* to the velocity of the particle m.

In either of these cases the result of (2–91) takes an especially simple form. The quantity $\mathbf{r} \times \mathbf{f}$ is called the *torque*, or *moment*, of the force $\mathbf{f}$ about B and is denoted by the symbol $\boldsymbol{\tau}_B$. The quantity $\mathbf{r} \times \mathbf{p}$ is called the *angular momentum*, or *moment of momentum*, of the particle about B and

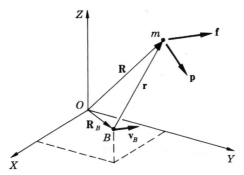

Fig. 2–35. Particle with momentum $\mathbf{p}$ subjected to force $\mathbf{f}$. Point B is moment center.

is denoted by the symbol $\mathbf{h}_B$. In terms of these, (2–91) takes the form

$$\boldsymbol{\tau}_B = \frac{d\mathbf{h}_B}{dt} \qquad (2\text{--}92)$$

whenever B is either a fixed point or a point which moves parallel to the particle. *The torque applied to a particle equals the time rate of change of the particle's angular momentum.*

Example 2–5. Consider the particle P of mass m shown in Fig. 2–36, attached to an inextensible but flexible and massless string that passes through a small hole at O. Except for the restraint provided by the string, the particle moves freely without friction in the XY plane. Initially, the string length OP is fixed and the particle moves in a circle of radius r_1 with peripheral speed v_1. Subsequently the string is pulled down through the hole, causing the particle to spiral inward until the string length is again fixed, with the particle moving in a smaller circle of radius r_2. The problem is to determine the peripheral speed v_2 with which the particle traverses the circle of radius r_2.

We begin the analysis by considering a general instant during the inward spiral in Fig. 2–37 when the particle is at radius r and angle θ. The only force in the XY plane acting on the particle is the string tension **f** shown in Fig. 2–37a.   The magnitude of the force is unknown, but it is directed toward the hole at O.   This means that there is no *torque* (about the fixed point O) exerted on the particle.   Therefore, according to (2–92),

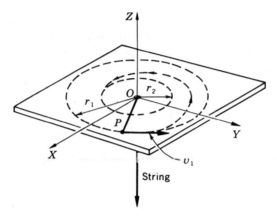

FIG. 2–36.   Example 2–5.   Particle P moves in circle of radius r_1 with speed v_1 until string is pulled down through hole at O.

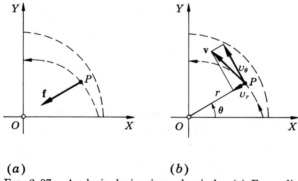

(*a*)   (*b*)

FIG. 2–37.   Analysis during inward spiral.   (*a*) Force diagram.   (*b*) Velocity diagram.

the angular momentum of the particle (with respect to O) must *remain constant* during the motion.

In Fig. 2–37b the velocity **v** at this same instant is shown tangent to the spiral.   The components of **v** are v_r along the radius and v_θ perpendicular to the radius.   The corresponding components of linear momentum are mv_r and mv_θ.   Note that the radial component does not contribute to the angular momentum with respect to O.   If $\mathbf{u}_z$ is the unit vector along OZ, the angular momentum about O is, simply,

$$\mathbf{h}_o = \mathbf{u}_z m r v_\theta. \tag{2-93}$$

For this to remain constant during the motion, the product $r v_\theta$ must remain constant, and in particular we must have

$$r_2 v_2 = r_1 v_1, \tag{2-94}$$

which is the desired result. If the radius is halved, the peripheral speed is doubled independently of the precise spiral followed when the radius is changed.

2–6   Kinetic state functions

In Sec. 1–3 the kinetic state functions T and T^* are defined for the special case of a particle moving along a straight line. The kinetic state of such a particle is fixed by giving either its momentum p or its velocity v. The *kinetic energy* $T(p)$ is the work done on the particle in increasing its momentum from 0 to p. The *kinetic coenergy* $T^*(v)$ is the Legendre transformation of $T(p)$. For the velocity-momentum relation (1–64) of special relativity, the complementary functions T and T^* are generally unequal in magnitude. For the newtonian velocity-momentum relation (1–62), the complementary functions are equal in magnitude, although they have different arguments.

The extension of these ideas to general three-dimensional motion is straightforward. The work done on a particle of mass m in giving it a momentum $\mathbf{p}$ with respect to an inertial reference frame is the *kinetic energy*,

$$T(\mathbf{p}) = \int_0^{\mathbf{p}} \mathbf{v} \cdot d\mathbf{p}, \tag{2-95}$$

independently of the velocity-momentum relation. For the newtonian relation of (2–86),

$$T = \int_0^{\mathbf{p}} \frac{1}{m} \mathbf{p} \cdot d\mathbf{p} = \frac{1}{2m} \mathbf{p} \cdot \mathbf{p} = \frac{p^2}{2m}. \tag{2-96}$$

The corresponding complementary function is the *kinetic coenergy*,

$$T^*(\mathbf{v}) = \mathbf{v} \cdot \mathbf{p} - T(\mathbf{p}) = \int_0^{\mathbf{v}} \mathbf{p} \cdot d\mathbf{v}, \tag{2-97}$$

independently of the velocity-momentum relation. For the newtonian relation of (2–86),

$$T^* = \tfrac{1}{2} m \mathbf{v} \cdot \mathbf{v} = \tfrac{1}{2} m v^2. \tag{2-98}$$

The above definitions apply to a single particle. For a system of

N particles the system kinetic state functions are defined to be the sums of the corresponding state functions of the individual particles.

$$T = \sum_{i=1}^{N} T_i$$
$$T^* = \sum_{i=1}^{N} T_i^*. \tag{2-99}$$

These system functions are related by the Legendre transformation

$$T^* = \sum_{i=1}^{N} \mathbf{v}_i \cdot \mathbf{p}_i - T, \tag{2-100}$$

where $\mathbf{v}_i$ and $\mathbf{p}_i$ are, respectively, the velocity and momentum of the ith particle.

The manner in which the kinetic coenergy T^* enters Hamilton's principle is described in Sec. 1–4. This application is developed at some length in Chaps. 2 to 4. The kinetic energy T appears here only in connection with conservation-of-energy statements. In Appendix D a complementary variational principle for dynamics is described in which the kinetic energy T enters the variational principle directly.

Example 2–6. To illustrate the evaluation of the kinetic coenergy of a simple system, we consider the flyball governor shown in Fig. 2–38. We assume, for simplicity, that the only significant mass elements in the system are the two flyballs, which we shall treat as particles[1] of mass m. The massless linkage constrains the flyballs to remain in the xz plane but does allow alterations in the angle θ. The kinetic state of the system is fixed by the velocities $\mathbf{v}_1$ and $\mathbf{v}_2$ of the masses. The kinetic coenergy T^* of the system is, according to (2–98) and (2–99),

$$T^* = \tfrac{1}{2}mv_1^2 + \tfrac{1}{2}mv_2^2. \tag{2-101}$$

This is a correct statement, but it may not be very useful until the velocity magnitudes v_1 and v_2 are expressed in terms of the significant geometric parameters of the system. In the present system these velocities are related to the rotational speed ω of the xyz frame with respect to the inertial frame XYZ and the angle θ of the linkage. To determine

[1] An extended mass can usually be represented in a model by a particle when the size of the mass is small in comparison with the radius of curvature of its trajectory. In Sec. 2–13 it is shown that there is always one point in an extended system of mass which has exactly the same trajectory as a single particle would have if the actual mass system were replaced by a single particle with the same total mass as that of the system. A single-particle model is, however, unable to represent the motions within an extended mass system. These motions can include rotations, expansions or contractions, vibrations, wave disturbances, etc. Models which can represent these motions are introduced in Chaps. 3 and 7.

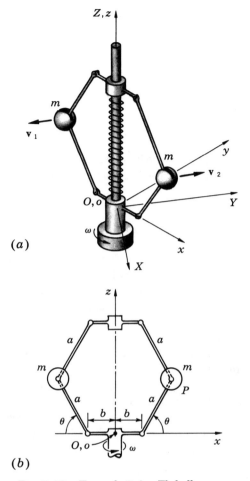

(a)

(b)

Fɪɢ. 2–38. Example 2–6. Flyball governor.

the velocity v_2 of the point labeled P in Fig. 2–38, we first write the position vector $\mathbf{R}_2$ of the point P with respect to 0.

$$\mathbf{R}_2 = (b + a\cos\theta)\mathbf{u}_x + a\sin\theta\,\mathbf{u}_z. \qquad (2\text{--}102)$$

The velocity v_2 is the time derivative of $\mathbf{R}_2$. In order to evaluate this derivative, we need the time derivatives of the unit vectors $\mathbf{u}_x$ and $\mathbf{u}_z$. Since the xyz frame has the angular velocity $\omega\mathbf{u}_z$, we have

$$\frac{d\mathbf{u}_x}{dt} = \omega \times \mathbf{u}_x = \omega\mathbf{u}_y$$

$$\frac{d\mathbf{u}_z}{dt} = \omega \times \mathbf{u}_z = 0, \qquad (2\text{--}103)$$

according to (2–43). Differentiating (2–102) with the aid of (2–103), we find

$$\mathbf{v}_2 = -\mathbf{u}_x a\dot{\theta} \sin \theta + \mathbf{u}_y(b + a \cos \theta)\omega + \mathbf{u}_z a\dot{\theta} \cos \theta \quad (2\text{–}104)$$

and

$$\mathbf{v}_2 \cdot \mathbf{v}_2 = v_2^2 = a^2\dot{\theta}^2 + (b + a \cos \theta)^2\omega^2. \quad (2\text{–}105)$$

Because of the symmetry of the linkage, $v_1^2 = v_2^2$, and thus (2–101) becomes

$$T^* = m[a^2\dot{\theta}^2 + (b + a \cos \theta)^2\omega^2]. \quad (2\text{–}106)$$

We shall make use of this relation in Example 2–10, where the equations of motion for the flyball governor are obtained by application of Hamilton's principle.

2–7 Work and potential energy

The *work increment* done by a force $\mathbf{f}_i$ which acts on a particle m_i when the particle undergoes an infinitesimal displacement $\delta\mathbf{R}_i$ is defined to be $\mathbf{f}_i \cdot \delta\mathbf{R}_i$. In the variational calculus associated with Hamilton's principle, the displacement $\delta\mathbf{R}_i$ is a displacement between two admissible trajectories which are under comparison at the same instant of time.   When implementing Hamilton's principle, it is necessary to account for the work increments of all the forces which act on all the mass particles within the system.   In Chap. 1 it was noted that for some forces it is possible to account for their work increments in terms of *potential-energy* changes.   We shall discuss this topic in greater detail in this section, where the following types of forces are considered: forces described by static vector fields, and forces exerted by idealized two-force elements such as springs and dashpots.

FORCE FIELDS

A force field is a region (usually three-dimensional) in which a suitable probe experiences a well-defined force $\mathbf{f}(\mathbf{R},\mathbf{v},t)$, which in general depends on the probe's position $\mathbf{R}$, on its velocity $\mathbf{v}$, and on the time t, as well as on certain other properties of the probe, such as its mass and electrical charge.   A good example is the electromagnetic-force field (2–87).   In this section we restrict our consideration to fields which, for a given probe, depend only on the probe's position; i.e., we consider only fields of the form $\mathbf{f}(\mathbf{R})$.

In Fig. 2–39 we show a mass particle m acting as a probe in a force field $\mathbf{f}(\mathbf{R})$. This means that when the particle is at the position $\mathbf{R}$, the force exerted *on* the particle *by* the field is $\mathbf{f}(\mathbf{R})$.   When Hamilton's principle is applied to a system which contains this particle, the work increment corresponding to the force exerted by the field is

$$\mathbf{f} \cdot \delta \mathbf{R}. \tag{2-107}$$

To study the properties of the field we examine the behavior of the integral

$$\int_{\mathbf{R}_0}^{\mathbf{R}} \mathbf{f} \cdot d\mathbf{R}, \tag{2-108}$$

which represents the net work done on the probe by the field as the probe is moved from a datum position $\mathbf{R}_0$ to a terminal position $\mathbf{R}$. For

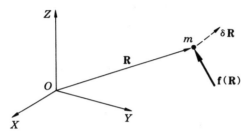

Fig. 2–39.   Force field $\mathbf{f}(\mathbf{R})$ acting on probe with mass m.

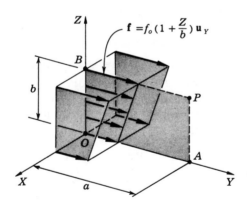

Fig. 2–40.   Force field which is uniquely determined by position of probe but for which the work integral depends on the path.

some force fields the integral (2–108) is always independent of the particular path followed in going from $\mathbf{R}_0$ to $\mathbf{R}$.   For other force fields it is not.   As an example of the latter type, consider the hypothetical field shown in Fig. 2–40, where the force field is always parallel to the Y axis and the force magnitude is independent of X and Y but varies linearly with Z.   Let the datum point $\mathbf{R}_0$ be the origin in Fig. 2–40.   Then the integral (2–108) has different values at the point P for the paths OAP and OBP (f_0a for OAP and $2f_0a$ for OBP).

If, on the other hand, the work integral (2–108) has the same value for every path connecting $\mathbf{R}_0$ and $\mathbf{R}$, then the net work done around any closed circuit which starts at $\mathbf{R}_0$, passes through $\mathbf{R}$, and returns to $\mathbf{R}_0$ will be zero. Conversely, if the net work done around any closed path in a region is zero, the work integral (2–107) from $\mathbf{R}_0$ to $\mathbf{R}$ cannot be different for two different paths (for if different results were obtained for paths 1 and 2, the integral around the closed path consisting of path 1 and the reverse of path 2 would not vanish). A force field $\mathbf{f}(\mathbf{R})$ is said to be *conservative* in a region if the net work done around *any* closed path in the region is zero.

A simple differential condition which a conservative field must satisfy can be obtained by using Stokes' theorem[1] to relate the work integral around a closed curve C to a surface integral over an arbitrary surface S bounded by C.

$$\oint_C \mathbf{f} \cdot d\mathbf{R} = \iint_S (\nabla \times \mathbf{f}) \cdot \mathbf{n} \, dS. \qquad (2\text{–}109)$$

In (2–109) dS is the surface element and $\mathbf{n}$ is the unit normal to the surface. If the field $\mathbf{f}(\mathbf{R})$ is conservative, (2–109) vanishes for every closed curve C, and hence for every surface S in the region. This implies that

$$\nabla \times \mathbf{f} = \text{curl } \mathbf{f} = 0 \qquad (2\text{–}110)$$

throughout the region. Conversely, if $\nabla \times \mathbf{f}$ vanishes in a region, so also does the line integral around any closed curve in the region. Thus a necessary and sufficient condition for a differentiable field to be conservative is that its curl vanish. Note that the force field of Fig. 2–40 violates (2–110). In this case $\mathbf{f} = f_0(1 + Z/b)\mathbf{u}_Y$ and

$$\nabla \times \mathbf{f} = \begin{vmatrix} \mathbf{u}_X & \mathbf{u}_Y & \mathbf{u}_Z \\ \dfrac{\partial}{\partial X} & \dfrac{\partial}{\partial Y} & \dfrac{\partial}{\partial Z} \\ 0 & f_0\left(1 + \dfrac{Z}{b}\right) & 0 \end{vmatrix} = -\frac{f_0}{b}\mathbf{u}_X. \qquad (2\text{–}111)$$

POTENTIAL ENERGY OF A FORCE FIELD

For a given probe in a force field $\mathbf{f}(\mathbf{R})$, it is only the probe's position $\mathbf{R}$ which determines the force experienced. In this context we say that the *state* of the probe is fixed by its position. For a conservative force field $\mathbf{f}(\mathbf{R})$, the work integral (2–108) is independent of path. For a fixed datum position $\mathbf{R}_0$, the integral depends only on the present position of

[1] See Eq. (C–22) in Appendix C.

the probe; i.e., the integral (2–108) is a *state function.* The state function most commonly used in connection with conservative force fields is the *potential energy*

$$V(\mathbf{R}) = - \int_{\mathbf{R}_0}^{\mathbf{R}} \mathbf{f} \cdot d\mathbf{R}, \tag{2–112}$$

which is the negative[1] of (2–108). Note that V is a scalar function of the state (i.e., the position $\mathbf{R}$) of the probe. The properties of the potential energy are that its negative gradient yields the force exerted by the force field on the probe,

$$-\boldsymbol{\nabla} V = \mathbf{f}, \tag{2–113}$$

and that the negative of its variation gives the work increment (2–107),

$$-\delta V = \mathbf{f} \cdot \delta \mathbf{R}. \tag{2–114}$$

Because of (2–114), when we apply Hamilton's principle we can account

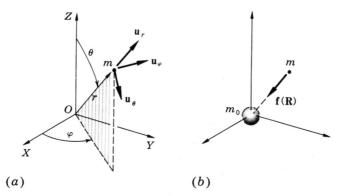

FIG. 2–41. (a) Spherical coordinates and (b) a central force field, in which a probe with mass m experiences an attractive force $\mathbf{f}$ toward the origin.

for the work increments of conservative forces acting on mass particles by using potential-energy decrements. The value of the potential energy (2–112) depends on the selection of the datum location $\mathbf{R}_0$, but the relations (2–113) and (2–114) are independent of the choice of datum.

Example 2–7. Gravity field. Consider the inverse-square central force field shown in Fig. 2–41b, with a probe of mass m. In terms of the spherical coordinates r, θ, φ shown in Fig. 2–41a, the position vector is $\mathbf{R} = r\mathbf{u}_r$ and the gravitational force attracting the probe toward the

[1] The sign of the potential energy is chosen so that when the field does positive work on the probe there is a *decrease* in the potential energy of the probe in the field.

origin is

$$\mathbf{f} = -\frac{Km_0 m}{r^2}\,\mathbf{u}_r, \qquad (2\text{--}115)$$

where K is the gravitational constant,[1] and m_0 is the mass of the attracting body at the origin.   It can be shown[2] that any central force field in which the magnitude of the force is a single-valued function of r alone is conservative, and hence that a potential energy (2–112) exists.   Taking the datum position $\mathbf{R}_0$ at infinite radius, inserting (2–115) into (2–112), and integrating along a radius, we obtain

$$V(r) = -\int_\infty^r \left(-\frac{Km_0 m}{r^2}\,\mathbf{u}_r\right)\cdot\mathbf{u}_r\,dr$$

$$= -\frac{Km_0 m}{r} \qquad (2\text{--}116)$$

for the potential energy of a probe of mass m in the gravity field.   This result is valid for all r if the attracting body at the origin is a particle. If the attracting body is a sphere of finite radius (e.g., the earth), then (2–115) and (2–116) are still valid[3] as long as the probe remains *outside* of the sphere.   The force (2–115) can be recovered from (2–116) by using (2–113) and (C–18).

$$\mathbf{f} = -\nabla V = -\mathbf{u}_r\frac{\partial}{\partial r}\left(-\frac{Km_0 m}{r}\right) = -\frac{Km_0 m}{r^2}\,\mathbf{u}_r. \qquad (2\text{--}117)$$

The potential energy (2–116) is sketched as a function of r in Fig. 2–42.

When we restrict our range of application of (2–116) to the near neighborhood of a sphere of radius r_0 (e.g., the surface of the earth), the following approximation is very useful.   Let $r = r_0 + z$, as indicated in Fig. 2–43.   Then we can expand (2–116) in powers of z/r_0 as follows:

$$V(r_0 + z) = -\frac{Km_0 m}{r_0}\,\frac{1}{1 + z/r_0}$$

$$= V(r_0)\left(1 - \frac{z}{r_0} + \frac{z^2}{r_0^2} - \cdots\right), \qquad (2\text{--}118)$$

from which we obtain the linearized approximation

$$V(r_0 + z) - V(r_0) = \frac{Km_0}{r_0^2}\,mz \qquad (2\text{--}119)$$

by neglecting terms of order $(z/r_0)^2$.

[1] $K = 6.670 \times 10^{-11}$ newton-m^2/kg^2, or 3.321×10^{-11} lbf-ft^2/lbm^2.   If m_0 represents the mass of the earth, the product Km_0 is 3.991×10^{14} newton-m^2/kg, or 4.380×10^{14} lbf-ft^2/lbm.

[2] See Probs. 2–20 and 2–21.

[3] See Prob. 2–56.

The left-hand side of (2–119) can be redefined as $V(z)$, which amounts to a shift of the datum position from $r = \infty$ to $r = r_0$. The right-hand side of (2–119) is simplified if we introduce the new constant

$$g = \frac{Km_0}{r_0^2}, \qquad (2\text{–}120)$$

which can be identified from (2–115) as the local acceleration of gravity

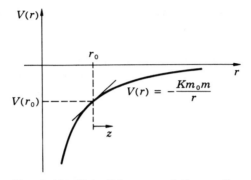

Fig. 2–42. Potential energy of the gravity field as a function of the radius and linearized approximation in the neighborhood of $r = r_0$.

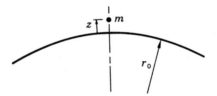

Fig. 2–43. Gravitational potential energy is approximately mgz when $z \ll r_0$.

at the position $r = r_0$. Thus, in place of (2–116), we may use the linearized approximation

$$V(z) = mgz \qquad (2\text{–}121)$$

for the potential energy of a mass m in the neighborhood of the datum $r = r_0$, where z is the elevation above the datum. The approximate relation (2–121) appears as the tangent to the true potential-energy curve in Fig. 2–42. Note that the approximate potential (2–121) implies that the vector force field acting on a probe m is

$$-\boldsymbol{\nabla} V = -\frac{\partial V}{\partial z}\,\mathbf{u}_z = -mg\mathbf{u}_z, \qquad (2\text{–}122)$$

which is independent of the probe's position. Such a field is said to be *uniform*. The approximation of the gravity field by the uniform field

(2–122) is nearly always adequate in earthbound engineering applications. Throughout this book we shall take the gravity field to be uniform unless specific mention is made to the contrary.

IDEAL TWO-FORCE ELEMENTS

A two-force element as shown in Fig. 2–44 is an element which has contact-force interactions with its environment at only two points, A and B. An *ideal* two-force element is assumed to be *massless*. Simple springs and dashpots are often treated as ideal two-force elements. Such massless elements would experience infinite accelerations if they were ever subjected to unbalanced force or torque. Thus, in order to maintain finite accelerations, it is necessary for the external forces acting on an ideal two-force member to be balanced at all times. This implies that

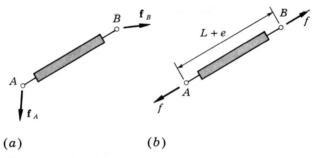

(a) (b)

FIG. 2–44. Two-force element acted upon by its environment at A and B. (a) General case. (b) Ideal case where forces acting on the element must be equal and opposite.

$\mathbf{f}_A$ and $\mathbf{f}_B$ in Fig. 2–44a must be equal and opposite and must act along the line joining A and B, as indicated in Fig. 2–44b, if the element is ideal. It should be noted that the forces shown in Fig. 2–44 represent the forces of the environment acting *on* the element. Generally, this is the most convenient viewpoint when discussing springs and dashpots. We shall, however, have occasion to shift our viewpoint when we consider the force exerted *by* such an element on an attached mass.

Let the element in Fig. 2–44b have a reference length L, and let its elongation be denoted by e. For certain types of force elements the force magnitudes f in Fig. 2–44b may be completely determined by the element itself, e.g., by the *elongation e* and the elongation *rate ė*. In such cases the equation relating f to the variables upon which it depends is called the *constitutive relation* for the element. Two simple examples are illustrated in Fig. 2–45. For an ideal spring the force f_s acting *on* the spring depends only on the elongation e_s. In many cases the constitutive relation is *linear*, at least for small elongations. The linear constitutive relation is

$$f_s = ke_s, \tag{2-123}$$

where k is called the *spring constant*. For an ideal dashpot the force f_d acting *on* the dashpot depends only on the elongation rate $v_d = \dot{e}_d$. For small rates the constitutive relation may have the linear form

$$f_d = bv_d, \tag{2-124}$$

where b is called the *dashpot constant*.

We next return to the general ideal two-force element of Fig. 2–44b and evaluate the net *work* done *on* the element *by* the environment during a change of configuration. Since the pair of external forces of magnitude f always act along the direction of AB, the net work done by the pair during an infinitesimal change of configuration which results in an elongation de is simply $f\,de$. The total work done *on* the element in

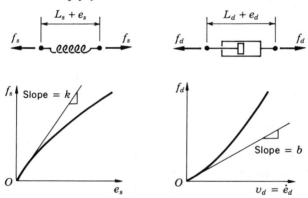

Fig. 2–45.   Constitutive relations for spring and dashpot.

bringing it from a datum configuration with $e = 0$ to a configuration with elongation e is given by the integral

$$\int_0^e f\,de. \tag{2-125}$$

In general, the work (2–125) depends on the path followed in bringing the element from the datum configuration to the final configuration. In the special case where the integral (2–125) is independent of path and depends only on its upper limit, the net work done on the element vanishes during any closed cycle which returns the element to its initial configuration. In this case the element is said to be *conservative*. An ideal two-force member is conservative if the magnitude f of the element force is a single-valued function of the elongation e, i.e., if the constitutive relation for the element is that of an ideal spring (see Fig. 2–45).

This requirement is simpler than the corresponding one for a three-dimensional conservative field, which includes the requirement of vanishing curl (2–110) because an ideal two-force element is essentially one-dimensional.

POTENTIAL ENERGY OF A FORCE ELEMENT

For a conservative ideal two-force element the work integral (2–125) is a function of the current *state* (i.e., the current elongation e of the element), and is called the potential energy $V(e)$ of the element,

$$V(e) = \int_0^e f \, de. \qquad (2\text{–}126)$$

In the special case of a *linear* spring with the constitutive relation (2–123), the potential energy is

$$V = \tfrac{1}{2}ke^2. \qquad (2\text{–}127)$$

The negative of the variation of (2–126),

$$-\delta V = -f \, \delta e, \qquad (2\text{–}128)$$

is the negative of the work done on the element, and thus represents the increment of work done *by* the element *on* its environment.

It is instructive to compare (2–126) with the corresponding definition (2–112) for the potential energy of a force field. In a conservative force *field* the potential energy is the *negative* of the work done *by* the field on an external agent, the probe, whereas for a conservative force *element* the potential energy is the work done *on* the element by an external agent. The two definitions are consistent. The appearance of a negative sign in (2–112) and the absence of one in (2–126) is due to the reversal in the significance of the forces involved. In dealing with force fields, we have chosen to emphasize the role of the force of the field acting on the probe (probably because our experience of living in the gravity field makes it easy for us to identify ourselves with the probe). In dealing with force elements, we have chosen to emphasize the role of the forces exerted by external agents on the element (probably because our experience in handling springs and dashpots makes it easy for us to identify ourselves with the external agent). In both cases the increment of work done *by* the field or *by* the element is given by a decrease in potential energy.

POTENTIAL ENERGY OF A SYSTEM

Consider a complex dynamic system containing a number of mass particles interconnected by conservative force elements and acted on by one or more conservative force fields. Let the individual potential energies of the separate elements and separate mass particles in the fields be denoted by V_i, with $i = 1, \ldots, n$, and define the *system potential energy* V as the sum of the individual energies,

$$V = \sum_{i=1}^{n} V_i. \qquad (2\text{–}129)$$

Because of (2–114) and (2–128), the total work done by all the conservative elements and fields under a variation of configuration is

$$-\delta V_1 - \delta V_2 - \cdots - \delta V_n = -\delta V. \qquad (2\text{–}130)$$

Thus the total work done by all the conservative forces is accounted for by the decrease in the system potential energy.

WORKLESS CONSTRAINTS

In modeling complex dynamic systems it is convenient to idealize the behavior of many subsidiary parts which serve to guide or constrain the motion of the significant moving masses.   Common idealizations in this

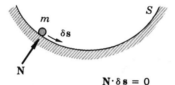

$$\mathbf{N} \cdot \delta \mathbf{s} = 0$$

FIG. 2–46.   Frictionless guiding surface S exerts normal force on m but does no work.

category include: rigid massless connecting links; massless and inextensible, but perfectly flexible, strings; frictionless pivots; massless levers; frictionless guiding surfaces and guiding channels; massless frictionless pulleys; etc.   These idealized devices may transmit or apply forces and may transmit or restrain motion, but as a result of the idealization, they do not do any net work under a variation of the system configuration.

For example, the frictionless surface S in Fig. 2–46 guides the motion of the mass particle m.   The surface can exert a normal reaction force $\mathbf{N}$ on m, but this force does no work under a displacement of the mass along the surface.

Another example is provided by the rigid massless lever on a frictionless pivot shown in Fig. 2–47.   Here it can be shown[1] that the assumption of a *massless* lever implies that the set of forces acting on the lever must be in *equilibrium* whenever the lever is at rest or moving with finite accelerations.   This is demonstrated by showing that nonequilibrium demands infinite accelerations.   With the lever in equilibrium, the net

[1] See Prob. 2–60.

work done by all the forces acting on it, during a variation of configuration $\delta\theta$, must be zero according to the principle of virtual work (see page 8). If there is no friction torque at the pivot, this means that the work done on the lever by f_1 must equal the work done by the lever against f_2. A massless lever thus serves to transmit work increments from one part of a system to another without loss. Another example of a device which transmits work is provided by the lazy-tongs mechanism

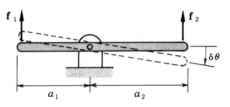

FIG. 2–47. Rigid massless lever on frictionless pivot. Work done by f_1 is equal to work
done against f_2.

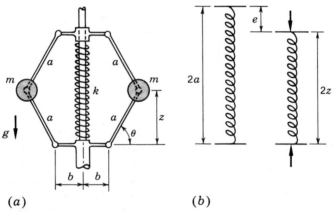

(a)   (b)

FIG. 2–48. Example 2–8. (a) Flyball governor. (b) Spring is
compressed a distance e in the configuration shown.

of Example 1–1. If the mechanism were assumed to be massless, then in the dynamic case the forces on it would still be in equilibrium and the work done on the linkage by f would just equal the work done by the linkage against W.

Example 2–8. To illustrate the foregoing ideas, we reconsider the flyball governor of Example 2–6 in Fig. 2–48 and seek to obtain the net work increment done on the masses in an infinitesimal variation of the angle θ. We assume that the linkage elements are rigid and massless and that there is no friction in the hinged joints. We take the spring to have an unstretched length $2a$ and a linear constitutive relation described by the spring constant k. A uniform gravitational field is assumed to act downward.

Under a geometrically admissible variation of configuration represented by $\delta\theta$, the elevation of the masses in the gravity field would be changed and the spring compression e would be altered. It is desired to obtain the work increment done by all forces acting on the masses in such a variation. The calculation is greatly simplified by the observation that no net work is done by any of the rigid massless linkage elements. The work done on the linkage by the spring is transmitted without loss to the masses. Thus the work increment done *on* the masses by all forces is just the sum of the work increments done *by* the spring and *by* the gravity field. These latter increments are conveniently obtained as potential-energy decrements. The potential energy of each mass in the gravity field is mgz, according to (2–121), and the potential energy of the spring is $\frac{1}{2}ke^2$, according to (2–127). On setting $e = 2(a - z)$, as indicated in Fig. 2–48b, we obtain the total potential energy of the system,

$$V = 2mgz + \tfrac{1}{2}k[2(a - z)]^2, \qquad (2\text{–}131)$$

in terms of the elevation z. In terms of θ this becomes

$$V = 2mga \sin \theta + 2ka^2(1 - \sin \theta)^2. \qquad (2\text{–}132)$$

If the angle θ undergoes a variation $\delta\theta$, the total work increment done *by* the spring and *by* the gravity field is the corresponding decrease in potential energy,

$$-\delta V = -2mga \cos \theta \, \delta\theta + 4ka^2(1 - \sin \theta) \cos \theta \, \delta\theta. \qquad (2\text{–}133)$$

This work increment is equal, as we have seen, to the total work done by all forces acting on both masses. We shall make use of (2–132) in Example 2–10, in which equations of motion for the flyball governor are derived.

2-8 Formulation of equations of motion

Dynamic analysis of an engineering system usually begins with the isolation and identification of an idealized model of the system. The next step is to give a precise mathematical statement of the dynamic behavior of the model. This is done by applying appropriate governing principles to the model to formulate differential equations of motion. For mechanical systems the governing requirements can be divided into three categories:

1. Geometric requirements, including kinematic relations
2. Dynamic-force requirements, including relations between forces and rates of change of momenta
3. Constitutive relations for forces in deformable elements and in fields and velocity-momentum relations for inertial elements

The *direct method* of formulating the equations of motion consists in stating each of the above requirements in the form of a mathematical equation and then using algebraic elimination to reduce the resulting set of equations. The alternative *variational method* consists in using Hamilton's principle to select out of the geometrically admissible motions those which also satisfy the dynamic relations. The variational indicator,

$$\text{V.I.} = \int_{t_1}^{t_2} \left[\delta(T^* - V) + \sum_i \mathbf{f}_i \cdot \delta \mathbf{R}_i \right] dt, \qquad (2\text{-}134)$$

is constructed for motions satisfying the requirements of categories 1 and 3 above. The requirements of category 2 are contained in Hamilton's principle, which states that the variational indicator (2–134) vanishes for arbitrary geometrically admissible variations in the neighborhood of a natural motion. To apply the variational method, one begins by establishing the variational indicator (2–134) for the system under investigation, and then uses the calculus of variations to obtain the desired equations of motion as the necessary conditions for the vanishing of the indicator.

In this section we shall illustrate the application of *both* methods to two examples. In the first example we give an unusually detailed treatment of a simple system, in an effort to provide clear understanding of the alternative methods. The second example is not so transparent. It provides an opportunity to compare the two methods when applied to systems whose complexity is more typical of that encountered in engineering practice.

Example 2–9. We consider in Fig. 2–49 a mass m which can move horizontally and which is connected to a foundation by a linear spring with spring constant k and a linear dashpot with dashpot constant b. It is desired to obtain the equation of motion of the mass under the influence of the prescribed force $f(t)$. In Fig. 2–50 the system is dissected and the governing requirements are stated. At the top the three elements are shown separately. The pertinent variables for each element are defined in the sketches on the left, and the governing requirements for that element are stated on the right. At the bottom the governing requirements having to do with the system are stated. For example, the pertinent variables for the spring are the elongation e_s and

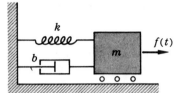

Fɪɢ. 2–49. Example 2–9. A mass supported by a spring and a dashpot and acted on by a prescribed force $f(t)$. The shaded wall and foundation are assumed to provide an inertial reference frame.

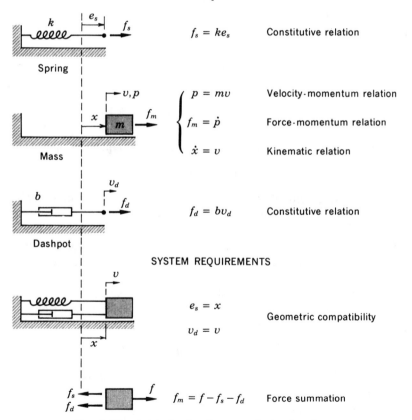

ELEMENT REQUIREMENTS

Spring

$f_s = ke_s$ Constitutive relation

Mass

$p = mv$ Velocity-momentum relation

$f_m = \dot{p}$ Force-momentum relation

$\dot{x} = v$ Kinematic relation

Dashpot

$f_d = bv_d$ Constitutive relation

SYSTEM REQUIREMENTS

$e_s = x$

$v_d = v$ Geometric compatibility

$f_m = f - f_s - f_d$ Force summation

Fɪɢ. 2–50.   Analysis of spring-mass-dashpot system.

the spring force f_s. They are related by the linear constitutive equation
(2–123). The system requirements at the bottom are of two types: the
geometric relations follow from the basic requirement that the system
remain connected during the motion, and the *force* requirement is that
the net force acting on the mass is the resultant of the prescribed external
force f and the internal interconnecting forces f_s and f_d.

Direct method. In Fig. 2–50 there are eight unknown variables,
$e_s, f_s, x, v, p, f_m, v_d, f_d$, and eight algebraic and differential relations connecting these variables. The direct method of analysis consists in laying out these eight equations which state all the governing requirements,
and then combining them, where possible, to achieve a more compact
mathematical representation. A systematic way to reduce the equations
of Fig. 2–50 is to start with the system force requirement[1] and then to
eliminate the force unknowns in favor of geometrical variables, which
can eventually all be expressed in terms of the single displacement x.
Thus we obtain, in turn,

$$f_m = f - f_s - f_d$$
$$\dot{p} = f - ke_s - bv_d$$
$$\frac{d}{dt}(mv) = f - kx - bv \tag{2–135}$$
$$m\ddot{x} = f - kx - b\dot{x},$$

which provide the single second-order differential equation

$$m\ddot{x} + b\dot{x} + kx = f(t) \tag{2–136}$$

for the displacement $x(t)$. Equation (2–136) is called an *equation of
motion* for the system of Fig. 2–49. Given a prescribed force history $f(t)$
and initial conditions on x and $\dot{x}$, the solution to (2–136) describes the
subsequent motion in terms of x. Once $x(t)$ is known, the time history
of any of the other variables can be obtained by substitution in the
relations listed in Fig. 2–50.

Variational method. To obtain the same equation of motion by
the variational method, we begin by constructing the variational indicator (2–134). The lagrangian for the system of Fig. 2–50 is

$$\mathcal{L} = T^* - V = \tfrac{1}{2}mv^2 - \tfrac{1}{2}ke_s^2, \tag{2–137}$$

according to (2–98) and (2–127). The lagrangian (2–137) accounts for
the kinetic coenergy of the mass and for the work done by the spring
force. The work done by the dashpot force f_d and by the prescribed

[1] Alternatively, one can start with the system geometric-compatibility requirements
and eliminate the geometrical variables in favor of force and momentum variables.
See Appendix D.

force f is accounted for by the work expression

$$\sum_i \mathbf{f}_i \cdot \delta \mathbf{R}_i = (-f_d + f)\, \delta x$$
$$= (-bv_d + f)\, \delta x. \qquad (2\text{--}138)$$

The variational indicator (2–134) thus takes the form

$$\text{V.I.} = \int_{t_1}^{t_2} [\delta(\tfrac{1}{2}mv^2 - \tfrac{1}{2}ke_s^2) + (-bv_d + f)\, \delta x]\, dt. \qquad (2\text{--}139)$$

According to Hamilton's principle, a natural motion of the system is a geometrically admissible motion for which (2–139) vanishes for arbitrary admissible variations. A simple way to ensure geometric admissibility for all motions considered is to use the geometric and kinematic relations of Fig. 2–50 to express all displacements and velocities in (2–139) in terms of the single (unknown) time history $x(t)$.

$$\text{V.I.} = \int_{t_1}^{t_2} [\delta(\tfrac{1}{2}m\dot{x}^2 - \tfrac{1}{2}kx^2) + (-b\dot{x} + f)\, \delta x]\, dt. \qquad (2\text{--}140)$$

Now we apply the calculus of variations to obtain the equation of motion which x must satisfy in order to make (2–140) vanish for arbitrary variations δx. First, we carry out the indicated variation in (2–140),

$$\text{V.I.} = \int_{t_1}^{t_2} (m\dot{x}\, \delta\dot{x} - kx\, \delta x - b\dot{x}\, \delta x + f\, \delta x)\, dt. \qquad (2\text{--}141)$$

Then we insert the commutativity relation $\delta\dot{x} = [d(\delta x)]/dt$ and integrate the first term by parts.

$$\text{V.I.} = m\dot{x}\, \delta x \Big|_{t_1}^{t_2} - \int_{t_1}^{t_2} (m\ddot{x} + b\dot{x} + kx - f)\, \delta x\, dt. \qquad (2\text{--}142)$$

According to the agreement implied in Hamilton's principle, that all variations vanish at the time limits t_1 and t_2, the term $m\dot{x}\, \delta x$ is zero at both limits. Then, if (2–142) is to vanish for arbitrary δx within the time interval from t_1 to t_2, it is necessary[1] that the contents of the parentheses within the integral be zero throughout the interval. Thus we conclude that the natural motion $x(t)$ must satisfy the differential equation

$$m\ddot{x} + b\dot{x} + kx = f(t). \qquad (2\text{--}143)$$

Note that (2–143), derived from Hamilton's principle, is identical with the equation of motion (2–136) obtained by the direct method.

Although the variational method appears to be quite different from the direct method, it is in fact simply an alternative procedure for bringing together the same information which is contained in the eight equations listed in Fig. 2–50. The velocity-momentum relation is tacitly employed in constructing the kinetic coenergy T^*, and the constitutive relation for

[1] See Appendix A.

the spring is tacitly employed in constructing the potential energy V. The constitutive relation for the dashpot is explicitly utilized in setting up the work expression (2–138). The geometric and kinematic relations of Fig. 2–50 are explicitly used in passing from (2–139) to (2–140). Finally, the form of the variational indicator and the accompanying variational principle are inherently equivalent to enforcing the system force requirement and the relation between force and rate of change of momentum.   The variational method thus makes implicit or explicit use of precisely the same governing requirements as the direct method.

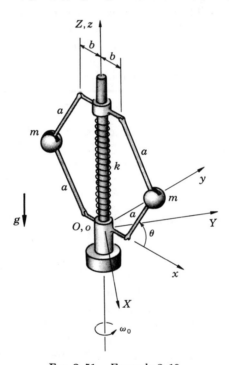

F̲ɪ̲ɢ̲. 2–51.   Example 2–10.

Since the two methods are equivalent at the level of basic physical content, the choice between them must be one of relative convenience. To some extent this depends on the experience and personal preferences of the analyst.   For a system as simple as that of the present example, there is certainly no advantage to the variational method.   If the manipulations of the calculus of variations are still unfamiliar to the reader, the variational method may seem even more clumsy than the direct method.   When, however, we consider more complex systems, in which the kinematic relations are more complicated and in which there are many workless constraints, we shall find that there are considerable economies of thought, as well as economies of manipulation, in the variational method.   The following example provides an opportunity to compare the two methods in a somewhat more complex system.

Example 2–10.   We return in Fig. 2–51 to the flyball governor already considered in Examples 2–6 and 2–8.   We shall obtain equations of motion which describe the motions of the flyballs for the case when the angular speed ω_0 of the governor is held constant.   As before, the linkage is taken to be massless and constrained to flex without friction in the xz plane.   The linear spring with constant k has a free length of $2a$.

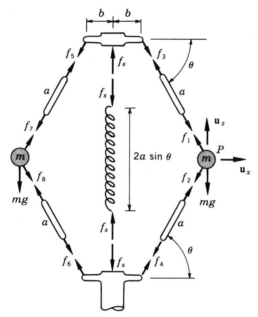

Fɪɢ. 2–52.   Force analysis in xz plane.

Direct method.   We begin by analyzing the forces in the system as shown in Fig. 2–52, where the force components in the xz plane acting on each element are represented.   It may be noted that there *are* force components normal to this plane, but since the linkage prohibits motion out of the plane, they have *no influence* on the equations of motion.   The out-of-plane forces would, however, be of importance in determining the loading on the members and on the joints.

Since the linkage members in Fig. 2–52 are massless, the forces acting on them must be balanced.   This requires, for example, that f_1 and f_3 act along the length of the link, and moreover that $f_3 = f_1$.   Applying the same requirement to the other links, plus making use of the system's symmetry, we find the relations

$$f_1 = f_3 = f_5 = f_7$$
$$f_2 = f_4 = f_6 = f_8,$$

(2–144)

so that all the linkage forces can be expressed in terms of f_1 and f_2. We then balance the forces on the sliding collar at the top to obtain

$$f_s = 2f_1 \sin \theta.$$

(2–145)

The spring force f_s is also given by the constitutive relation,

$$f_s = k(2a - 2a \sin \theta).$$

(2–146)

The net force in the xz plane acting on the mass particle at P in Fig. 2–52 is

$$\mathbf{f}_m = -(f_1 + f_2)\cos \theta \, \mathbf{u}_x + [(f_1 - f_2) \sin \theta - mg]\mathbf{u}_z.$$

(2–147)

Next we evaluate the momentum $\mathbf{p}$ of the mass particle at P. Here we utilize the result (2–104) obtained in Example 2–6 for the velocity of the mass particle to write

$$\mathbf{p} = m[-\mathbf{u}_x a\dot\theta \sin \theta + \mathbf{u}_y(b + a \cos \theta)\omega_0 + \mathbf{u}_z a\dot\theta \cos \theta].$$

(2–148)

In order to apply Newton's law, $\mathbf{f}_m = d\mathbf{p}/dt$, it is necessary to take the time derivative of (2–148). Since the xyz frame has angular velocity $\mathbf{u}_z\omega_0$, the derivatives of the unit vectors are

$$\frac{d\mathbf{u}_x}{dt} = \mathbf{u}_y\omega_0 \qquad \frac{d\mathbf{u}_y}{dt} = -\mathbf{u}_x\omega_0 \qquad \frac{d\mathbf{u}_z}{dt} = 0.$$

(2–149)

On using (2–149), we find the total rate of change of the momentum $\mathbf{p}$ to be

$$\frac{d\mathbf{p}}{dt} = m\{-\mathbf{u}_x[a\ddot\theta \sin \theta + a\dot\theta^2 \cos \theta + (b + a \cos \theta)\omega_0^2] \\ - \mathbf{u}_y(2a\omega_0\dot\theta \sin \theta) + \mathbf{u}_z(a\ddot\theta \cos \theta - a\dot\theta^2 \sin \theta)\}.$$

(2–150)

Now applying Newton's law, we equate the x and z components of Eqs. (2–147) and (2–150) to obtain

$$-(f_1 + f_2) \cos \theta = -m[a\ddot\theta \sin \theta + a\dot\theta^2 \cos \theta \\ + (b + a \cos \theta)\omega_0^2]$$

(2–151)

$$(f_1 - f_2) \sin \theta - mg = m[a\ddot\theta \cos \theta - a\dot\theta^2 \sin \theta].$$

At this stage we have completed the mathematical statement of the governing requirements and have four equations in (2–145), (2–146), and (2–151) for the four unknowns f_1, f_2, f_s, and θ. A single equation of motion can be obtained by eliminating the three forces in favor of the angle θ.   After considerable algebraic manipulation we find

$$a\ddot{\theta} + \omega_0^2(b + a \cos \theta) \sin \theta - 2a \frac{k}{m} (1 - \sin \theta) \cos \theta + g \cos \theta = 0$$

$$(2\text{–}152)$$

as a single second-order differential equation of motion.   This illustrates the application of the direct method to formulate dynamic equations of motions.   Note the importance of the forces f_1, f_2, and f_s in the intermediate analysis, although they do not appear in the final equation of motion.   Note also the kinematical complexity in passing from $\mathbf{p}$ to $d\mathbf{p}/dt$.

Variational method.   Returning to Fig. 2–51, we note that since the rotational speed ω_0 is *fixed*, the only admissible variations are those corresponding to a variation in the angle θ.   There are no nonconservative forces which do work under a variation $\delta\theta$; so the general variational indicator (2–134) reduces to

$$\text{V.I.} = \int_{t_1}^{t_2} \delta(T^* - V) \, dt. \tag{2–153}$$

For this system, the kinetic coenergy T^* is given by (2–106) in Example 2–6 and the potential energy V is given by (2–132) in Example 2–8, so that we have

$$\text{V.I.} = \int_{t_1}^{t_2} \delta\{m[a^2\dot{\theta}^2 + (b + a \cos \theta)^2\omega_0^2]$$
$$- [2mga \sin \theta + 2ka^2(1 - \sin \theta)^2]\} \, dt. \tag{2–154}$$

In applying Hamilton's principle, it is necessary to consider geometrically admissible motions and geometrically admissible variations.   In (2–154) the geometrical requirements have already been met by the device of expressing all geometrical and kinematical variables in terms of $\theta(t)$, which represents an arbitrary admissible time history of motion.   Admissible variations are represented by $\delta\theta$.

At this stage the governing physical considerations have all been attended to, and it only remains to draw out the equations of motion from (2–154) by the routine application of the calculus of variations.   Of all admissible motions a natural motion is distinguished by the requirement that the variational indicator vanish for arbitrary admissible variations.   To begin, we carry out the indicated variation in (2–154),

$$\text{V.I.} = \int_{t_1}^{t_2} 2m \left[a^2\dot{\theta} \, \delta\dot{\theta} - (b + a \cos \theta)\omega_0^2 a \sin \theta \, \delta\theta - ga \cos \theta \, \delta\theta \right.$$
$$\left. + 2 \frac{k}{m} a^2(1 - \sin \theta) \cos \theta \, \delta\theta \right] dt. \tag{2–155}$$

Next we insert the commutativity relation $\delta\dot{\theta} = [d(\delta\theta)]/dt$ and integrate the first term by parts to get

$$\text{V.I.} = 2ma^2\dot\theta\,\delta\theta\,\Big|_{t_1}^{t_2} - 2ma\int_{t_1}^{t_2}\Big[a\ddot\theta + \omega_0^2(b + a\cos\theta)\sin\theta$$
$$+ g\cos\theta - 2\frac{k}{m}a(1 - \sin\theta)\cos\theta\Big]\delta\theta\,dt. \quad (2\text{--}156)$$

Now the first term on the right vanishes at the limits because of the agreement that variations shall vanish at t_1 and t_2. The necessary condition for the remaining integral to vanish for arbitrary $\delta\theta$ is that the contents of the square brackets should be zero at every instant throughout the interval from t_1 to t_2. On setting the contents of the brackets equal to zero, we get the following equation, which must be satisfied by the natural motion:

$$a\ddot\theta + \omega_0^2(b + a\cos\theta)\sin\theta + g\cos\theta - 2\frac{k}{m}a(1 - \sin\theta)\cos\theta = 0.$$
$$(2\text{--}157)$$

This is identical with the equation of motion (2–152) obtained by the direct method. Notice that in the variational method the difficult steps are confined to the identification of the contributions to the state functions T^* and V and to the representation of these contributions in forms which facilitate ensuring geometric admissibility. The subsequent manipulation is of a standardized routine nature. In Sec. 2–11 it will be seen that in many cases this manipulation can be streamlined even further.

2–9   Techniques for assuring admissible variations

An essential part of the variational approach is embodied in the requirement that the variations considered must be geometrically admissible. The most common technique for assuring admissible variations is to express all kinematic quantities in terms of geometric variables which can be varied arbitrarily without violating the geometric constraints. This technique is developed systematically in Sec. 2–10. In this section we illustrate two alternative techniques which are occasionally useful and which provide additional insight into the admissibility requirements. The two techniques are described in Example 2–11 below. In the first procedure the admissibility requirements are introduced after the variational operations have been performed on the variational indicator. In the second procedure the admissibility requirements are introduced directly into the variational indicator by means of the Lagrange-multiplier technique.

Example 2–11. The system in Fig. 2–53 consists of a mass m supported in a uniform gravity field by a flexible massless inextensible string which runs from a fixed support at A through frictionless massless pulleys to a linear spring with spring constant k.   Let the extension of the spring from its unstretched length be denoted by x, and let the displacement of the mass measured from the position corresponding to $x = 0$ be denoted by y.

In a variation of the configuration of the system of Fig. 2–53, work will be done by the spring and by the gravity field, but no net work is done by the string or the pulleys which serve to enforce geometric constraints. Therefore, if the work of the spring and the gravity field are accounted for

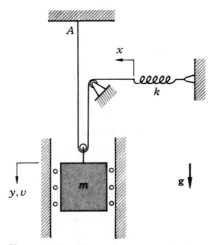

FIG. 2–53. Example 2–11. Mass-spring-pulley system in gravity field.

in the potential energy V, no additional work expressions are required in the variational indicator.

$$\text{V.I.} = \int_{t_1}^{t_2} \delta(T^* - V)\,dt$$

$$= \int_{t_1}^{t_2} \delta(\tfrac{1}{2}mv^2 - \tfrac{1}{2}kx^2 + mgy)\,dt. \qquad (2\text{–}158)$$

We next perform the indicated variation,

$$\text{V.I.} = \int_{t_1}^{t_2} (mv\,\delta v - kx\,\delta x + mg\,\delta y)\,dt. \qquad (2\text{–}159)$$

At this stage the variational indicator is expressed in terms of the three variables v, x, and y and their variations.   We have not yet introduced the geometric admissibility requirements.

For an admissible motion it is necessary that the velocity v and displacement y of the mass satisfy the kinematic relation

$$v = \dot{y}. \qquad (2\text{–}160)$$

Furthermore, the inextensibility of the string connecting the spring and the mass requires that

$$x = 2y. \tag{2-161}$$

Applying the variational operator to (2–160) and (2–161), we obtain the following relations between the variations which must be satisfied in an admissible variation:

$$\delta v = \delta \dot{y} = \frac{d}{dt}(\delta y) \tag{2-162}$$

$$\delta x = 2\delta y.$$

Thus all the variations in (2–159) can be expressed in terms of δy. If we insert (2–162) into (2–159) and perform an integration by parts, we obtain

$$\text{V.I.} = mv\, \delta y \Big|_{t_1}^{t_2} - \int_{t_1}^{t_2} \left[\frac{d}{dt}(mv) + 2kx - mg \right] \delta y\, dt. \tag{2-163}$$

Since variations vanish by agreement at t_1 and t_2, the first term on the right is zero, and the necessary condition for the variational indicator to vanish for arbitrary δy is that the contents of the square brackets vanish at each instant within the interval from t_1 to t_2. This yields the equation of motion

$$m\dot{v} + 2kx = mg, \tag{2-164}$$

in terms of the variables v and x. This equation can be expressed in terms of the single variable y by introducing the admissibility conditions (2–160) and (2–161):

$$m\ddot{y} + 4ky = mg. \tag{2-165}$$

Note that (2–164) can be interpreted as a direct application of Newton's law to the mass of Fig. 2–53. The forces acting to increase the momentum of the mass are its weight mg and (in the opposite sense) the two string tensions, each of which has the magnitude kx.

Lagrange-multiplier technique

To illustrate the use of Lagrange multipliers, we return to Fig. 2–53, the variational indicator (2–158), and the geometric admissibility conditions (2–160) and (2–161). For simplicity, let us introduce (2–160) directly into (2–158) so that we have the variational indicator

$$\text{V.I.} = \int_{t_1}^{t_2} \delta(\tfrac{1}{2}m\dot{y}^2 - \tfrac{1}{2}kx^2 + mgy)\, dt, \tag{2-166}$$

together with the remaining geometric constraint,

$$x = 2y. \tag{2-167}$$

The Lagrange-multiplier technique[1] for introducing the constraint (2–167) into (2–166) is to consider the extended variational indicator,

$$\text{V.I.} = \int_{t_1}^{t_2} [\delta(\tfrac{1}{2}m\dot{y}^2 - \tfrac{1}{2}kx^2 + mgy) + \lambda\delta(x - 2y)]\, dt, \quad (2\text{–}168)$$

where $\lambda(t)$ is an undetermined function called *a Lagrange multiplier*. Note that (2–168) reduces to (2–166) for arbitrary λ if the constraint (2–167) is satisfied. We now apply the calculus of variations to (2–168) in the usual way, treating x and y as independent variables. After an integration by parts we find

$$\text{V.I.} = m\dot{y}\,\delta y\Big|_{t_1}^{t_2} - \int_{t_1}^{t_2} [(m\ddot{y} - mg + 2\lambda)\,\delta y + (kx - \lambda)\,\delta x]\, dt, \quad (2\text{–}169)$$

where the first term on the right vanishes by agreement and the necessary conditions for the vanishing of the integral, for arbitrary independent variations δx and δy, are the two equations

$$m\ddot{y} - mg + 2\lambda = 0 \quad \text{and} \quad kx - \lambda = 0, \quad (2\text{–}170)$$

which must be satisfied at each instant throughout the interval from t_1 to t_2. These two equations, together with the geometric constraint (2–167), are equivalent to the equation of motion (2–165) obtained previously. To show this, we first eliminate the Lagrange multiplier from (2–170) (by adding 2 times the second equation to the first equation),

$$m\ddot{y} - mg + 2kx = 0, \quad (2\text{–}171)$$

and then use (2–167) to eliminate x in favor of y.

$$m\ddot{y} - mg + 4ky = 0. \quad (2\text{–}172)$$

Note that (2–172) is equivalent to (2–165).

INTERPRETATION OF THE LAGRANGE MULTIPLIER

The Lagrange multiplier $\lambda(t)$ in the preceding analysis can be identified as the force required to maintain the geometric constraint (2–167). This interpretation is suggested by the analogy between the extended variational indicator (2–168) and the general variational indicator for Hamilton's principle (2–134), which includes a work expression. To pursue this interpretation, let us consider the modified system shown in Fig. 2–54. This system is identical with the original system, except that a deformable element E has been inserted in the previously inextensible string. Hamilton's principle can be applied to the modified system provided that the work done by the element E is included in the work

[1] See Appendix A.

expression of the variational indicator (2–134). If we denote the tensile force in the element E by λ, the work increment done by the element under independent variations of x and y is

$$\lambda\delta(x - 2y), \tag{2-173}$$

since the contraction in length of the element is $x - 2y$. Note that this work increment is just the term that has been added in the extended variational integral (2–168). Thus (2–168) can be interpreted as the variational indicator for Hamilton's principle applied to the system of Fig. 2–54. The necessary conditions (2–170) for the vanishing of the indicator are the equations of motion for the system of Fig. 2–54. These

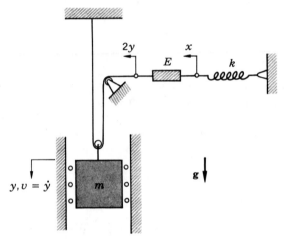

Fig. 2–54. System of Fig. 2–53 with the addition of a fictitious deformable element E.

equations apply whatever the nature of the element E; e.g., E could be a spring or a dashpot. To proceed further, it is necessary to have a particular constitutive relation for the element E, i.e., a description of the relation between the tension λ and the deformation of the element. In the actual system of Fig. 2–52, the element E is simply a length of inextensible string. Its constitutive relation requires that there be no deformation, whatever the value of λ. This requirement is embodied in the geometric constraint (2–167).

The preceding interpretation shows that when the Lagrange-multiplier technique is employed to enforce a geometric constraint, the multiplier itself can be interpreted as the effective-force component which would do work if the constraint were relaxed.

2–10 Generalized variables

In this section we examine the problem of selecting appropriate variables to describe the dynamic behavior of a system. We shall see that the application of the variational approach is considerably simplified whenever a set of independent geometrical variables can be found which can be varied independently and which permit representation of any admissible configuration of the system. The requirements of geometric admissibility are automatically met by the employment of such sets of variables. We begin with the concept of generalized coordinates. Point coordinates (e.g., cartesian coordinates, spherical coordinates) are used to locate a point with respect to a reference frame. *Generalized coordinates* are used to locate a dynamic *system* with respect to a reference frame. A set of generalized coordinates may include cartesian coordinates or spherical coordinates, but may also include convenient lengths or angles which

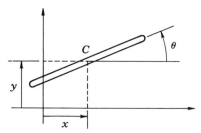

Fig. 2–55. Generalized coordinates
for rigid rod moving in the plane are
x, y, and θ.

have no association with conventional coordinate systems. For example, consider the rigid rod moving in the plane shown in Fig. 2–55. A possible set of generalized coordinates for locating the rod are the x and y coordinates of the center C of the rod and the angle θ between the rod and the x axis. Another example is shown in Fig. 2–56, where m_1 and m_2 move parallel to the horizontal x axis, and m_3 swings as a pendulum in the plane of the sketch. A possible set of generalized coordinates are the x displacements of m_1 and m_2, each measured from the corresponding equilibrium position, and the angle θ between the pendulum arm and the vertical.

A set of generalized coordinates is said to be *complete* if the values of the coordinates corresponding to an arbitrary geometrically admissible configuration of the system are sufficient to fix the location of *all* parts of the system. As a simple example, the cartesian coordinates (x,y,z) are a complete set of coordinates for a point free to move in three dimensions, but the set (x,y) is incomplete.

A set of generalized coordinates is said to be *independent* if, when all but any one of the coordinates is fixed, there still remains a continuous range of values for that one coordinate which corresponds to a range of admissible system configurations. As a simple example, the spherical coordinates (r,θ,φ) are an independent set of coordinates for a point free to move in three dimensions, but for the same point the set (r,θ,φ,x,y,z) is not independent, since the values of the cartesian coordinates are frozen as soon as the spherical coordinates are fixed.

In addition to generalized coordinates which describe geometrically admissible configurations of a dynamic system, we can introduce generalized infinitesimal *variational variables* to describe admissible variations. In most cases the variations of the generalized coordinates are the most convenient generalized variational variables. For example, in Fig. 2–55, the variations δx, δy, and $\delta \theta$ are a set of variational variables which can

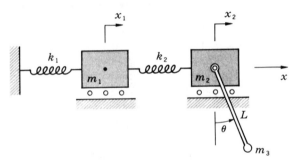

Fig. 2–56. Generalized coordinates for system are x_1, x_2, and θ.

be used to describe variations in the configuration of the rod.   Similarly, in Fig. 2–56, the variations δx_1, δx_2, and $\delta \theta$ are a set of variational variables which can be used to describe variations in the configuration of the system.

The concepts of completeness and independence introduced for generalized coordinates can also be applied in conjunction with infinitesimal variational variables.   A set of variational variables is said to be *complete* if the values of the variational variables corresponding to an admissible variation of configuration are sufficient to fix the variations in position of *all* parts of the system.   A set of variational variables is said to be *independent* if, when all but any one of the variational variables is fixed, there still remains a continuous range of values for that one variational variable which corresponds to a range of admissible variations in the configuration of the system.   The number of independent variational variables in a complete set of infinitesimal variational variables is called the number of

degrees of freedom that the system has in the given configuration.   For
example, the rod in Fig. 2–55 has three degrees of freedom, as does the
system in Fig. 2–56.   For a wide class of systems the number of degrees of
freedom is identical with the number of independent coordinates in a
complete set of generalized coordinates, but this is *not always* the case.
The relation between these numbers depends on the nature of the constraints in the system.

GEOMETRIC CONSTRAINTS

A geometric constraint is any requirement which acts to reduce the
number of degrees of freedom of a system.   These requirements can take
many forms.   For example, the requirement that the rod in Fig. 2–55 is
to remain in the plane of the sketch is a geometric constraint.   Similarly,
the requirement that the masses in Fig. 2–56 are to move along the x axis
is a geometric constraint.

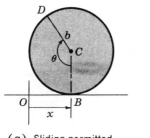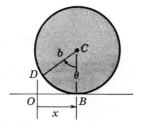

(*a*) Sliding permitted   (*b*) No slip at B

FIG. 2–57.   Rolling constraint.   (*a*) When sliding is permitted, x and θ are independent.   (*b*) When disk rolls without
slip, $x = b\theta$.

As another example consider the *rolling constraint* illustrated in Fig.
2–57.   The disk of radius b is constrained to remain in the plane of the
sketch and in contact with the horizontal roadbed.   The coordinates
x and θ constitute a complete set of generalized coordinates in both cases.
In case (*a*) it is supposed that the disk is free to roll or slide, so that x and
θ are also independent.   In case (*b*) an additional constraint requires
that the disk roll *without slipping*, so that any change in θ requires a
change in x, and vice versa.   If the line CD is chosen so that D coincides
with O when the x coordinate of the center C is zero, then the relation
between x and θ can be written

$$x = b\theta. \tag{2-174}$$

This relation is an analytical statement of the rolling constraint.   Note
that it relates generalized coordinates of a complete set, but is independent

of time and does not contain infinitesimal variational variables. Note also that (2–174) implies the following restriction on the infinitesimal variations:

$$\delta x = b \, \delta\theta. \tag{2–175}$$

A different kind of constraint is encountered in problems of *forced motion*. A simple example is illustrated by the two cases shown in Fig. 2–58. In both cases the two masses are constrained to move along the x axis. The coordinates x_1 and x_2 constitute a complete set. In case (a)

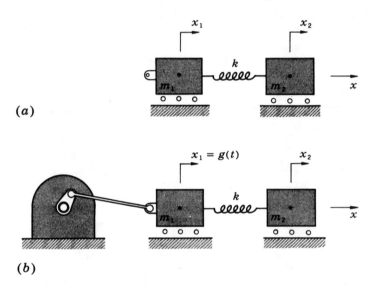

Fig. 2–58. Time-varying constraint. (a) x_1 and x_2 are independent. (b) x_1 is forced to follow a prescribed history $x_1 = g(t)$ by external drive.

the coordinates are also independent. In case (b) an external drive is connected to m_1. Instead of studying the dynamics of the masses *and* the drive, we make the idealization that the drive constrains m_1 to execute a prescribed time history $x_1 = g(t)$ [where the function $g(t)$ depends on the properties of the drive and on its schedule of operation] independently of the motion of m_2. This means that, as far as the system of m_1 and m_2 goes, we have introduced the *time-varying constraint*

$$x_1 = g(t), \tag{2–176}$$

and there remains only the single independent coordinate x_2. Note that (2–176) is an equation relating generalized coordinates and time but does not contain infinitesimal variational variables. Note also that (2–176) implies the following restriction on the infinitesimal variations:

$$\delta x_1 = 0. \tag{2–177}$$

A geometric constraint which can be expressed analytically as an equation relating generalized coordinates and time is said to be *holonomic*. All the constraints illustrated thus far have been holonomic. The most important property of a holonomic constraint is that there is a one-to-one correspondence between a restriction on the generalized coordinates and a restriction on the infinitesimal variations. This correspondence is illustrated by the pair (2–174) and (2–175) and by the pair (2–176) and (2–177). If all the constraints in a system are holonomic, this correspondence implies that the number of independent generalized coordinates in a complete set is the *same* as the number of degrees of freedom.

NONHOLONOMIC CONSTRAINTS

Any geometric constraint which cannot be expressed analytically as an equation relating generalized coordinates and time is *nonholonomic*. The

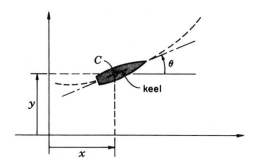

FIG. 2–59.   Example 2–12.   Action of keel may
be considered as a nonholonomic constraint.

most important type of nonholonomic constraint[1] is one which can only be expressed as an equation relating infinitesimal variations to generalized coordinates. In systems which have such constraints, the number of degrees of freedom is *less* than the number of independent generalized variables in a complete set.

Example 2–12.   In Fig. 2–59 we show a boat on a body of water whose surface is in the plane of the sketch. The boat has a keel which offers little resistance to longitudinal translation but offers very great resistance to transverse translation. Let us idealize this situation by considering a model in which transverse motions are completely prohibited by a geometric constraint. In such a model the translation of the point C on the

---

[1] For illustrations see, in addition to Example 2–12, Probs. 2–31 and 2–32.

boat must always be parallel to the instantaneous heading of the keel. We do, however, admit that the heading angle θ may vary during the motion.

If we consider only the two-dimensional motion of the boat in the plane, the generalized coordinates x, y, and θ constitute a complete set. The variations δx, δy, and $\delta \theta$ are a complete set of infinitesimal variational variables. These variations are not, however, independent. The requirement that any translation must be in the heading direction implies the constraining relation

$$\delta y - \delta x \tan \theta = 0. \tag{2–178}$$

Thus there are only two independent infinitesimal variations, and the boat has only *two degrees of freedom*.

The *three* generalized coordinates x, y, and θ are, however, *independent*. To demonstrate this it is necessary to show that if any two of them are fixed, there still remains a range of values for the third which corresponds to a range of admissible configurations. If x and y are fixed in Fig. 2–59,

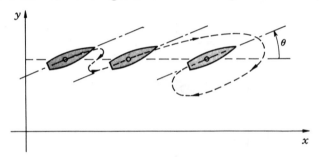

FIG. 2–60.   Configurations with fixed values of y and θ but differing values of x.   Possible admissible paths between configurations are sketched.

the angle θ can be varied continuously without violating the constraint. If y and θ are fixed, there also remains a continuous range of admissible values of x, as indicated in Fig. 2–60, even though it is no longer possible to pass from one admissible configuration to another by simply varying x alone.   A similar argument applies when x and θ are fixed.   Thus, as a result of the constraint represented by (2–178), the system has *three* independent coordinates but only *two* degrees of freedom.

To demonstrate that the constraint here is nonholonomic, it is necessary to show that it is not possible to express the constraint in the form

$$h(x,y,\theta,t) = 0, \tag{2–179}$$

where h is an arbitrary function of the arguments listed.   If a relation of the form (2–179) did exist, its variation δh would have the form

$$X\ \delta x + Y\ \delta y + \Theta\ \delta\theta = 0, \qquad (2\text{–}180)$$

where
$$X = \frac{\partial h}{\partial x} \qquad Y = \frac{\partial h}{\partial y} \qquad \Theta = \frac{\partial h}{\partial \theta}. \qquad (2\text{–}181)$$

Because of (2–181) it will be necessary for the coefficients of (2–180) to satisfy the *integrability requirements*[1]

$$\frac{\partial Y}{\partial x} = \frac{\partial X}{\partial y} \qquad \frac{\partial Y}{\partial \theta} = \frac{\partial \Theta}{\partial y} \qquad \frac{\partial \Theta}{\partial x} = \frac{\partial X}{\partial \theta}. \qquad (2\text{–}182)$$

Now the constraint relation (2–178) can be put in the form of (2–180), with coefficients

$$X = -\tan\theta \qquad Y = 1 \qquad \Theta = 0, \qquad (2\text{–}183)$$

but these coefficients do *not* satisfy the integrability requirements (2–182).

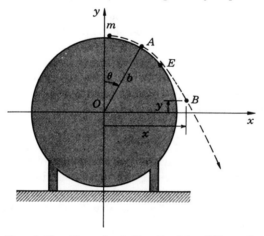

FIG. 2–61. Example 2–13. Particle sliding off cylinder. At A, m has *one* degree of freedom, $\delta\theta$; at B, m has *two* degrees of freedom, δx and δy.

Therefore the constraint here cannot be expressed in the holonomic form (2–179), and thus must be nonholonomic.

Example 2–13. To illustrate another type of nonholonomic constraint, consider the problem of the mass particle m in Fig. 2–61, which is constrained to remain in the plane of the sketch and which can slide on the surface of a fixed cylinder of radius b *or* move freely in the plane of the sketch when it is off the cylinder. The cartesian coordinates x and y of the particle constitute a complete set of generalized coordinates in this case. The requirement that the particle stay on the cylinder *or* on the

[1] See page 16.

outside of the cylinder can be expressed analytically in the following form:

$$x^2 + y^2 \geq b^2. \qquad (2\text{–}184)$$

This relation involves only the coordinates, but since it involves an inequality rather than an equality, it must be considered to represent a *nonholonomic constraint*.

In dealing with a constraint of this type it is convenient to consider that the entire motion consists of two regimes. In the first regime the particle is taken to be *constrained* to slide *on* the cylinder. Here the single generalized coordinate θ is sufficient and the particle has *one* degree of freedom, $\delta\theta$. In the second regime the particle is *off* the cylinder. Here two generalized coordinates are required to form a complete set and the particle has *two* degrees of freedom. The entire trajectory of motion is then made up of a first-regime trajectory up to some point E, together with a second-regime trajectory beyond the point E. In general, the location of the point E where the particle leaves the cylinder cannot be determined in advance, but must be determined[1] as part of the solution.

The nonholonomic constraint (2–184) may be considered to be *piecewise holonomic*. As long as a trajectory remains entirely in either regime, the constraint is holonomic, but when a trajectory includes portions in both regimes, there is a sudden change (during the course of the motion) from one holonomic constraint to another.

GENERALIZED HOLONOMIC SYSTEM

If all the constraints in a dynamic system are holonomic, the system is said to be holonomic. By using generalized coordinates it is possible to develop a generalized representation which applies to holonomic systems of arbitrary complexity. In the generalized representation it is convenient to introduce generalized *velocities*, generalized *forces*, and generalized *momenta* associated with the generalized coordinates.

Consider a holonomic system of N mass particles m_i $(i = 1, \ldots, N)$, as indicated in Fig. 2–62. Let

$$\eta_1, \eta_2, \ldots, \eta_m \qquad (2\text{–}185)$$

be a *complete* set of generalized coordinates. A possible choice here might be the cartesian coordinates for every particle. In this case $m = 3N$. It may be that some of the constraints have already been observed in establishing the coordinates (2–185), in which case $m < 3N$. For example, if a certain particle is constrained to remain in the XY plane, the Z coordinate for that particle could be omitted in (2–185). If all the constraints have not been observed, the set (2–185) is not independent and

[1] See Prob. 2–45.

there are p holonomic constraints of the form

$$h_1(\eta_1, \eta_2, \ldots, \eta_m, t) = 0$$
$$h_2(\eta_1, \eta_2, \ldots, \eta_m, t) = 0$$
$$\cdots \cdots \cdots \cdots \cdots \cdots \cdots \qquad (2\text{--}186)$$
$$h_p(\eta_1, \eta_2, \ldots, \eta_m, t) = 0.$$

Examples of constraints of this form are provided by the rolling constraint (2–174) and the time-varying constraint (2–176).

Because of the p constraints among the m coordinates, there are only $n = m - p$ independent coordinates, and the system has n degrees of freedom. Occasionally, it is convenient to employ the redundant set

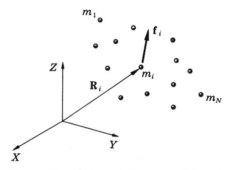

Fig. 2–62. System of N mass particles subjected to holonomic constraints (not shown). System has n degrees of freedom with $1 \leq n \leq 3N$.

(2–185) and to include the constraint relations (2–186) by means of the techniques described in Sec. 2–9. In most cases it is more convenient to select from the start a set of generalized coordinates,

$$\xi_1, \xi_2, \ldots, \xi_n, \qquad (2\text{--}187)$$

which is *complete* and *independent* and for which the corresponding set of variations,

$$\delta\xi_1, \delta\xi_2, \ldots, \delta\xi_n, \qquad (2\text{--}188)$$

is also a *complete* and *independent* set. Any admissible configuration of the system can be represented by the coordinates (2–187), and any admissible variation can be represented by the set (2–188). The analysis of holonomic systems is generally simplest in terms of such coordinates.

The position vector $\mathbf{R}_i$ of any particle m_i in Fig. 2–62 is fixed by the coordinates (2–187) and (if there are time-varying constraints) by the time t,

$$\mathbf{R}_i = \mathbf{R}_i(\xi_1, \xi_2, \ldots, \xi_n, t). \tag{2–189}$$

The velocity $\mathbf{v}_i$ of this particle is

$$\mathbf{v}_i = \frac{d\mathbf{R}_i}{dt} = \sum_{j=1}^{n} \frac{\partial \mathbf{R}_i}{\partial \xi_j} \dot{\xi}_j + \frac{\partial \mathbf{R}_i}{\partial t}. \tag{2–190}$$

The time derivatives $\dot{\xi}_j$ of the generalized coordinates are called *generalized velocities*. Equation (2–190) indicates that the actual particle velocities are usually complicated combinations of the generalized velocities. The final term on the right of (2–190) represents the effect (if any) of an externally imposed time-varying constraint.

In the variational approach we consider neighboring admissible configurations at the *same* instant of time. The variation of the position vector $\mathbf{R}_i$ of the ith particle in Fig. 2–62 is expressed in terms of the variations of the generalized coordinates by the variation of (2–189),

$$\delta\mathbf{R}_i = \sum_{j=1}^{n} \frac{\partial \mathbf{R}_i}{\partial \xi_j} \delta\xi_j. \tag{2–191}$$

GENERALIZED FORCES

We now suppose that on each particle m_i in Fig. 2–62 there acts a force $\mathbf{f}_i$. Let us evaluate the total work increment done by this set of forces under an admissible variation of the system's configuration. Using (2–191), we have

$$\sum_{i=1}^{N} \mathbf{f}_i \cdot \delta\mathbf{R}_i = \sum_{i=1}^{N} \sum_{j=1}^{n} \mathbf{f}_i \cdot \frac{\partial \mathbf{R}_i}{\partial \xi_j} \delta\xi_j$$

$$= \sum_{j=1}^{n} \left(\sum_{i=1}^{N} \mathbf{f}_i \cdot \frac{\partial \mathbf{R}_i}{\partial \xi_j} \right) \delta\xi_j. \tag{2–192}$$

Each parentheses in (2–192) is a combination of components of the actual forces taken along directions $\delta\mathbf{R}_i$ which result from incrementing a single generalized coordinate ξ_j. We call these combinations *generalized forces* and denote them by the symbols $\Xi_j, j = 1, 2, \ldots, n$:

$$\Xi_j = \sum_{i=1}^{N} \mathbf{f}_i \cdot \frac{\partial \mathbf{R}_i}{\partial \xi_j}. \tag{2–193}$$

Introducing these symbols in (2–192), we write

$$\sum_{i=1}^{N} \mathbf{f}_i \cdot \delta\mathbf{R}_i = \sum_{j=1}^{n} \Xi_j \, \delta\xi_j. \tag{2–194}$$

The total work of the actual force system acting through the displacements $\delta \mathbf{R}_i$ equals the total work of the generalized forces acting through the corresponding generalized displacements $\delta \xi_j$. Since each term of (2-194) represents work the dimensions of Ξ_j depend on those of ξ_j. If ξ_j has the dimension of *length*, then Ξ_j has the dimension of *force*; if ξ_j represents an *angle*, then Ξ_j has the dimension of *torque*.

When it is necessary to obtain the generalized force Ξ_j corresponding to a certain set of actual forces $\mathbf{f}_i$, one can evaluate the Ξ_j according to (2-193), or one can make use of (2-194) and the fact that the variations (2-188) are independent, as follows: To obtain Ξ_1, for example, consider the variation in which all the $\delta \xi_j$ vanish except for $\delta \xi_1$. Study the geometry of the system to determine the corresponding $\delta \mathbf{R}_i$ and evaluate the left side of (2-194). Since the right side is simply $\Xi_1 \, \delta \xi_1$, the generalized force Ξ_1 follows on dividing both sides by $\delta \xi_1$.

The nature of the actual set of forces $\mathbf{f}_i$ and the corresponding generalized forces Ξ_j can be arbitrary; i.e., generalized forces can be used to describe any of the following sets of forces: all forces acting, external forces, internal forces, conservative forces, nonconservative forces, or any other well-defined set of forces. In this book generalized forces Ξ_j are most commonly used to describe the nonconservative forces acting in a system.

GENERALIZED MOMENTA

For a single mass particle moving along a straight line the kinetic coenergy T^* has the property that its derivative with respect to the velocity v of the particle yields the momentum p of the particle,

$$\frac{\partial T^*}{\partial v} = \frac{\partial}{\partial v} \left(\int_0^v p \, dv \right) = p. \tag{2-195}$$

For the holonomic system sketched in Fig. 2-62, the kinetic coenergy

$$T^* = \sum_{i=1}^{N} \int_0^{v_i} \mathbf{p}_i \cdot d\mathbf{v}_i \tag{2-196}$$

is a function of the generalized coordinates ξ_i, the generalized velocities $\dot{\xi}_i$, and the time t, as can be seen if (2-190) is inserted in (2-196). The derivative of the kinetic coenergy (2-196) with respect to the generalized velocity $\dot{\xi}_k$ is called the *generalized momentum* π_k, corresponding to the coordinate ξ_k. By using (2-190) we find

$$\pi_k = \frac{\partial T^*}{\partial \dot{\xi}_k} = \sum_{i=1}^{N} \frac{\partial T^*}{\partial \mathbf{v}_i} \cdot \frac{\partial \mathbf{v}_i}{\partial \dot{\xi}_k} = \sum_{i=1}^{N} \mathbf{p}_i \cdot \frac{\partial \mathbf{v}_i}{\partial \dot{\xi}_k} = \sum_{i=1}^{N} \mathbf{p}_i \cdot \frac{\partial \mathbf{R}_i}{\partial \xi_k}. \tag{2-197}$$

The generalized momentum π_k is thus a combination of the components of the actual momenta $\mathbf{p}_i$ along the directions of the displacements $\delta\mathbf{R}_i$ which result from an increment $\delta\xi_k$ in the generalized coordinate ξ_k.

Whenever there are no time-varying constraints in a holonomic system there is an interesting identity connecting the actual and the generalized momenta. This identity is obtained by expressing the sum of the kinetic state functions (2–100) in terms of generalized variables. When time-varying constraints are absent, (2–190) reduces to

$$\mathbf{v}_i = \sum_{j=1}^{n} \frac{\partial\mathbf{R}_i}{\partial\xi_j}\,\dot{\xi}_j. \tag{2–198}$$

When (2–198) is substituted in (2–100), we obtain

$$T + T^* = \sum_{i=1}^{N} \mathbf{v}_i \cdot \mathbf{p}_i = \sum_{j=1}^{n} \dot{\xi}_j \left(\sum_{i=1}^{N} \frac{\partial\mathbf{R}_i}{\partial\xi_j} \cdot \mathbf{p}_i \right) = \sum_{j=1}^{n} \dot{\xi}_j \pi_j, \tag{2–199}$$

on interchanging the order of summation and introducing the definition (2–197) of generalized momentum. Thus the sum $T + T^*$ for a system with time-invariant constraints can be obtained by adding the products of the actual velocities $\mathbf{v}_i$ and the corresponding momenta $\mathbf{p}_i$ or by adding the products of the generalized velocities $\dot{\xi}_j$ and corresponding generalized momenta π_j. Note the parallelism between the definition (2–193) for generalized force and the definition (2–197) for generalized momentum. Note also the parallelism between (2–194) and (2–199). The former states that the sum of the *actual* force-displacement products equals the sum of the *generalized* force-displacement products. The latter states that the sum of the *actual* velocity-momentum products equals the sum of the *generalized* velocity-momentum products.

Example 2–14. To illustrate the relations between the *actual* velocities, forces, and momenta in a system and the *generalized* forces, velocities, and momenta corresponding to a particular choice of generalized coordinates, we consider the spherical pendulum shown in Fig. 2–63. A single mass particle m is suspended by a rigid massless rod of length a from a frictionless ball joint at the origin. This is a holonomic system with two degrees of freedom and with no externally forced time-varying constraints. The angles φ and θ indicated in Fig. 2–63 are an independent, complete set of generalized coordinates: $\xi_1 = \theta$, $\xi_2 = \varphi$. The position vector $\mathbf{R}$ for the mass particle is

$$\mathbf{R} = a\sin\theta\cos\varphi\,\mathbf{u}_X + a\sin\theta\sin\varphi\,\mathbf{u}_Y - a\cos\theta\,\mathbf{u}_Z, \tag{2–200}$$

which is an illustration of (2–189). By differentiation the velocity is

$$\mathbf{v} = \frac{d\mathbf{R}}{dt} = a(\cos\theta\cos\varphi\,\dot{\theta} - \sin\theta\sin\varphi\,\dot{\varphi})\mathbf{u}_X$$

$$+ a\,(\cos\theta\sin\varphi\,\dot{\theta} + \sin\theta\cos\varphi\,\dot{\varphi})\mathbf{u}_Y + a\sin\theta\,\dot{\theta}\mathbf{u}_Z. \tag{2–201}$$

The actual velocity is thus a complicated combination of the *generalized velocities* $\xi_1 = \theta$ and $\xi_2 = \dot{\varphi}$. For comparison with the general result (2–190), we can regroup the terms in (2–201) as follows:

$$\mathbf{v} = (a \cos \theta \cos \varphi \, \mathbf{u}_X + a \cos \theta \sin \varphi \, \mathbf{u}_Y + a \sin \theta \, \mathbf{u}_Z)\dot{\theta}$$
$$+ (-a \sin \theta \sin \varphi \, \mathbf{u}_X + a \sin \theta \cos \varphi \, \mathbf{u}_Y)\dot{\varphi}. \quad (2\text{–}202)$$

To obtain the *generalized forces* corresponding to the actual force $\mathbf{f} = f_0 \mathbf{u}_X$, we use (2–194) as follows: For a variation in which $\xi_1 = \theta$ sustains an increment $\delta\theta$ while $\delta\varphi = 0$, the work done by $\mathbf{f}$ is the product of f_0 and the X component of the displacement $\delta\mathbf{R}$. This component is

$$a \cos \theta \cos \varphi \, \delta\theta, \quad (2\text{–}203)$$

as may be seen directly from the geometry of Fig. 2–63 or by applying

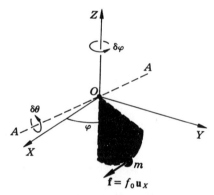

FIG. 2–63. Example 2–14. Spherical pendulum with force acting parallel to X axis.

the variational operator to (2–200). Substituting in (2–194), we have

$$\Xi_1 \, \delta\xi_1 = f_0 a \cos \theta \cos \varphi \, \delta\theta. \quad (2\text{–}204)$$

Now by simply canceling $\delta\xi_1 = \delta\theta$ from both sides of (2–204), we obtain

$$\Xi_1 = f_0 a \cos \theta \cos \varphi. \quad (2\text{–}205)$$

A similar calculation based on the variation in which $\xi_2 = \varphi$ sustains an increment $\delta\varphi$, while $\delta\theta = 0$ yields

$$\Xi_2 = -f_0 a \sin \theta \sin \varphi. \quad (2\text{–}206)$$

Note that since ξ_1 and ξ_2 are angles, the generalized forces have the dimensions of torque. Note also that the generalized force Ξ_1 corresponding to $\xi_1 = \theta$ can be interpreted as the moment of $\mathbf{f}$ about the

axis AA, which is the axis for the infinitesimal rotation $\delta\theta$.   The generalized force Ξ_2 corresponding to $\xi_2 = \varphi$ can be interpreted as the moment
of $\mathbf{f}$ about the Z axis, which is the axis for the infinitesimal rotation $\delta\varphi$.

To obtain the *generalized momenta* we begin by evaluating the kinetic
coenergy in terms of the generalized coordinates.   If we assume the
newtonian relation (2–86), the actual momentum $\mathbf{p}$ of the mass particle
is simply $m\mathbf{v}$, where $\mathbf{v}$ is given by (2–201).   The kinetic coenergy is

$$T^* = \tfrac{1}{2}mv^2 = \tfrac{1}{2}ma^2(\dot\theta^2 + \dot\varphi^2 \sin^2 \theta). \qquad (2\text{–}207)$$

Then, according to (2–197), we calculate

$$\pi_1 = \frac{\partial T^*}{\partial \dot\theta} = ma^2\dot\theta$$

$$\qquad (2\text{–}208)$$

$$\pi_2 = \frac{\partial T^*}{\partial \dot\varphi} = ma^2\dot\varphi \sin^2 \theta.$$

Observe that the generalized momentum π_1 corresponding to $\xi_1 = \theta$ can
be interpreted as the particle's angular-momentum component about the
axis AA (the axis for $\delta\theta$) and that the generalized momentum π_2 corresponding to $\xi_2 = \varphi$ can be interpreted as the angular-momentum component about the Z axis (the axis for $\delta\varphi$).

The identity (2–199) for systems with time-invariant constraints
reduces in the present case to

$$\mathbf{v} \cdot \mathbf{p} = \dot\xi_1\pi_1 + \dot\xi_2\pi_2. \qquad (2\text{–}209)$$

By using (2–202) and (2–208), the reader can readily verify the validity
of (2–209).

2–11   Hamilton's principle

In Sec. 1–4 the variational indicator for Hamilton's principle is derived
for a particular system with a single degree of freedom.   The extension
there to more complex systems is stated without proof.   It is now possible to give a generalized derivation of *Hamilton's principle* for holonomic
systems of arbitrary complexity.   In this section we present such a derivation, and by applying the calculus of variations to the variational indicator, we derive the corresponding generalized equations of motion known
as *Lagrange's equations*.   In many holonomic systems the simplest route
to the equations of motion is via Lagrange's equations.

We consider the general holonomic system of N mass particles m_i with
n degrees of freedom sketched in Fig. 2–62 and described in the preceding section.   The dynamic equations relating forces and momenta for

this system are

$$\mathbf{f}_i - \frac{d\mathbf{p}_i}{dt} = 0 \qquad i = 1, 2, \ldots, N, \qquad (2\text{-}210)$$

where $\mathbf{f}_i$ is the resultant force acting *on* the particle m_i, and $\mathbf{p}_i$ is the momentum of the particle. The scalar variational indicator

$$\sum_{i=1}^{N} \left(\mathbf{f}_i - \frac{d\mathbf{p}_i}{dt} \right) \cdot \delta\mathbf{R}_i \qquad (2\text{-}211)$$

vanishes for arbitrary variations if every equation of the set (2–210) is satisfied. The converse is usually untrue because the $\delta\mathbf{R}_i$ are generally not independent. To see what *can* be learned when (2–211) vanishes for arbitrary admissible variations, we introduce in (2–211) the generalized coordinates ξ_j of (2–187) and use (2–191) to obtain

$$\sum_{j=1}^{n} \left[\sum_{i=1}^{N} \left(\mathbf{f}_i - \frac{d\mathbf{p}_i}{dt} \right) \cdot \frac{\partial\mathbf{R}_i}{\partial\xi_j} \right] \delta\xi_j. \qquad (2\text{-}212)$$

Since the $\delta\xi_j$ are independent, we see that the necessary conditions for the indicator to vanish for arbitrary admissible variations are a set of n generalized force-momentum relations,

$$\sum_{i=1}^{N} \left(\mathbf{f}_i - \frac{d\mathbf{p}_i}{dt} \right) \cdot \frac{\partial\mathbf{R}_i}{\partial\xi_j} = 0 \qquad j = 1, \ldots, n. \qquad (2\text{-}213)$$

Each equation in (2–213) is a combination of components of the equations (2–210), where the components are taken along the directions of displacements $\delta\mathbf{R}_i$, which result from an increment $\delta\xi_j$ of a single generalized variable ξ_j. If each of the N particles is unconstrained in three dimensions, then $n = 3N$ and the set (2–213) is completely equivalent to the set (2–210). If, however, there are constraints, then $n < 3N$ and the set (2–213) does not include those combinations of components of (2–210) which apply to directions of motions prohibited by the constraints. The set (2–213) does, however, provide n independent force-momentum relations, so that when combined with the geometric relations (2–189) and with the constitutive relations and velocity-momentum relations, they lead to n equations of motion for the n generalized coordinates.

Hamilton's principle provides an alternative route to an equivalent set of n equations of motion. To derive the principle we use the constitutive and the velocity-momentum relations to transform the variational indicator (2–211) into a more convenient form. We consider first the expression

$$\sum_{i=1}^{N} \mathbf{f}_i \cdot \delta\mathbf{R}_i, \qquad (2\text{-}214)$$

which represents the work done by all forces acting on all masses within the system. In general, this work can be divided into a contribution from conservative elements and fields plus a contribution from nonconservative forces. Let the potential energy of the conservative forces be $V(\xi_1, \ldots, \xi_n)$, and let the *generalized* forces corresponding to the *nonconservative* forces be denoted by Ξ_j. Then, according to (2–114) and (2–193), we can represent (2–214) as follows:

$$\sum_{i=1}^{N} \mathbf{f}_i \cdot \delta \mathbf{R}_i = -\delta V + \sum_{j=1}^{n} \Xi_j \, \delta \xi_j. \qquad (2\text{–}215)$$

The total work done equals the decrease in potential energy of the conservative elements plus an explicit evaluation of the work increments of the nonconservative forces.

Substituting (2–215) into (2–211), we obtain

$$-\delta V + \sum_{j=1}^{n} \Xi_j \, \delta \xi_j - \sum_{i=1}^{N} \frac{d\mathbf{p}_i}{dt} \cdot \delta \mathbf{R}_i \qquad (2\text{–}216)$$

for the force-momentum variational indicator. We next consider the momentum terms in (2–216). Introducing the identity responsible for integration by parts, we have

$$-\sum_{i=1}^{N} \frac{d\mathbf{p}_i}{dt} \cdot \delta \mathbf{R}_i = -\sum_{i=1}^{N} \frac{d}{dt} (\mathbf{p}_i \cdot \delta \mathbf{R}_i) + \sum_{i=1}^{N} \mathbf{p}_i \cdot \frac{d}{dt} (\delta \mathbf{R}_i). \quad (2\text{–}217)$$

The last term on the right of (2–217) is identical with the variation of the kinetic coenergy,

$$\delta T^* = \delta \left(\sum_{i=1}^{N} \int \mathbf{p}_i \cdot d\mathbf{v}_i \right) = \sum_{i=1}^{N} \mathbf{p}_i \cdot \delta \mathbf{v}_i, \qquad (2\text{–}218)$$

because of the commutativity relation

$$\delta \mathbf{v}_i = \delta \left(\frac{d\mathbf{R}_i}{dt} \right) = \frac{d}{dt} (\delta \mathbf{R}_i). \qquad (2\text{–}219)$$

Thus, in the place of (2–217), we can write

$$-\sum_{i=1}^{N} \frac{d\mathbf{p}_i}{dt} \cdot \delta \mathbf{R}_i = -\sum_{i=1}^{N} \frac{d}{dt} (\mathbf{p}_i \cdot \delta \mathbf{R}_i) + \delta T^*. \qquad (2\text{–}220)$$

After insertion of (2–220) into (2–216), the variational indicator takes the form

$$\delta T^* - \delta V + \sum_{j=1}^{n} \Xi_j \, \delta \xi_j - \sum_{i=1}^{N} \frac{d}{dt} (\mathbf{p}_i \cdot \delta \mathbf{R}_i). \qquad (2\text{–}221)$$

The indicator (2–221) vanishes for arbitrary admissible variations at any instant for which the dynamic equations of (2–210) are satisfied. Conversely, if at any instant the indicator vanishes for arbitrary admissible variations, the n generalized force-momentum relations of (2–213) must be satisfied at that instant.

The final transformation of the indicator is accomplished by integrating (2–221) with respect to time over an interval from $t = t_1$ to $t = t_2$, which extends beyond the period of interest in the analysis. The last term on the right of (2–221) drops out at the limits, provided we agree to consider that all variations $\delta \mathbf{R}_i$ vanish at $t = t_1$ and $t = t_2$. The resulting variational indicator has the following form:

$$\text{V.I.} = \int_{t_1}^{t_2} \left[\delta(T^* - V) + \sum_{j=1}^{n} \Xi_j \, \delta \xi_j \right] dt. \qquad (2\text{--}222)$$

Admissible variations are represented by the n independent $\delta \xi_j$. We have agreed that these are to vanish at t_1 and t_2, but they are otherwise arbitrary functions of time within the interval from t_1 to t_2. If the dynamic equations (2–210) are satisfied for *every* instant of time between t_1 and t_2, then (2–222) vanishes for arbitrary-admissible-variation histories. Conversely, if (2–222) vanishes for arbitrary admissible variations, the n generalized force-momentum relations of (2–213) must be satisfied at every instant within the interval from t_1 to t_2. This constitutes a derivation of *Hamilton's principle* for a general holonomic system with n degrees of freedom: *An admissible motion of the system between specified configurations at t_1 and t_2 is a natural motion if, and only if, the variational indicator (2–222) vanishes for arbitrary admissible variations.*

LAGRANGE'S EQUATIONS

We next apply the calculus of variations to the generalized variational indicator (2–222) to obtain the necessary conditions for the indicator to vanish for arbitrary variations. Let us introduce the symbol $\mathcal{L}$ for the *lagrangian function* (or simply the *lagrangian*),

$$\mathcal{L} = T^* - V. \qquad (2\text{--}223)$$

In general, V is a function of the ξ_j while T^* is a function of the ξ_j, the $\dot{\xi}_j$, and[1] the time t. The lagrangian is therefore a function of these same variables,

$$\mathcal{L} = \mathcal{L}(\xi_1, \xi_2, \ldots, \xi_n, \dot{\xi}_1, \dot{\xi}_2, \ldots, \dot{\xi}_n, t), \qquad (2\text{--}224)$$

and its variation is

$$\delta \mathcal{L} = \sum_{j=1}^{n} \left(\frac{\partial \mathcal{L}}{\partial \xi_j} \delta \xi_j + \frac{\partial \mathcal{L}}{\partial \dot{\xi}_j} \delta \dot{\xi}_j \right). \qquad (2\text{--}225)$$

[1] The time t enters T^* explicitly only when there are time-varying constraints.

If we substitute (2–225) into (2–222) and insert the commutativity relation $\delta\dot{\xi}_j = d(\delta\xi_j)/dt$, the variational indicator becomes

$$\text{V.I.} = \int_{t_1}^{t_2} \sum_{j=1}^{n} \left[\frac{\partial\mathcal{L}}{\partial\dot{\xi}_j}\frac{d}{dt}(\delta\xi_j) + \frac{\partial\mathcal{L}}{\partial\xi_j}\delta\xi_j + \Xi_j\,\delta\xi_j \right] dt. \qquad (2\text{--}226)$$

Integrating the first terms in the integrand by parts and using the agreement that the $\delta\xi_j$ vanish at $t = t_1$ and $t = t_2$, we find

$$\text{V.I.} = -\int_{t_1}^{t_2} \sum_{j=1}^{n} \left[\frac{d}{dt}\left(\frac{\partial\mathcal{L}}{\partial\dot{\xi}_j}\right) - \frac{\partial\mathcal{L}}{\partial\xi_j} - \Xi_j \right] \delta\xi_j\,dt. \qquad (2\text{--}227)$$

The necessary conditions for (2–227) to vanish for arbitrary $\delta\xi_j$ are that the n equations,

$$\frac{d}{dt}\left(\frac{\partial\mathcal{L}}{\partial\dot{\xi}_j}\right) - \frac{\partial\mathcal{L}}{\partial\xi_j} = \Xi_j \qquad j = 1, 2, \ldots, n, \qquad (2\text{--}228)$$

must be satisfied at each instant within the interval from t_1 to t_2. The equations in (2–228) are called *Lagrange's equations*. They are the equations of motion of the system in terms of the generalized coordinates ξ_j. They are equivalent to the generalized force-momentum relations (2–213) after the constitutive equations and velocity-momentum relations have been used to eliminate forces and momenta in favor of geometrical variables, and after all geometrical variables have been expressed in terms of the generalized coordinates. For many holonomic systems the most direct route to the equations of motion is furnished by construction of Lagrange's equations. The following steps are required:

1. Establish a complete set of independent generalized coordinates ξ_j.
2. Identify generalized nonconservative forces Ξ_j (if any).
3. Construct the lagrangian (2–223).
4. Substitute in Lagrange's equations (2–228).

Example 2–15.   An idealized double pendulum is shown in Fig. 2–64. The two mass particles are attached to rigid massless links connected by frictionless joints. The motion is constrained to remain in the plane of the figure.

The significant elements are the mass particles and the gravity field. Geometric admissibility can be achieved by the choice of the complete and independent set of generalized coordinates θ_1 and θ_2 indicated. There are no nonconservative elements in the system, so that all the dynamic properties of the system reside in the lagrangian,

$$\mathcal{L} = T^* - V = \tfrac{1}{2}m_1v_1^2 + \tfrac{1}{2}m_2v_2^2 - m_1gz_1 - m_2gz_2. \qquad (2\text{--}229)$$

In order to express this in terms of the generalized coordinates, we use the geometric relations

$$z_1 = a(1 - \cos \theta_1)$$
$$z_2 = b(1 - \cos \theta_1) + c(1 - \cos \theta_2),$$

(2–230)

from Fig. 2–64a, and the kinematic relations

$$v_1^2 = a^2\dot\theta_1^2$$
$$v_2^2 = b^2\dot\theta_1^2 + c^2\dot\theta_2^2 + 2bc\dot\theta_1\dot\theta_2 \cos (\theta_2 - \theta_1),$$

(2–231)

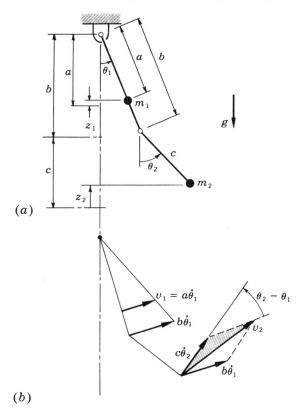

(a)

(b)

Fig. 2–64.   Example 2–15.   Double pendulum.   (a) System and coordinates.   (b) Velocity analysis.

from Fig. 2–64b.   The second of (2–231) is the law of cosines applied to the shaded triangle in Fig. 2–64b.   The lagrangian (2–229) then becomes

$$\mathcal{L} = \tfrac{1}{2}(m_1a^2 + m_2b^2)\dot\theta_1^2 + m_2bc\dot\theta_1\dot\theta_2 \cos (\theta_2 - \theta_1)$$
$$+ \tfrac{1}{2}m_2c^2\dot\theta_2^2 - (m_1a + m_2b)g(1 - \cos \theta_1) - m_2cg(1 - \cos \theta_2).$$ (2–232)

At this point the physical considerations have been completed. Substitution in Lagrange's equations (2–228) requires only routine mathematical operations. The final equations of motion are a pair of coupled nonlinear differential equations for $\theta_1(t)$ and $\theta_2(t)$:

$$\frac{d}{dt}\left[(m_1a^2 + m_2b^2)\dot{\theta}_1 + m_2bc\dot{\theta}_2 \cos (\theta_2 - \theta_1)\right]$$

$$- m_2bc\dot{\theta}_1\dot{\theta}_2 \sin (\theta_2 - \theta_1) + (m_1a + m_2b)g \sin \theta_1 = 0$$

$$\frac{d}{dt}\left[m_2bc\dot{\theta}_1 \cos (\theta_2 - \theta_1) + m_2c^2\dot{\theta}_2\right]$$

$$+ m_2bc\dot{\theta}_1\dot{\theta}_2 \sin (\theta_2 - \theta_1) + m_2cg \sin \theta_2 = 0.$$

(2–233)

LINEARIZATION OF THE EQUATIONS OF MOTION

An approximate picture of the dynamic behavior of the double pendulum in the neighborhood of $\theta_1 = 0$, $\theta_2 = 0$, can be obtained from a set of *linear* differential equations. One procedure for obtaining the linear equations is first to obtain the complete nonlinear equations (2–233) and then to apply a systematic linearization. An alternative procedure is to introduce suitable approximations into the lagrangian (2–232) so that Lagrange's equations are automatically linear.

Following the first procedure, we replace the trigonometric functions in (2–233) by the series expansions

$$\sin \theta = \theta - \frac{\theta^3}{3!} + \frac{\theta^5}{5!} - \cdots$$

$$\cos \theta = 1 - \frac{\theta^2}{2!} + \frac{\theta^4}{4!} - \cdots,$$

(2–234)

and then, when the equations are expanded, we retain only those terms which are *linear* in the small angles θ_1, θ_2 and their derivatives $\dot{\theta}_1$, $\dot{\theta}_2$; i.e., we neglect all squares, products, and higher powers of these small quantities. The resulting linearized equations are

$$\begin{bmatrix} m_1a^2 + m_2b^2 & m_2bc \\ m_2bc & m_2c^2 \end{bmatrix} \begin{Bmatrix} \ddot{\theta}_1 \\ \ddot{\theta}_2 \end{Bmatrix} + \begin{bmatrix} (m_1a + m_2b)g & 0 \\ 0 & m_2cg \end{bmatrix} \begin{Bmatrix} \theta_1 \\ \theta_2 \end{Bmatrix} = \begin{Bmatrix} 0 \\ 0 \end{Bmatrix}.$$

(2–235)

These are still coupled second-order differential equations, but they are *linear*, with *constant coefficients*.

Following the alternative procedure, we return to the lagrangian (2–232), insert the series expansions (2–234), and then retain only those terms which are *quadratic* in the small quantities θ_1, θ_2, $\dot{\theta}_1$, $\dot{\theta}_2$; i.e., we retain squares and products of two variables, but we neglect all higher powers and products.

$$\mathcal{L} = \tfrac{1}{2}(m_1 a^2 + m_2 b^2)\dot{\theta}_1^2 + m_2 b c \dot{\theta}_1 \dot{\theta}_2$$
$$+ \tfrac{1}{2} m_2 c^2 \dot{\theta}_2^2 - \tfrac{1}{2}(m_1 a + m_2 b)g\theta_1^2 - \tfrac{1}{2} m_2 c g \theta_2^2. \quad (2\text{–}236)$$

When the quadratic lagrangian (2–236) is inserted in Lagrange's equations (2–228), the linear equations (2–235) are obtained. The alternative procedure is particularly convenient when only the linearized equations are required and when it is possible to obtain the quadratic lagrangian directly from the sketch of the system. To this end it is useful to note that when v is a first-order small quantity, then $\tfrac{1}{2}mv^2$ is a second-order small quantity, and hence a linear approximation to v will provide a quadratic approximation to $\tfrac{1}{2}mv^2$. Thus, in Fig. 2–64b, a linear approxi-

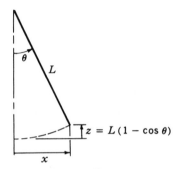

FIG. 2–65. Correct to second order in θ, the displacement z can be approximated by $\tfrac{1}{2}L\theta^2$ or by $x^2/2L$.

mation to v_2 is $b\dot{\theta}_1 + c\dot{\theta}_2$, so that a quadratic approximation to T^* is

$$\tfrac{1}{2}m_1 b^2 \dot{\theta}_1^2 + \tfrac{1}{2}m_2(b\dot{\theta}_1 + c\dot{\theta}_2)^2. \quad (2\text{–}237)$$

Note that (2–237) is identical with the kinetic-coenergy contribution to (2–236).

In the case of a gravitational potential-energy term mgz, it is necessary to employ a quadratic approximation to z in order to maintain a quadratic approximation. Convenient quadratic approximations for pendulum configurations are displayed in Fig. 2–65. Note that the potential-energy contributions to the quadratic lagrangian (2–236) follow directly from Fig. 2–64a when the approximations of Fig. 2–65 are available.

2–12   Conservation laws

Under certain special circumstances the equations of motion of a dynamic system may possess standardized integrals which can be recognized in advance. These standardized integrals are statements of conservation of energy or conservation of momentum. In this section we consider circumstances under which these integrals occur for holonomic systems of N particles which have n degrees of freedom.

CONSERVATION OF ENERGY

We shall derive the energy integral for the class of systems in which there are no nonconservative forces and for which there are no time-varying constraints. For a system of this class the lagrangian (2-224) is a function of n independent generalized coordinates ξ_j and the corresponding n generalized velocities $\dot{\xi}_j$. The equations of motion are Lagrange's equations (2-228), with $\Xi_j = 0$,

$$\frac{d}{dt}\left(\frac{\partial \mathcal{L}}{\partial \dot{\xi}_j}\right) - \frac{\partial \mathcal{L}}{\partial \xi_j} = 0 \qquad j = 1, \ldots, n. \tag{2-238}$$

We now demonstrate that one integral of (2-238) can always be found. This integral amounts to a statement that the total mechanical energy $T + V$ is conserved during the motion of the system. Our demonstration is independent of the velocity-momentum relations for the system.

Let us multiply the jth equation in (2-238) by the generalized velocity $\dot{\xi}_j$ and then sum over j to get

$$\sum_{j=1}^{n}\left[\dot{\xi}_j \frac{d}{dt}\left(\frac{\partial \mathcal{L}}{\partial \dot{\xi}_j}\right) - \dot{\xi}_j \frac{\partial \mathcal{L}}{\partial \xi_j}\right] = 0. \tag{2-239}$$

The first terms on the left are then transformed by using the identity that underlies integration by parts.

$$\sum_{j=1}^{n}\left[\frac{d}{dt}\left(\dot{\xi}_j \frac{\partial \mathcal{L}}{\partial \dot{\xi}_j}\right) - \frac{\partial \mathcal{L}}{\partial \dot{\xi}_j}\ddot{\xi}_j - \dot{\xi}_j \frac{\partial \mathcal{L}}{\partial \xi_j}\right] = 0. \tag{2-240}$$

This result can now be integrated with respect to time. The first terms on the left are explicitly time derivatives, and the remaining terms represent the total time derivative of the lagrangian since

$$\frac{d\mathcal{L}}{dt} = \sum_{j=1}^{n}\frac{\partial \mathcal{L}}{\partial \dot{\xi}_j}\frac{d}{dt}(\dot{\xi}_j) + \sum_{j=1}^{n}\frac{\partial \mathcal{L}}{\partial \xi_j}\frac{d}{dt}(\xi_j). \tag{2-241}$$

Therefore the integral of (2-240) is

$$\sum_{j=1}^{n}\dot{\xi}_j\frac{\partial \mathcal{L}}{\partial \dot{\xi}_j} - \mathcal{L} = E, \tag{2-242}$$

where E is a constant of integration having the dimensions of work or energy. Now, since $\mathcal{L} = T^* - V$ and the potential energy V is independent of the velocities $\dot{\xi}_j$, we have

$$\frac{\partial \mathcal{L}}{\partial \dot{\xi}_j} = \frac{\partial T^*}{\partial \dot{\xi}_j} = \pi_j, \qquad (2\text{–}243)$$

and thus we recognize that the summation in (2–242) is precisely that of (2–199). The integral (2–242) then becomes

$$T + T^* - \mathcal{L} = E$$
$$T + T^* - (T^* - V) = E \qquad (2\text{–}244)$$
$$T + V = E.$$

This result expresses the *law of conservation of mechanical energy:* The sum of the instantaneous kinetic energy and the instantaneous potential energy remains constant throughout the motion. A motion of a dynamic system for which (2–244) holds is called a *conservative motion.* If all natural motions of a system are conservative, the system is said to be a *conservative system.* The foregoing derivation has shown that sufficient conditions for a holonomic system to be conservative are that there be no nonconservative forces and no time-varying constraints.

The conservation law (2–244) applies independently of the particular velocity-momentum relations assumed. For systems with nonlinear velocity-momentum relations or nonlinear constitutive relations it is important to note that the state functions which appear in the conservation law are energies (unstarred) and not coenergies (starred). For linear systems the coenergies are equal in magnitude to the corresponding energies. In particular, for the newtonian case where $T^* = T$, a useful corollary of (2–244) is

$$T^* + V = E. \qquad (2\text{–}245)$$

The integral (2–245) is a single first integral of the equations of motion which relates the coordinates ξ_j and the velocities $\dot{\xi}_j$. In conservative systems with n degrees of freedom it is necessary to obtain $n - 1$ additional independent first integrals in order to completely fix the velocities as functions of coordinates; however, in systems with only one degree of freedom the energy integral is sufficient by itself to fix the velocity $\dot{\xi}_1$ as a function of the single generalized coordinate ξ_1.

Example 2–16. A mass particle m moves without friction on a horizontal plane restrained by a massless inextensible string which wraps around a fixed cylindrical peg of radius r. The problem is to determine how the magnitude v of the velocity varies as the particle spirals inward, under the assumption that the string remains taut and simply wraps itself around the peg.

This problem may be solved immediately if it is recognized that the system is conservative.   The position of a particle in a plane can be fixed by two generalized coordinates, e.g., the coordinates L and θ in Fig. 2–66b. The string wrapping around the peg provides a holonomic-constraint relation between L and θ.   If L_0 is the unwrapped length when $\theta = 0$, then

$$L = L_0 - r\theta. \qquad (2\text{–}246)$$

This constraint is independent of time.   Furthermore, there are no

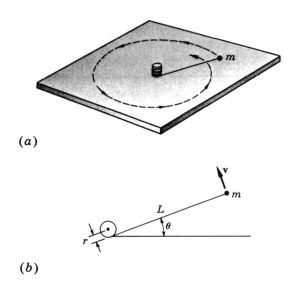

(a)

(b)

FIG. 2–66. Example 2–16. Particle moving in a plane restrained by a string which wraps around a fixed peg. (a) Perspective view. (b) Top view.

nonconservative forces which do work during the motion.   There is no friction, and the tension in the string does no work.   This last can be seen by considering an admissible variation of configuration described by $\delta\theta$. The admissible displacements are at right angles to the line of action of the string tension.   Thus the system is conservative, and the total mechanical energy is conserved.   Since there is no change in potential energy during the motion, the kinetic energy T must remain constant. This implies that the kinetic coenergy T^* also remains constant, and hence that the magnitude v of the velocity remains *unchanged* as the particle spirals inward.

It is instructive to compare this example with Example 2–5, where the inward spiral is caused by pulling the string through a central hole.   In that case angular momentum is conserved but the kinetic energy is increased; work is done by the force pulling the string through the hole. In the present example energy is conserved but the angular momentum is decreased; the string tension exerts a retarding torque about the center of the peg.

CONSERVATION OF GENERALIZED MOMENTUM

We return to the general holonomic system with n degrees of freedom and Lagrange's equations (2–228); i.e.,

$$\frac{d}{dt}\left(\frac{\partial \mathcal{L}}{\partial \dot{\xi}_j}\right) - \frac{\partial \mathcal{L}}{\partial \xi_j} = \Xi_j \qquad j = 1, 2, \ldots, n. \qquad (2\text{–}247)$$

In general, the lagrangian depends on all n of the coordinates ξ_j and their derivatives $\dot{\xi}_j$.   There are, however, cases in which one (or more) of the coordinates ξ_k is absent from the lagrangian although the corresponding velocity $\dot{\xi}_k$ is present.   If at the same time the corresponding nonconservative generalized force Ξ_k is also absent, the kth equation integrates immediately into

$$\frac{\partial \mathcal{L}}{\partial \dot{\xi}_k} = \pi_k = \text{constant}, \qquad (2\text{–}248)$$

i.e., the *generalized momentum* corresponding to a particular coordinate ξ_k *is conserved* whenever (1) the coordinate ξ_k does not enter the lagrangian explicitly, and (2) the corresponding generalized force Ξ_k is zero.   Under these circumstances the coordinate ξ_k is said to be a *cyclic*, or *ignorable*, coordinate.

Example 2–17.   To illustrate the use of the conservation laws, we consider the spherical pendulum of Fig. 2–67.   A mass particle m is suspended without friction from a fixed pivot O by a rigid massless arm of length L.   The system is holonomic, with two degrees of freedom.   We shall take the spherical coordinates θ and φ as generalized coordinates. The kinetic coenergy (assuming the newtonian velocity-momentum relation) is

$$T^* = \tfrac{1}{2}mL^2(\dot{\theta}^2 + \dot{\varphi}^2 \sin^2 \theta), \qquad (2\text{–}249)$$

and the potential energy with respect to the datum plane $Z = 0$ is

$$V = -mgL \cos \theta. \qquad (2\text{–}250)$$

There are no nonconservative forces acting, and the pendulum constraint is independent of time.

We begin by noting that while both θ and $\dot{\theta}$ appear in the lagrangian $T^* - V$, the coordinate φ does not enter except through its derivative $\dot{\varphi}$.

Since $\Xi_\varphi \equiv 0$, the coordinate φ is cyclic, and the corresponding generalized momentum is conserved; i.e.,

$$\pi_\varphi = \frac{\partial T^*}{\partial \varphi} = mL^2\dot{\varphi}\sin^2\theta = \pi_0, \qquad (2\text{-}251)$$

where π_0 is a constant which can be evaluated if the initial conditions of the motion are known. Note that (2-251) states that the angular-momentum component of the particle about the Z axis remains constant throughout the motion. This is a first integral of one of the equations of motion.

A second independent integral can be obtained here (still without

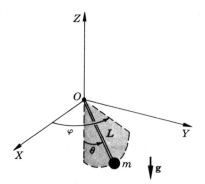

Fig. 2-67. Example 2-17. Spherical pendulum in gravity field.

formulating the actual[1] equations of motion) because the system also conserves energy. Substituting in (2-245), we have

$$\tfrac{1}{2}mL^2(\dot{\theta}^2 + \dot{\varphi}^2\sin^2\theta) + mgL\cos\theta = E, \qquad (2\text{-}252)$$

where E is a constant which can be evaluated if the initial coordinates and velocities are known. The two integrals (2-251) and (2-252) are sufficient to determine both velocities, $\dot{\theta}$ and $\dot{\varphi}$, in terms of the coordinate θ and the constants of integration π_0 and E.

$$\dot{\theta}^2 = \frac{2E}{mL^2} + \frac{2g}{L}\cos\theta - \left(\frac{\pi_0}{mL^2\sin\theta}\right)^2$$

$$\dot{\varphi} = \frac{\pi_0}{mL^2\sin^2\theta}. \qquad (2\text{-}253)$$

Note that the velocities are independent of the cyclic coordinate φ.

[1] See Prob. 2-46.

2-13   Momentum principles for a system of particles

For a single mass particle the force applied to the particle equals the time rate of change of the momentum of the particle; see (2–88). Also, the torque applied to the particle with respect to a point B equals the time rate of change of the particle's angular momentum with respect to B for any fixed point B and for certain moving points B; see (2–92). In the present section we develop corresponding linear- and angular-momentum principles for a system of particles. These principles often permit useful

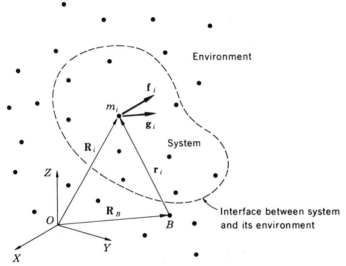

FIG. 2-68.   System of N particles of fixed identity $m_1, m_2, \ldots , m_N$. Point B is moment center.

deductions about the behavior of a group of particles without requiring detailed analysis of the motions of the constituent particles.

In Fig. 2–68 a system of N particles which interact with each other and with their environment is sketched. We suppose that these particles can be identified and that the same set of N particles constitutes the system at all times. Let f_i represent the resultant force of the external environment acting on the particle m_i, and let g_i represent the resultant force acting on m_i due to its interaction with all the other particles in the system; i.e., we assume that the total force acting on each particle m_i in the system can be decomposed into an *external* force f_i and an *internal* force g_i.

It is not necessary to know the precise nature of the internal-force system. To obtain the system momentum principles, the least stringent assumption that has to be made concerning the internal forces is the assumption that at all times the total set of internal forces is an *equilibrium* set, i.e., a set of forces g_i such that the vector sum of all the g_i vanishes,

$$\sum_{i=1}^{N} g_i = 0, \qquad (2\text{-}254)$$

and the vector sum of the moments of all the g_i about an arbitrary point B vanishes,

$$\sum_{i=1}^{N} r_i \times g_i = 0. \qquad (2\text{-}255)$$

At relative velocities which are small in comparison with the velocity of light,[1] the known interatomic and intermolecular forces are all of this type.

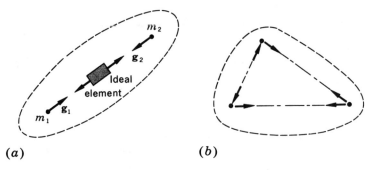

(a) $\qquad\qquad\qquad\qquad$ (b)

Fig. 2-69. Self-equilibrating internal force systems (a) due to ideal force elements, (b) due to mutual pairwise interactions which are individually in equilibrium.

In macroscopic models the internal forces due to massless connecting links and ideal force elements are always in equilibrium, as indicated in Fig. 2-69a. If all the internal forces occur in equal and opposite pairs, as indicated in Fig. 2-69b, the complete force system will be an equilibrium set.

In order to derive the momentum principles it is not, however, necessary to postulate anything about the nature of individual interactions. The only requirement is that the entire set of internal forces should be an equilibrium set. For example, the assumption that internal forces in

[1] The forces between charged particles moving at large relative velocities are *not* instantaneously in equilibrium. To extend the momentum principles to this case it is necessary to assign additional momentum to the electromagnetic field surrounding a particle. See R. P. Feynman, R. B. Leighton, and M. Sands, "The Feynman Lectures on Physics," Addison-Wesley Publishing Co., Reading, Mass., 1964, vol. 2, pp. 26-5 and 27-9.

a solid body occur only as central-force interactions between pairs of molecules is too restrictive to include the models currently employed in solid-state physics.[1] Models based on pairwise central-force interactions predict constraining relations between elastic constants which are seldom observed in real materials (e.g., in the isotropic case every elastic solid should have a Poisson's ratio of $\frac{1}{4}$ according to the pairwise central-force model). An alternative requirement for equilibrium of the internal-force system is that the total virtual work of the internal forces should vanish,

$$\sum_{i=1}^{N} \mathbf{g}_i \cdot \delta\mathbf{R}_i = 0, \tag{2-256}$$

for arbitrary *rigid-body* variations in the position of the system.[2]

LINEAR-MOMENTUM PRINCIPLE

To obtain the linear-momentum principle for the system of Fig. 2–68, we write the momentum principle (2–88) for each particle m_i of the system and then sum over i.

$$\sum_{i=1}^{N} (\mathbf{f}_i + \mathbf{g}_i) = \sum_{i=1}^{N} \frac{d\mathbf{p}_i}{dt}. \tag{2-257}$$

Substituting (2–254) and interchanging the order of summation and differentiation on the right, we get

$$\sum_{i=1}^{N} \mathbf{f}_i = \frac{d}{dt}\left(\sum_{i=1}^{N} \mathbf{p}_i\right). \tag{2-258}$$

This equation for the entire system is similar to (2–88) for a single particle. To emphasize this similarity, we define the total external force $\mathbf{F}$ acting on the system and the total linear momentum $\mathbf{P}$ of the system as follows:

$$\mathbf{F} = \sum_{i=1}^{N} \mathbf{f}_i \quad \text{and} \quad \mathbf{P} = \sum_{i=1}^{N} \mathbf{p}_i. \tag{2-259}$$

In terms of these, (2–258) becomes

$$\mathbf{F} = \frac{d\mathbf{P}}{dt}, \tag{2-260}$$

---

[1] F. Seitz and D. Turnbull (eds.), "Solid State Physics," Academic Press Inc., New York, 1958, vol. 7. See H. B. Huntington, The Elastic Constants of Crystals, pp. 213–351.

[2] See Prob. 2–48.

which is the linear-momentum principle for a system: *The resultant external force acting on a system equals the time rate of change of the system's linear momentum with respect to an inertial frame.*

When the newtonian velocity-momentum relation (2–86) applies, the system momentum **P** can be interpreted in terms of the motion of the centroid, or mass center. In Fig. 2–68 let the total mass of the system be denoted by M; i.e.,

$$M = \sum_{i=1}^{N} m_i. \tag{2-261}$$

Then the *centroid* C is defined by the position vector $\mathbf{R}_c$, where

$$M\mathbf{R}_c = \sum_{i=1}^{N} m_i\mathbf{R}_i. \tag{2-262}$$

The centroid is not necessarily a material particle of the system. It is simply a point whose position is a weighted average of the positions of the N particles of the system. As the particles move about, the location of the centroid will generally change. Let us differentiate (2–262) with respect to time and introduce the newtonian velocity-momentum relation (2–86).

$$M \frac{d\mathbf{R}_c}{dt} = \sum_{i=1}^{N} m_i \frac{d\mathbf{R}_i}{dt}$$

$$M\mathbf{v}_c = \sum_{i=1}^{N} m_i\mathbf{v}_i = \sum_{i=1}^{N} \mathbf{p}_i = \mathbf{P}. \tag{2-263}$$

Thus, in newtonian mechanics, the total linear momentum of a system is the product of the total mass M and the velocity of the centroid. When (2–263) is introduced into (2–260), we find

$$\mathbf{F} = M \frac{d\mathbf{v}_c}{dt} = M\mathbf{a}_c, \tag{2-264}$$

which is a system relation analogous to Newton's second law for a single particle in the form (2–89). The centroid of a system moves like a particle having total mass M and which is acted on by the total external force T. This result is widely used in applications; e.g., a space vehicle with astronauts moving around within it is a complex dynamic system. The orbit of the vehicle's centroid can, however, be predicted with precision from analyses of the motion of a single particle.

ANGULAR-MOMENTUM PRINCIPLE

To obtain the angular-momentum principle for the system of Fig. 2–68, we write the linear-momentum principle (2–88) for each particle m_i of the system, take the moment of this relation with respect to point B, and then sum over i.

$$\sum_{i=1}^{N} \mathbf{r}_i \times (\mathbf{f}_i + \mathbf{g}_i) = \sum_{i=1}^{N} \mathbf{r}_i \times \frac{d\mathbf{p}_i}{dt}. \qquad (2\text{--}265)$$

As in the derivation of the angular-momentum principle (2–92) for a single particle, we shall find that the final results are simplified if we consider a special class of moment centers B. At the outset, however, we take B to be an arbitrary point with position vector $\mathbf{R}_B$ and velocity $\mathbf{v}_B = d\mathbf{R}_B/dt$. We then introduce the requirement (2–255) for equilibrium of the internal-force system and expand the right-hand side of (2–265) to obtain

$$\sum_{i=1}^{N} \mathbf{r}_i \times \mathbf{f}_i = \sum_{i=1}^{N} \frac{d}{dt}(\mathbf{r}_i \times \mathbf{p}_i) - \sum_{i=1}^{N} \frac{d\mathbf{r}_i}{dt} \times \mathbf{p}_i$$

$$= \frac{d}{dt}\left(\sum_{i=1}^{N} \mathbf{r}_i \times \mathbf{p}_i\right) - \sum_{i=1}^{N} (\mathbf{v}_i - \mathbf{v}_B) \times \mathbf{p}_i. \qquad (2\text{--}266)$$

Since the momentum $\mathbf{p}_i$ is parallel to the velocity $\mathbf{v}_i$, the terms $\mathbf{v}_i \times \mathbf{p}_i$ will all vanish. The remaining terms in the final summation on the right can be transformed as follows:

$$\sum_{i=1}^{N} \mathbf{v}_B \times \mathbf{p}_i = \mathbf{v}_B \times \sum_{i=1}^{N} \mathbf{p}_i = \mathbf{v}_B \times \mathbf{P}, \qquad (2\text{--}267)$$

where $\mathbf{P}$ is the total linear momentum (2–259). We then introduce the following notation for the other sums in (2–266): Let $\boldsymbol{\tau}_B$ represent the total torque about B of the external forces acting on the system,

$$\boldsymbol{\tau}_B = \sum_{i=1}^{N} \mathbf{r}_i \times \mathbf{f}_i, \qquad (2\text{--}268)$$

and let $\mathbf{H}_B$ represent the total angular momentum of the system with respect to B,

$$\mathbf{H}_B = \sum_{i=1}^{N} \mathbf{r}_i \times \mathbf{p}_i. \qquad (2\text{--}269)$$

On inserting (2–267), (2–268), and (2–269) into (2–266), we obtain

$$\boldsymbol{\tau}_B = \frac{d}{dt}\mathbf{H}_B + \mathbf{v}_B \times \mathbf{P}, \qquad (2\text{--}270)$$

which is an angular-momentum principle valid for any moment center B. This principle takes a simpler form if we restrict the choice of moment centers B to those points for which the term $v_B \times P$ is zero. These special moment centers are

1. Any point B which has zero velocity with respect to the inertial reference frame $OXYZ$.
2. Any point B whose velocity v_B is parallel to the linear momentum $P = Mv_c$. The most important case in this category is where the moment center B is taken to coincide with the centroid C of the system.

If B *is* one of these special moment centers, the angular-momentum principle (2–270) takes the same general form as the linear-momentum principle,

$$\tau_B = \frac{d}{dt} H_B. \tag{2-271}$$

The total external torque acting on a system of particles equals the time rate of change of the system's total angular momentum.

PROBLEMS

2–1. Express the position vectors of points A and B in terms of the length L of the rod AB and the time history $\theta(t)$.

(a) Obtain the velocities of A and B.
(b) Obtain the accelerations of A and B.

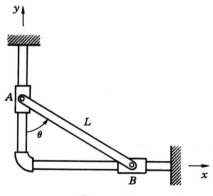

PROB. 2–1

2-2. Express the position of the point P in terms of the unit vectors $\mathbf{u}_r$, $\mathbf{u}_\theta$, and $\mathbf{u}_z$ of cylindrical coordinates. Determine the time derivatives of the unit vectors.

(*a*) Express the velocity of P in terms of cylindrical coordinates.
(*b*) Express the acceleration of P in cylindrical coordinates.

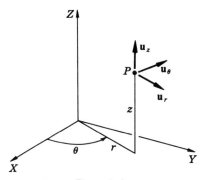

PROB. 2-2

2-3. A point P is moving across a disk in a straight line MN with a constant speed v_0 relative to the disk. The disk is spinning with a constant angular speed ω_0.

(*a*) Find an expression for the velocity of P relative to the nonrotating frame $OXYZ$.
(*b*) Find an expression for the acceleration of P with respect to $OXYZ$. Identify the various components.

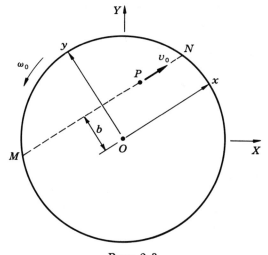

PROB. 2-3

2-4. A point P moves with uniform speed v_0 along a great circle of a sphere of radius a which rotates with uniform angular speed ω_0 about the Z axis. Answer the following questions for the instant when P passes through a point on the sphere with latitude (angle from equator) 60°, with a heading $\alpha = 30°$ east of north.

(a)   Find the velocity of P with respect to the nonrotating frame $OXYZ$.
(b)   Find the acceleration of P with respect to $OXYZ$.

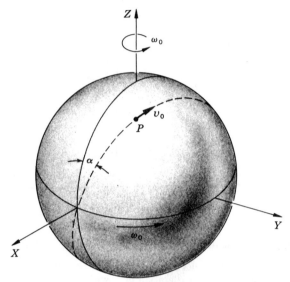

PROB. 2-4

2-5. A disk rolls without slip on a horizontal surface with (variable) angular speed ω. The point P is fixed to the disk as shown.

(a)   Find the velocity of P.
(b)   Find the acceleration of P.

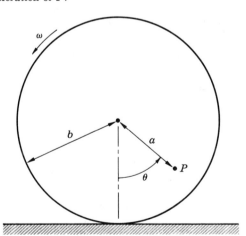

PROB. 2-5

2–6. A disk of radius a rolls without slip on a fixed cylindrical surface of radius b. Express ω in terms of $d\theta/dt$.

(a) Find the velocity of the center of the disk in terms of θ and its derivative.
(b) Find the acceleration of the center of the disk in terms of θ and its derivatives.

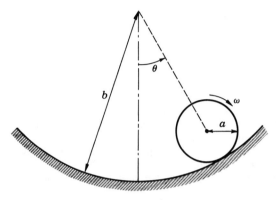

PROB. 2–6

2–7. Parts of a machine are shown, with all the driving mechanisms omitted for simplicity. Find the magnitude of the velocity of the point A on the moving belt. Show the velocity components on a vector sketch.

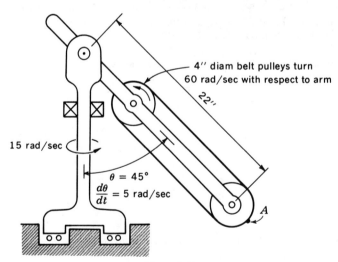

PROB. 2–7

2–8. Consider the crank, connecting rod, and piston of Fig. 2–23. Show that the position vector of P can be written

$$\mathbf{R}_P = (a \cos \theta + b \cos \varphi)\mathbf{u}_X.$$

(a) Verify that the time derivative of $\mathbf{R}_P$ yields the same velocity as (2–84).

(b) Resolve (2–85) into components parallel to OX and OY, and thereby obtain an analytical expression for the acceleration of P.

(c) Verify that the second derivative of $\mathbf{R}_P$, above, yields the same acceleration as (b).

2–9. Write out the nine equations indicated in (2–22) for the rotation matrix of (2–17). These nine equations divide into three groups of three:

(i)   Those corresponding to the main diagonal of the unit matrix

(ii) Those corresponding to the three zeros above the main diagonal of the unit matrix

(iii) Those corresponding to the three zeros below the main diagonal

(a) Show that the set (iii) is identical with the set (ii).

(b) Show that the set (i) can be interpreted as the requirements that the vectors $\mathbf{u}_x$, $\mathbf{u}_y$, and $\mathbf{u}_z$ have *unit* length.

(c) Show that the set (ii) can be interpreted as the requirements that the vectors $\mathbf{u}_x$, $\mathbf{u}_y$, and $\mathbf{u}_z$ are mutually orthogonal.

2–10. Show that (2–23) can be represented in matrix notation as

$$\{\mathbf{U}\}^t \begin{Bmatrix} x \\ y \\ z \end{Bmatrix} = \{\mathbf{U}'\}^t \begin{Bmatrix} x' \\ y' \\ z' \end{Bmatrix}.$$

Use (2–20) to eliminate $\{\mathbf{U}\}$ and thereby verify (2–24).

2–11. Let $[A]$ and $[B]$ be arbitrary rotation matrices; i.e., $[A][A]^t = [B][B]^t = [1]$, and $\det [A] = \det [B] = 1$. Prove that their product $[C] = [A][B]$ is also a rotation matrix; i.e., show that $[C][C]^t = [1]$ and that $\det [C] = 1$.

2–12. Given the rotation matrix $[C]$ of (2–17).

(a) Show that $[C] - [C]^t$ is a skew-symmetric matrix,

$$[C] - [C]^t = \begin{bmatrix} 0 & N & -M \\ -N & 0 & L \\ M & -L & 0 \end{bmatrix}.$$

(b) By using (2–29), show that the direction cosines of the axis of rotation are given by

$$l = \frac{L}{R} \qquad m = \frac{M}{R} \qquad n = \frac{N}{R},$$

where $R^2 = L^2 + M^2 + N^2$.

(c) Show that the sine of the angle of rotation is given by

$$\sin \theta = \tfrac{1}{2}R$$

and that the cosine of the angle of rotation is given by

$$\cos \theta = \tfrac{1}{2}(c_{xx'} + c_{yy'} + c_{zz'} - 1).$$

2–13. Verify that the rotation matrix

$$[C_z] = \begin{bmatrix} \cos\theta_z & \sin\theta_z & 0 \\ -\sin\theta_z & \cos\theta_z & 0 \\ 0 & 0 & 1 \end{bmatrix}$$

corresponds to a rotation about the z axis through an angle θ_z. Construct the rotation matrices $[C_x]$ and $[C_y]$ which correspond to rotations about the x and y axes through angles θ_x and θ_y, respectively.

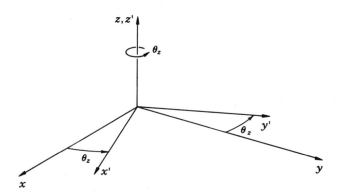

PROB. 2–13

2–14. Consider the following sequence of rotations:

(i) The $x'y'z'$ triad is rotated from coincidence with the fixed xyz triad through the angle θ_x about the x axis and then fixed.

(ii) The $x''y''z''$ triad is rotated from coincidence with the $x'y'z'$ triad through the angle $\theta_{y'}$ about the y' axis and then fixed.

(iii) Finally, the $x'''y'''z'''$ triad is rotated from coincidence with the $x''y''z''$ triad through the angle $\theta_{z''}$ about the z'' axis.

Construct the rotation matrix which represents the resultant single rotation from the unprimed triad to the triple-primed triad.

2–15. Let θ_1 represent the acute angle whose tangent is $\frac{3}{4}$, and let θ_2 represent the acute angle whose tangent is $\frac{4}{3}$. Consider the sequence of rotations in which

(i) The $x'y'z'$ triad is rotated from coincidence with the fixed xyz triad through the angle θ_1 about the y axis and then fixed.

(ii) The $x''y''z''$ triad is rotated from coincidence with the $x'y'z'$ triad through the angle θ_2 about the z' axis.

(a) Obtain the rotation matrix $[C]$ which represents the single rotation which would carry the $x''y''z''$ triad from coincidence with the xyz triad into the final position obtained above.

(b) Locate the axis of this rotation and find the magnitude of the angle of rotation.

2–16. Show that when the angles θ_x, $\theta_{y'}$, and $\theta_{z''}$ in Prob. 2–14 are taken as small angles $\Delta\theta_x$, $\Delta\theta_y$, and $\Delta\theta_z$, respectively, the individual rotations in Prob. 2–14 can be represented approximately by the matrices

$$
\begin{bmatrix} 1 & 0 & 0 \\ 0 & 1 & \Delta\theta_x \\ 0 & -\Delta\theta_x & 1 \end{bmatrix}
\begin{bmatrix} 1 & 0 & -\Delta\theta_y \\ 0 & 1 & 0 \\ \Delta\theta_y & 0 & 1 \end{bmatrix}
\begin{bmatrix} 1 & \Delta\theta_z & 0 \\ -\Delta\theta_z & 1 & 0 \\ 0 & 0 & 1 \end{bmatrix},
$$

with error terms proportional to the squares of the small angles. Verify that the result of compounding these three rotations, taken *in any order*, is a single rotation which can be represented by (2–33).

2–17. Consider the two small rotations represented by the matrices

$$
\begin{bmatrix} 1 & \Delta\theta_z & -\Delta\theta_y \\ -\Delta\theta_z & 1 & \Delta\theta_x \\ \Delta\theta_y & -\Delta\theta_x & 1 \end{bmatrix}
\begin{bmatrix} 1 & \Delta\varphi_z & -\Delta\varphi_y \\ -\Delta\varphi_z & 1 & \Delta\varphi_x \\ \Delta\varphi_y & -\Delta\varphi_x & 1 \end{bmatrix}.
$$

Show that the single rotation which results from compounding these, in either order, can be represented approximately by the matrix

$$
\begin{bmatrix} 1 & \Delta\theta_z + \Delta\varphi_z & -(\Delta\theta_y + \Delta\varphi_y) \\ -(\Delta\theta_z + \Delta\varphi_z) & 1 & \Delta\theta_x + \Delta\varphi_x \\ \Delta\theta_y + \Delta\varphi_y & -(\Delta\theta_x + \Delta\varphi_x) & 1 \end{bmatrix},
$$

where the error terms are proportional to the sums of the squares of the small angles involved.

2–18. Consider a three-dimensional velocity space in which the cartesian coordinates of each point represent the velocity components of a particle. Consider the function $T^*(\mathbf{v})$ defined by (2–97) as a scalar field in this space. If $\boldsymbol{\nabla}_v$ represents the vector differential operator for this space, show that

$$
\boldsymbol{\nabla}_v T^* = \mathbf{p};
$$

i.e., the gradient of the kinetic coenergy of a particle in velocity space is the momentum of the particle.

2–19. Given a force field with axial symmetry about the axis of Z such that $\mathbf{f} = f(r)\mathbf{u}_\theta$. Determine the form of the function $f(r)$ in order for the field to be conservative in any region which does not include the Z axis.

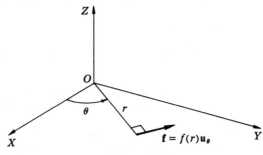

PROB. 2–19

2-20. Consider a two-dimensional central-force field in which the force on a probe is in the xy plane and always directed along a line joining the origin to the probe. If the magnitude of the force is a function only of the distance between the probe and the origin, show that the force field is conservative in any region which does not include the origin.

2-21. Consider a three-dimensional central-force field in which the force on a probe is always directed along a line joining the origin to the probe. If the magnitude of the force is a function only of the distance between the probe and the origin, show that the force field is conservative in any region which does not include the origin.

2-22. A prescribed force $\mathbf{f}(t)$ is applied to a point P in a system by an external *force source*. It is supposed that the time history of $\mathbf{f}(t)$ is maintained by the source independently of the motion of point P. Show that the work done by $\mathbf{f}(t)$ during a variation in configuration is given by $-\delta V$ if V is the *pseudo-potential* function

$$V = -\mathbf{f}(t) \cdot \mathbf{R},$$

where $\mathbf{R}$ is the position vector for the point P. Thus, although a prescribed force $\mathbf{f}(t)$ is not generally conservative, its work increment can be computed as a decrement of a pseudo potential.

2-23. The flexible inextensible massless string passes around a massless frictionless pulley attached to the block of mass m. Show that when a force f is applied to the

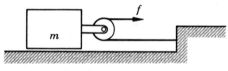

PROB. 2-23

string there is a force $2f$ applied to the block. Show that under a geometrically admissible variation in configuration, the work done on the block by the pulley system is *equal* to the work done by f.

2-24. A mass particle m is attached to a massless inextensible string going to large masses M, as shown. Find the potential energy V of the entire system, due to the

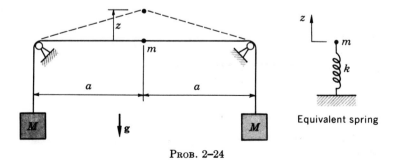

Equivalent spring

PROB. 2-24

154 *Dynamics of mechanical systems*

uniform gravity field, as a function of z. Show that for small values of z the effect of the large masses is statically equivalent to a linear spring. Determine the stiffness k of the equivalent linear spring in terms of M, a, and g.

2-25. A rigid massless lever ABC supports a mass m and is restrained by two springs. Analyze this system in a fashion analogous to that displayed in Fig. 2-50.

(a) Obtain a single differential equation of motion by the direct method.
(b) Construct T^* and V, and apply Hamilton's principle to obtain the equation of motion.

Follow a procedure analogous to that leading to (2-143). Assume that the inclination of the lever remains *small*.

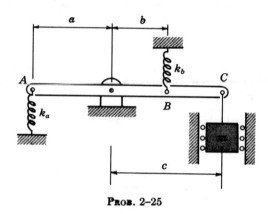

PROB. 2-25

2-26. Obtain equations of motion for the system of two masses on a horizontal surface

(a) By the direct method
(b) By the use of Hamilton's principle

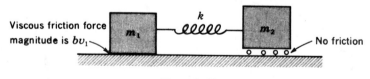

PROB. 2-26

2-27. Obtain equations of motion for the three masses connected by massless ropes and pulleys

(a) By the direct method
(b) By the use of Hamilton's principle

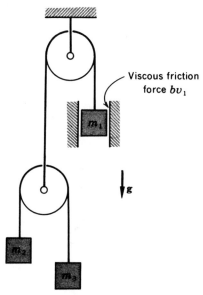

Viscous friction
force bv_1

PROB. 2-27

2-28. Consider a single mass particle moving in three dimensions subject to the force

$$\mathbf{f} = f_r\mathbf{u}_r + f_\theta\mathbf{u}_\theta + f_\varphi\mathbf{u}_\varphi,$$

where r, θ, and φ are the spherical coordinates of Fig. 2-22. Evaluate T^* in these same coordinates, and apply Hamilton's principle to obtain the equivalent of Newton's law (2-89) in spherical coordinates. Compare the acceleration components, so derived, with (2-78).

2-29. A mass particle slides down a fixed inclined plane, without friction. Use x and y as coordinates and incorporate the geometric-constraint relation by means of a Lagrange multiplier. Apply Hamilton's principle to obtain equations of motion analogous to (2-170).

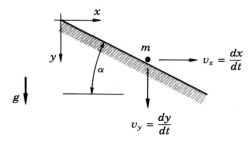

PROB. 2-29

2-30. The disk of radius a, shown in the figure, is constrained, so that it always remains in contact with the horizontal plane.   The plane of the disk is always vertical; however, the disk is free to spin about its vertical diameter, and the disk can roll without slipping on the horizontal plane about an axis perpendicular to its own plane.

(a) Establish a complete and independent set of generalized coordinates for the disk.
(b) How many degrees of freedom does the disk have?
(c) Formulate analytical expressions for any nonholonomic constraints.

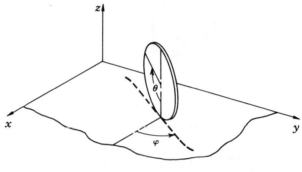

PROB. 2–30

2-31. A rigid sphere of radius a is free to roll without slipping on a horizontal plane.

(a) How many generalized coordinates are required to form a complete set of independent coordinates?
(b) How many degrees of freedom are there?

2-32. A single gas molecule is bouncing about within a rigid rectangular enclosure. Discuss the nature of the geometric constraint on the molecule's motion.

2-33. A particle P is moving in three-dimensional space under the influence of a force which acts parallel to the axis of Y.   Taking the spherical coordinates r, θ, and φ as generalized coordinates, find the corresponding generalized forces.

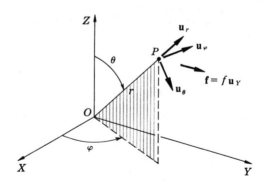

PROB. 2–33

2-34. The pendulum shown consists of two mass particles attached to a rigid mass-less frame which is constrained to pivot about O, in the plane of the sketch. Neglecting friction, obtain the equation of motion for the pendulum.

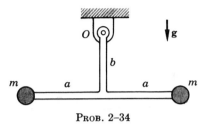

PROB. 2-34

2-35. Consider the system of Fig. 2-56 under the assumption that the pendulum arm connecting m_2 and m_3 is massless.

(a) Construct T^* and V for the system in terms of the coordinates shown.
(b) Obtain the differential equations of motion.

2-36. Assume that the disk in Prob. 2-5 is massless but that it carries a particle of mass m at P. Find the equation of motion under the assumption that a uniform gravitational field acts vertically downward.

2-37. The system shown consists of two mass particles and three identical linear springs, constrained to remain in the plane of the sketch. When $f = 0$, the equilibrium configuration of the springs is an equilateral triangle of side a.

(a) Define a set of generalized coordinates for analyzing small motions of the system under the influence of the prescribed force $f(t)$.
(b) What are the corresponding generalized forces?
(c) Construct the lagrangian.
(d) Obtain the equations of motion.

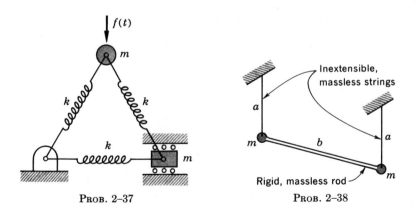

PROB. 2-37   PROB. 2-38

2-38. Consider the *small* motions of the bifilar pendulum shown.

(a) Select a complete set of independent generalized coordinates.
(b) How many degrees of freedom does the system have?

(c) Evaluate the lagrangian.

(d) Obtain the equations of motion.

2-39. A massless rigid shaft AB rotates in frictionless bearings. A mass m slides without friction on a rigid horizontal arm CD, and is restrained by a linear spring of unstrained length a.

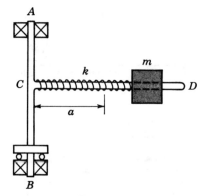

PROB. 2–39

(a) Select a complete set of independent generalized coordinates.

(b) How many degrees of freedom does the system have?

(c) Construct the lagrangian.

(d) Obtain the equations of motion.

2-40. Consider the single-degree-of-freedom system consisting of the mass m, supported by a linear spring, in the uniform gravity field. Construct the lagrangian and obtain the equations of motion for the system

(a) In terms of the elongation e of the spring from its unstrained length a

(b) In terms of the displacement ξ of the mass from its equilibrium position

What role does gravity play in the oscillatory motion of the mass?

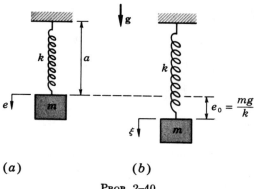

(a)   (b)

PROB. 2–40

2-41. The pendulum shown consists of a mass particle m which moves in the plane of the sketch subject to the constraint provided by the inextensible massless string which is connected to the linear spring.

(a) Select a complete set of independent coordinates.
(b) Are the constraints holonomic?
(c) Construct the lagrangian.
(d) Obtain the equations of motion.   Take the length of the string, from the supports at A to the mass particle m, to be a when the mass is at rest in equilibrium.

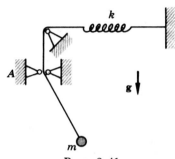

PROB. 2-41

2-42. Two pendulums constrained to swing in the same plane are joined by an ideal linear spring whose unstrained length is b.

(a) Select a complete set of independent generalized coordinates.
(b) Construct the lagrangian.
(c) Obtain *linearized* equations of *small* motion.

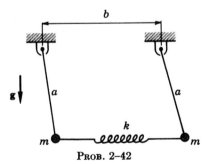

PROB. 2-42

2-43. Four equal mass particles are attached to the corners of a rigid massless rectangular frame supported by the linear springs, as shown.   Assume that the horizontal springs are not preloaded.

(a) Select a complete set of independent generalized coordinates for describing *small* motions of the system.

(b) Construct the lagrangian.

(c) Obtain the equations of motion.

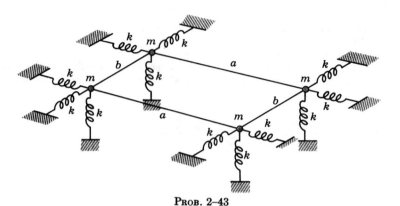

PROB. 2–43

2–44. The pendulum is constrained to remain in the plane of the sketch as the string is pulled through the hole at the uniform speed v_0. Use Hamilton's principle to find the force f as a function of v_0, a, θ, $\dot{\theta}$, and $\ddot{\theta}$.

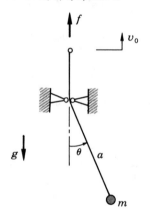

PROB. 2–44

2–45. Consider the particle in Fig. 2–61 sliding (without friction) off the cylinder under the influence of gravity.

(a) Analyze the motion in the first regime and show that if the particle starts from rest at $\theta = 0$, its angular speed during the first regime is given by

$$\left(\frac{d\theta}{dt}\right)^2 = 2\frac{g}{b}(1 - \cos\theta).$$

(b) Determine the condition for locating the point E where the particle leaves the cylinder. Evaluate θ_E.

2–46. Obtain the differential equations of motion for the spherical pendulum of Fig. 2–67 by applying Lagrange's equations to (2–249) and (2–250). Try to obtain the results (2–253) by integrating the equations of motion.

2–47. A set of n forces $\mathbf{f}_i$ $(i = 1, \ldots, n)$ is said to be in *equilibrium* if

$$\sum_{i=1}^{n} \mathbf{f}_i = 0 \quad \text{and} \quad \sum_{i=1}^{n} \mathbf{r}_i \times \mathbf{f}_i = 0,$$

where the $\mathbf{r}_i$ are position vectors from a common origin to the points P_i where the forces $\mathbf{f}_i$ act. Show that if the $\mathbf{R}_i$ are position vectors from any other origin, then

$$\sum_{i=1}^{n} \mathbf{R}_i \times \mathbf{f}_i = 0$$

whenever the set of forces is in equilibrium.

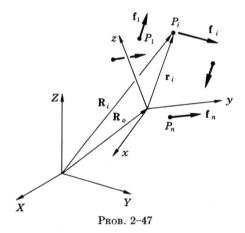

PROB. 2–47

2–48. Let the points P_i in Prob. 2–47 be rigidly fixed in the xyz frame. A general infinitesimal variation in the position of the xyz frame with respect to the XYZ frame can be represented by a translation $\delta\mathbf{R}_0$ and a rotation $\delta\theta$. Show that the total virtual work of the set of forces $\mathbf{f}_i$ under such a variation can be put in the form

$$\sum_{i=1}^{n} \mathbf{f}_i \cdot \delta\mathbf{R}_i = \left(\sum_{i=1}^{n} \mathbf{f}_i \right) \cdot \delta\mathbf{R}_0 + \left(\sum_{i=1}^{n} \mathbf{r}_i \times \mathbf{f}_i \right) \cdot \delta\theta,$$

and hence that the virtual work of a set of forces under arbitrary rigid-body variations of position will vanish if, and only if, the forces are in equilibrium.

2–49. Prove that the total *linear momentum* of an isolated system of interacting particles is *conserved*. Isolation in this context means that there are no external forces acting on the system.

2–50. Prove that the total *angular momentum* of an isolated system of interacting particles with respect to any fixed point is *conserved*.

2-51. Consider two stars A and B which attract one another in an otherwise empty universe. At a given instant, A and B are in the positions shown and are moving with the indicated velocities. At a later time, star A passes through the origin.

(a) When this occurs where is star B?
(b) If at this time the velocity of A is directed along the axis of X, what is the magnitude and direction of the velocity of B?

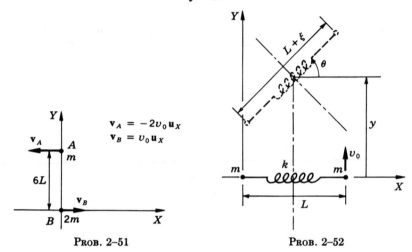

$$\mathbf{v}_A = -2v_0\,\mathbf{u}_X$$
$$\mathbf{v}_B = v_0\,\mathbf{u}_X$$

PROB. 2–51   PROB. 2–52

2-52. A system consists of a pair of mass particles, each of mass m, joined by a massless linear spring of natural length L. At $t = 0$ the system is released, in a force-free environment, from the configuration shown, with the left-hand mass at rest and the right-hand mass moving with the velocity $v_0 \mathbf{u}_Y$. Formulate equations for the subsequent motion in terms of the variables ξ, y, and θ. Describe qualitatively the general character of the motion.

2-53. Consider the behavior of a pendulum mounted in an airplane which is executing a steady banked turn of radius b while flying at a speed v_0. Obtain an equation for the motion of the pendulum in the vertical plane through the center of the turn. What is the equilibrium position of the pendulum as viewed by an observer in the airplane?

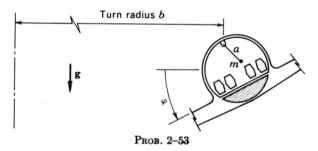

PROB. 2–53

2-54. Consider a system of N particles m_i in which each particle is subject to a force $\mathbf{f}_i = m_i\mathbf{g}$, where $\mathbf{g}$ is the constant acceleration vector of a *uniform gravity field*.

(a) Show that for the purposes of the linear- and angular-momentum theorems, this set of forces can be replaced by a single force $M\mathbf{g}$ which passes through the centroid; i.e., for a *uniform* gravity field the *center of gravity* coincides with *center of mass*.

(b) Show that the potential energy of the system is given by Mgz_c.

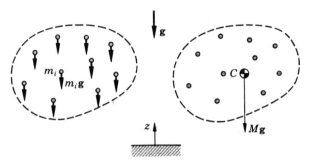

PROB. 2-54

2-55. Consider a uniform thin spherical shell of radius a and total mass m_0. It is required to obtain the gravitational potential energy $V(r)$ of a probe of mass m at a distance r from the center of the shell, where $r > a$.

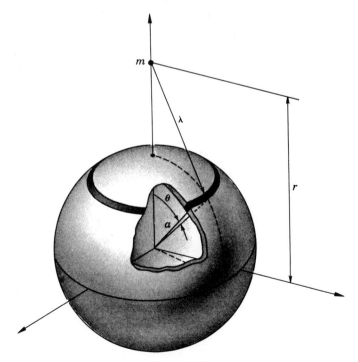

PROB. 2-55

(a) First show that the potential due to a ring of the shell between θ and $\theta + d\theta$ is

$$dV = -\frac{Km_0m}{2\lambda}\sin\theta\,d\theta,$$

where $\lambda^2 = r^2 + a^2 - 2ra\cos\theta$.

(b) Then integrate the result of (a) by eliminating θ in favor of λ to obtain

$$V = -\frac{Km_0m}{r}.$$

2-56. Use the result of Prob. 2–55 to show that the gravitational force (2–115) and the gravitational potential (2–116) apply whenever the probe is outside of any spherical attracting body which has a spherically symmetric mass distribution.   Note that the mass density in the attracting body *can* change with radius without altering these conclusions.

2-57. Repeat Prob. 2–55 for the case where the probe is inside the shell ($r < a$). What gravitational force does the probe experience when it is inside the shell?

2-58. Use the results of Probs. 2–55 and 2–57 to obtain the gravitational force and gravitational potential for a probe of mass m as a function of r when the probe is in the interior of a homogeneous sphere of mass M and radius b.

2-59. Imagine that a hole is drilled straight through the earth from the North Pole to the South Pole.   Neglecting air resistance, what would happen to a coin dropped into the hole?   Assuming the earth to be a uniform sphere of radius $b = 3{,}960$ miles, estimate the time interval between reappearances of the coin.

2-60. Consider a set of *massless* particles P_i acted on by external forces $\mathbf{f}_i$, and by internal forces $\mathbf{g}_i$ which satisfy (2–254) and (2–255).   Use the linear- and angular-momentum principles to show that the external forces must be a set of forces in *equilibrium* for all motions of the particles P_i which have finite accelerations.

CHAPTER THREE

Dynamical properties of a rigid body

In Chap. 2, mass particles are used as models for inertia elements. The dynamical properties of linear momentum, angular momentum, kinetic energy, and kinetic coenergy are introduced for individual particles and for systems of particles. In this chapter we develop the corresponding properties for an extended rigid body considered as a single inertia element. Because the motion of a rigid body can involve rotation as well as translation, the dynamical properties of a rigid body are more complex than those of a particle. The central aim of the chapter is to develop the concepts of angular momentum and kinetic coenergy for a rigid body and to obtain efficient procedures for their calculation. Once these tools are available, we return in Chap. 4 to the study of dynamics of mechanical systems. There we use the results of this chapter to study the dynamic behavior of systems whose principal inertial elements are rigid bodies.

3-1   Rigid bodies

In studying the gross dynamic motions of extended bodies which undergo relatively little internal deformation (e.g., flywheels and connecting rods), it is convenient to introduce the idealization of perfectly rigid bodies. A rigid body can be considered to be a distribution of mass rigidly fixed to a rigid frame. As indicated in Fig. 3-1, it is sometimes convenient to consider a *continuous* distribution of rigid mass, while for other purposes it is useful to consider a distribution of a very large number of *discrete* mass particles rigidly bound together. We shall consider that an integral over a continuous mass distribution is a natural extension of a summation over a very large number of distinct mass particles. In either case the essential requirement of rigidity is that the distance between any two elements of mass within the body must remain constant.

The concept of rigidity has a simple unambiguous meaning only within

165

the framework of nonrelativistic mechanics. Our treatment here is restricted to classical newtonian mechanics. In particular, from this point on, we consider only the newtonian velocity-momentum relation (2-86). This implies that the complementary kinetic state functions T and T^* are always equal in magnitude. For formal reasons we shall generally maintain the notation T^* for functions of velocity and T for functions of momentum, although it is occasionally convenient, as in Eq. (3-35), to use a mixed notation.

The problem of describing the position of a rigid body is synonymous with the problem of describing the position of a rigid frame (see Sec. 2-2). Since each element of mass in a rigid body can be taken as rigidly attached to a frame, the entire body can be considered to be located as soon as one rectangular frame, fixed in the body, has been located. The position of a

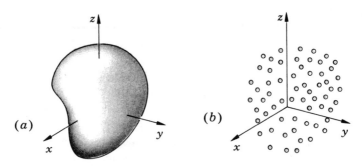

FIG. 3-1. A rigid body may be considered to be either (a) a continuous distribution of mass rigidly fixed to a frame or (b) a very large number of mass particles rigidly attached to a frame.

rectangular frame, with respect to a reference frame, can be fixed (see Fig. 2-5) by giving the *position* of a single point and the angle of *rotation* between the frames. An unconstrained rigid body has *six* independent degrees of freedom: three independent translation components and three independent rotation components.

The velocities of all points in a rigid body are determined once the angular velocity ω of the body is known and the linear velocity of any one point in the body is known. Thus, in Fig. 3-2, if the frame $Oxyz$ and the arbitrary point P are both fixed in the same rigid body, then, since

$$\mathbf{R} = \mathbf{R}_o + \mathbf{r}, \tag{3-1}$$

the velocity of point P, with respect to the reference frame, is

$$\mathbf{v} = \mathbf{v}_o + \omega \times \mathbf{r}, \tag{3-2}$$

according to (2-55).

In order to capitalize on the momentum principles developed in Sec. 2-13 for systems of particles, we shall initially consider a rigid body to have the structure indicated in Fig. 3-1b. We consider a large number N of mass particles $m_i(i = 1, \ldots, N)$ rigidly attached to a rectangular frame xyz. Such a frame is called a *body-coordinate system*. The dynamical properties of a rigid body are readily developed by assuming that the

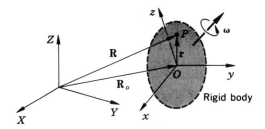

FIG. 3-2. The velocity of any point in a rigid body is determined by the velocity of a single point in the body and by the angular velocity ω of the rigid body.

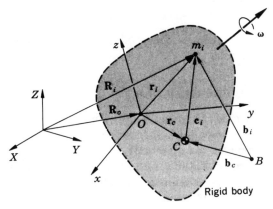

FIG. 3-3. Rigid body represented by frame xyz and typical mass particle m_i. Point C is mass center, and point B is arbitrary moment center.

momentum principles (2-260) and (2-271) apply to a rigidly interconnected system of particles. In Fig. 3-3 a typical particle m_i is sketched within a rigid body attached to a body-coordinate system $Oxyz$. The motion of the rigid body with respect to the inertial reference frame XYZ is described by the displacement vector $\mathbf{R}_o(t)$ and the angular velocity $\omega(t)$. The *mass center*, or *centroid*, C defined by (2-262) is indicated in

Fig. 3–3.   If M is the total mass of the rigid body,

$$M = \sum_{i=1}^{N} m_i, \tag{3-3}$$

it follows from (2–262) that for an arbitrary center B,

$$M\mathbf{b}_c = \sum_{i=1}^{N} m_i \mathbf{b}_i. \tag{3-4}$$

In particular, if O is taken as the center,

$$M\mathbf{r}_c = \sum_{i=1}^{N} m_i \mathbf{r}_i, \tag{3-5}$$

while if the centroid itself is taken as the center,

$$\sum_{i=1}^{N} m_i \mathbf{c}_i = 0. \tag{3-6}$$

The *linear momentum* $\mathbf{P}$ of a rigid body is the vector sum of the linear momenta of the individual particles which make up the rigid body.   The simplest representation of $\mathbf{P}$ is that given by (2–263),

$$\mathbf{P} = M\mathbf{v}_c; \tag{3-7}$$

i.e., the linear momentum is the product of the total mass times the velocity of the centroid.

3–2   Angular momentum of a rigid body

The angular momentum of a rigid body with respect to a moment center B is the sum of the moments about B of the momenta of all the individual particles which make up the rigid body.   The angular momentum can be defined for any moment center B, although the relation between the external torque and the angular momentum takes the simple form of (2–271) for only a restricted set of moment centers.   In this section we develop expressions for the angular momentum of a rigid body with respect to an *arbitrary* moment center B; i.e., we study angular momentum for its own sake and postpone the considerations involved in the torque–angular momentum relation until Sec. 3–5.

The angular momentum of the rigid body indicated in Fig. 3–3, with respect to the *arbitrary* moment center B, is

$$\mathbf{H}_B = \sum_{i=1}^{N} \mathbf{b}_i \times m_i \mathbf{v}_i, \tag{3-8}$$

where $\mathbf{b}_i$ is the vector from the moment center to the ith particle, and

$m_i\mathbf{v}_i$ is the momentum of that particle. The general expression (3–8) takes many special forms, depending on the particular choice of the moment center B. For example, if we let B coincide with the mass center C, the angular momentum with respect to the centroid is

$$\mathbf{H}_c = \sum_{i=1}^{N} \mathbf{c}_i \times m_i\mathbf{v}_i, \tag{3-9}$$

where $\mathbf{c}_i$ is the position vector of the ith particle with respect to the centroid. By setting

$$\mathbf{v}_i = \mathbf{v}_c + \boldsymbol{\omega} \times \mathbf{c}_i \tag{3-10}$$

in (3–9), we obtain

$$\mathbf{H}_c = \sum_{i=1}^{N} m_i\mathbf{c}_i \times \mathbf{v}_c + \sum_{i=1}^{N} m_i\mathbf{c}_i \times (\boldsymbol{\omega} \times \mathbf{c}_i)$$

$$= 0 + \sum_{i=1}^{N} m_i\mathbf{c}_i \times (\boldsymbol{\omega} \times \mathbf{c}_i), \tag{3-11}$$

on using (3–6). Note that the centroidal angular momentum is independent of the linear velocity of the centroid. In Sec. 3–3 the properties of sums of vector triple products of the form of (3–11) are examined in detail.

Returning now to the general expression (3–8) for the angular momentum of a rigid body with respect to an arbitrary moment center B, we give a convenient representation of $\mathbf{H}_B$ in terms of the centroidal angular momentum. Noting from Fig. 3–3 that

$$\mathbf{b}_i = \mathbf{b}_c + \mathbf{c}_i, \tag{3-12}$$

we expand (3–8) as follows:

$$\mathbf{H}_B = \sum_{i=1}^{N} (\mathbf{b}_c + \mathbf{c}_i) \times m_i\mathbf{v}_i$$

$$= \mathbf{b}_c \times \sum_{i=1}^{N} m_i\mathbf{v}_i + \sum_{i=1}^{N} \mathbf{c}_i \times m_i\mathbf{v}_i. \tag{3-13}$$

The first summation on the right of (3–13) represents the linear momentum $\mathbf{P}$ of the body, and the second summation is just the centroidal angular momentum (3–9), so that (3–13) becomes

$$\mathbf{H}_B = \mathbf{b}_c \times \mathbf{P} + \mathbf{H}_c. \tag{3-14}$$

The angular momentum of a rigid[1] body about an arbitrary point B is the sum of the centroidal angular momentum $\mathbf{H}_c$ plus the moment of the linear momentum $\mathbf{P}$ with respect to B when $\mathbf{P}$ is considered to act at the centroid. This result is illustrated in Fig. 3–4. A direct consequence of (3–14) is that if the linear momentum $\mathbf{P}$ of a rigid body is known, then

[1] The derivation of (3–14) applies equally well to a nonrigid system of particles.

once the angular momentum with respect to any one center has been determined, the angular momentum with respect to any other center can immediately be obtained.[1]

A moment center which is convenient in many practical applications is a point which is fixed in the rigid body *and also fixed in inertial space.* In gyroscopic devices this simultaneous fixing is accomplished by suspending the rigid body through a set of gimbal rings. A spherical ball joint accomplishes the same thing (over a limited range of positions). In a rigid rotor which revolves in stationary bearings, every point of the rotor which lies on the axis of rotation determined by the bearing centers is fixed in the rotor and is also stationary. To deal with applications of this sort it is useful to derive an explicit expression for the angular

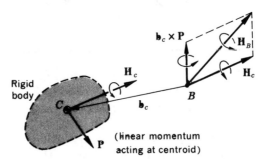

FIG. 3–4. Angular momentum of rigid body about arbitrary point B is sum of centroidal angular momentum $\mathbf{H}_c$ and moment of linear momentum $\mathbf{P}$, about B.

momentum of a rigid body with respect to a point O, which is fixed in the body and which is also motionless.

In Fig. 3–3 let us assume that the point O (the origin of the body coordinate system $Oxyz$) has no velocity; i.e., $\mathbf{v}_o \equiv 0$. Then the velocity of the typical mass particle m_i, as given by (3–2), reduces to

$$\mathbf{v}_i = \boldsymbol{\omega} \times \mathbf{r}_i, \tag{3–15}$$

and the angular momentum (3–8) becomes, when the moment center B is taken to coincide with O,

$$\mathbf{H}_o = \sum_{i=1}^{N} m_i \mathbf{r}_i \times (\boldsymbol{\omega} \times \mathbf{r}_i). \tag{3–16}$$

Note the similarity in the structure of (3–11) and (3–16). The vector triple products have the same form: The only difference is in the location of the moment centers. In (3–11) the position vectors $\mathbf{c}_i$ emanate from

[1] See Prob. 3–3.

the centroid, and in (3–16) the position vectors r_i emanate from a point O which is fixed in the rigid body *and* fixed in inertial space.

3-3 The inertia tensor

To proceed further with the analysis of rigid-body angular momentum, we study the structure of sums having the form of (3–11) and (3–16). Since the following developments are identical in the two cases, we shall take (3–16) as our prototype sum, but we shall temporarily refrain from attaching any special significance to the location of the origin employed. The situation we deal with is indicated in Fig. 3–5. The rigid body is made up of many mass particles m_i, fixed to a body-coordinate system

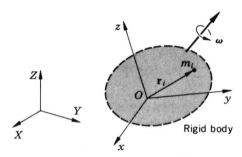

Fig. 3–5. Rigid body fixed to xyz frame undergoing angular velocity ω. Typical particle m_i is indicated for evaluation of sum $\displaystyle\sum_{i=1}^{N} m_i \mathbf{r}_i \times (\boldsymbol{\omega} \times \mathbf{r}_i)$.

$Oxyz$ which has angular velocity ω. We must evaluate the sum (3–16), where the r_i are position vectors to the m_i from the origin (which for the purposes of this section is a completely arbitrary point fixed in the body).

We begin the transformation of (3–16) by making use of the identity[1]

$$\mathbf{a} \times (\mathbf{b} \times \mathbf{c}) = (\mathbf{a} \cdot \mathbf{c})\mathbf{b} - (\mathbf{b} \cdot \mathbf{a})\mathbf{c} \qquad (3\text{–}17)$$

to obtain

$$\mathbf{H}_o = \sum_{i=1}^{N} m_i [r_i^2 \boldsymbol{\omega} - (\boldsymbol{\omega} \cdot \mathbf{r}_i)\mathbf{r}_i]. \qquad (3\text{–}18)$$

Further progress is made by switching from vector to matrix notation. If we introduce the x, y, and z components of the vectors in (3–18),

$$\mathbf{H}_o = H_x \mathbf{u}_x + H_y \mathbf{u}_y + H_z \mathbf{u}_z$$

$$\boldsymbol{\omega} = \omega_x \mathbf{u}_x + \omega_y \mathbf{u}_y + \omega_z \mathbf{u}_z \qquad (3\text{–}19)$$

$$\mathbf{r}_i = x_i \mathbf{u}_x + y_i \mathbf{u}_y + z_i \mathbf{u}_z,$$

[1] See Eq. (C–4).

the vector equation separates into three component equations,

$$H_x = [\Sigma m_i(r_i^2 - x_i^2)]\omega_x + [-\Sigma m_i x_i y_i]\omega_y + [-\Sigma m_i x_i z_i]\omega_z$$
$$H_y = [-\Sigma m_i y_i x_i]\omega_x + [\Sigma m_i(r_i^2 - y_i^2)]\omega_y + [-\Sigma m_i y_i z_i]\omega_z \quad (3\text{-}20)$$
$$H_z = [-\Sigma m_i z_i x_i]\omega_x + [-\Sigma m_i z_i y_i]\omega_y + [\Sigma m_i(r_i^2 - z_i^2)]\omega_z.$$

The sums in square brackets in (3-20) represent various second moments of the mass distribution of the rigid body. We call the terms on the main diagonal *moments of inertia* and represent them by the symbols I_{xx}, I_{yy}, and I_{zz}, where

$$I_{xx} = \sum_{i=1}^{N} m_i(r_i^2 - x_i^2) = \sum_{i=1}^{N} m_i(y_i^2 + z_i^2)$$
$$I_{yy} = \sum_{i=1}^{N} m_i(r_i^2 - y_i^2) = \sum_{i=1}^{N} m_i(z_i^2 + x_i^2) \quad (3\text{-}21)$$
$$I_{zz} = \sum_{i=1}^{N} m_i(r_i^2 - z_i^2) = \sum_{i=1}^{N} m_i(x_i^2 + y_i^2).$$

The remaining terms are called *products of inertia* and are represented by the symbols I_{xy}, I_{yz}, etc., where

$$I_{xy} = - \sum_{i=1}^{N} m_i x_i y_i \qquad I_{yz} = - \sum_{i=1}^{N} m_i y_i z_i \qquad \text{etc.} \quad (3\text{-}22)$$

With the abbreviations (3-21) and (3-22), the equations in (3-20) take the form

$$H_x = I_{xx}\omega_x + I_{xy}\omega_y + I_{xz}\omega_z$$
$$H_y = I_{yx}\omega_x + I_{yy}\omega_y + I_{yz}\omega_z \quad (3\text{-}23)$$
$$H_z = I_{zx}\omega_x + I_{zy}\omega_y + I_{zz}\omega_z.$$

We now introduce matrix notation[1] with the column matrices

$$\{H\}_o = \begin{Bmatrix} H_x \\ H_y \\ H_z \end{Bmatrix} \qquad \{\omega\} = \begin{Bmatrix} \omega_x \\ \omega_y \\ \omega_z \end{Bmatrix} \quad (3\text{-}24)$$

and the square matrix

$$[I]_o = \begin{bmatrix} I_{xx} & I_{xy} & I_{xz} \\ I_{yx} & I_{yy} & I_{yz} \\ I_{zx} & I_{zy} & I_{zz} \end{bmatrix}. \quad (3\text{-}25)$$

The three scalar equations (3-23) may then be written compactly as the single matrix equation

$$\{H\}_o = [I]_o\{\omega\}. \quad (3\text{-}26)$$

[1] See Appendix B.

The column matrix $\{H\}_o$ contains the components of the vector sum $\mathbf{H}_o$ of (3-16). Note that the vector $\mathbf{H}_o$ is independent of the orientation of the xyz frame originating at O in Fig. 3-5, whereas the elements of the matrix $\{H\}_o$ do depend on the orientation of the coordinate system. If a different frame $Ox'y'z'$ is selected, the vector $\mathbf{H}_o$ remains invariant but the column $\{H'\}_o$ will have different elements, given by the transformation law (2-24). In the same way the column matrix $\{\omega\}$ in (3-26) contains the components of the angular-velocity vector ω. At a given instant the vector ω is invariant with respect to changes in the orientation of the frame $Oxyz$ *within* the body, but the elements of the column matrix $\{\omega\}$ will change according to (2-24).

The square inertia matrix $[I]_o$ in (3-25) contains elements which are second moments of the mass distribution *with respect to a particular coordinate system*. If a different frame $Ox'y'z'$ is selected, the resulting inertia matrix $[I']_o$ will have different elements. The transformation law for inertia matrices is discussed in Sec. 3-7.

The physically invariant quantity underlying a *column* matrix and the transformation law (2-24) is called a *vector*. In an analogous fashion the physically invariant description of the second moments of a particular mass distribution with respect to a particular origin, which underlies a *square inertia matrix* and its transformation law, is called an *inertia tensor*. The nine elements of the matrix $[I]_o$ of (3-25) are said to be the *components of the inertia tensor* in the x, y, z coordinate system. Other tensor quantities which have the same structure as the inertia tensor are *stress* at a point and *strain* at a point.[1]

An important property of the inertia tensor is that its components always form a symmetrical matrix; i.e.,

$$I_{xy} = I_{yx} \qquad I_{yz} = I_{zy} \qquad I_{zx} = I_{xz}. \tag{3-27}$$

This result follows immediately from the definitions (3-22) of the products of inertia.

The developments in this section leading to the inertia tensor have been for an arbitrary origin O fixed in the rigid body; i.e., to any point of a rigid body there corresponds an inertia tensor (the relations between the components of inertia tensors corresponding to different origins are discussed in Sec. 3-6). At any origin O in a rigid body a coordinate system can be chosen, and the inertia matrix $[I]_o$ can be evaluated with the aid of (3-21) and (3-22). Furthermore, the column matrix $\{H\}_o$ constructed (from the inertia matrix and the angular-velocity column) according to (3-26) will represent the angular momentum of the body with respect to O provided that the point O has certain special charac-

[1] See, for example, S. H. Crandall and N. C. Dahl (eds.), "An Introduction to the Mechanics of Solids," McGraw-Hill Book Company, New York, 1959, p. 151.

teristics.   From (3–11) and (3–16) we see that $\{H\}_o$ is the total angular momentum when the origin is either

1.   A point fixed in inertial space, or
2.   The center of mass of the rigid body.

For most other origins the column $\{H\}_o$ defined by (3–26) does not represent[1] the complete angular momentum of the body with respect to O. In our treatment of rigid-body dynamics we shall generally reserve the subscript o to denote case 1, and we shall use the subscript c to denote case 2.

When the moment center is a point which falls in either case 1 or 2, the relation (3–26) states compactly the relation between angular momentum, the mass distribution of the rigid body, and the angular velocity of the motion.   The vector direction of the angular momentum determined by (3–26) is *not* in general parallel to the vector direction of the angular velocity; i.e., angular momentum is not usually aligned with the axis of rotation.   This fact is responsible for much of the complexity of rigid-body dynamics.

Example 3–1.   Given the uniform rigid cube shown in Fig. 3–6.   It is desired to obtain the components of the inertia tensor for the origin at a corner of the cube.   It is also desired to obtain the components of the angular-momentum vector $\mathbf{H}_o$ when the origin is at rest and the axis of the angular velocity $\boldsymbol{\omega}$ is along one edge of the cube.

In Fig. 3–6 the xyz coordinate frame has been taken with the axes along the edges of the cube.   The angular velocity is directed along the z axis.   The components of the inertia tensor are obtained by evaluating (3–21) and (3–22).   The model we have been using for a rigid body has consisted of a very large number N of discrete mass particles, as indicated in Fig. 3–1b.   In this example it is more convenient to consider the cube as a continuous distribution of mass and to replace the summations of (3–21) and (3–22) by integrations.   If the total mass M of the cube is uniformly distributed, the mass in a differential volume $dx\,dy\,dz$ is

$$dm = \frac{M}{L^3}\,dx\,dy\,dz. \tag{3–28}$$

To obtain the moment of inertia I_{zz} we construct the following integral in place of the sum in (3–21):

$$I_{zz} = \int\!\!\int\!\!\int_0^L (y^2 + z^2)\,\frac{M}{L^3}\,dx\,dy\,dz = \tfrac{2}{3}ML^2. \tag{3–29}$$

Similarly, for the product of inertia I_{zy} we construct the following integral

[1] See Prob. 3–4.

in place of the sum in (3–22):

$$I_{xy} = - \iiint\limits_{0}^{L} xy \frac{M}{L^3}\, dx\, dy\, dz = -\tfrac{1}{4}ML^2. \tag{3-30}$$

The remaining components of the inertia tensor follow from symmetry, and the resulting inertia matrix is

$$[I]_o = ML^2 \begin{bmatrix} \tfrac{2}{3} & -\tfrac{1}{4} & -\tfrac{1}{4} \\ -\tfrac{1}{4} & \tfrac{2}{3} & -\tfrac{1}{4} \\ -\tfrac{1}{4} & -\tfrac{1}{4} & \tfrac{2}{3} \end{bmatrix}. \tag{3-31}$$

Now, if the magnitude of the angular velocity is ω, the components of the

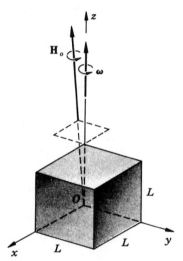

FIG. 3–6.   Example 3–1.   Angular momentum $\mathbf{H}_o$ of cube is not parallel to angular velocity ω.

angular-momentum vector $\mathbf{H}_o$ are given by (3–26).

$$\begin{Bmatrix} H_x \\ H_y \\ H_z \end{Bmatrix} = ML^2 \begin{bmatrix} \tfrac{2}{3} & -\tfrac{1}{4} & -\tfrac{1}{4} \\ -\tfrac{1}{4} & \tfrac{2}{3} & -\tfrac{1}{4} \\ -\tfrac{1}{4} & -\tfrac{1}{4} & \tfrac{2}{3} \end{bmatrix} \begin{Bmatrix} 0 \\ 0 \\ \omega \end{Bmatrix} = ML^2\omega \begin{Bmatrix} -\tfrac{1}{4} \\ -\tfrac{1}{4} \\ \tfrac{2}{3} \end{Bmatrix}. \tag{3-32}$$

The orientation of the resulting vector $\mathbf{H}_o$ is sketched in Fig. 3–6.

3–4   Kinetic coenergy of a rigid body

We return to the model of Fig. 3–1b in which a rigid body is pictured as a very large number of mass particles rigidly linked together.   In Fig. 3–7 a rigid body is indicated and one typical mass particle m_i is displayed. The kinetic coenergy of the rigid body is the sum of the coenergies of the individual particles,

$$T^* = \sum_{i=1}^{N} \tfrac{1}{2} m_i \mathbf{v}_i \cdot \mathbf{v}_i. \tag{3-33}$$

This can be given several interesting forms[1] by expanding $\mathbf{v}_i$ in different ways. For example, let us assume that there is a point O in the rigid

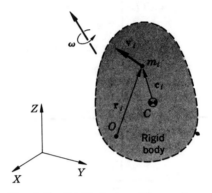

FIG. 3-7. Rigid body with angular velocity $\boldsymbol{\omega}$. Linear velocity of point O is zero. C is centroid.

body which is fixed with respect to the inertial reference; i.e., $\mathbf{v}_o \equiv 0$. Then $\mathbf{v}_i$ is given by (3-15), and (3-33) becomes

$$T^* = \tfrac{1}{2} \sum_{i=1}^{N} m_i (\boldsymbol{\omega} \times \mathbf{r}_i) \cdot (\boldsymbol{\omega} \times \mathbf{r}_i)$$

$$= \tfrac{1}{2} \boldsymbol{\omega} \cdot \sum_{i=1}^{N} m_i \mathbf{r}_i \times (\boldsymbol{\omega} \times \mathbf{r}_i) \tag{3-34}$$

on interchanging the order of dot and cross products. The summation remaining in (3-34) is the angular momentum $\mathbf{H}_o$ given by (3-16), so that

$$T^* = \tfrac{1}{2} \boldsymbol{\omega} \cdot \mathbf{H}_o. \tag{3-35}$$

An alternative expansion is obtained by utilizing the center of mass C with velocity $\mathbf{v}_c$. In this case,

$$\mathbf{v}_i = \mathbf{v}_c + \boldsymbol{\omega} \times \mathbf{c}_i, \tag{3-36}$$

and (3-33) becomes

$$T^* = \tfrac{1}{2} \sum_{i=1}^{N} m_i [\mathbf{v}_c \cdot \mathbf{v}_c + 2 \mathbf{v}_c \cdot \boldsymbol{\omega} \times \mathbf{c}_i + (\boldsymbol{\omega} \times \mathbf{c}_i) \cdot (\boldsymbol{\omega} \times \mathbf{c}_i)]. \tag{3-37}$$

[1] See Prob. 3-5.

By introducing (3–3), (3–6), and (3–11), we reduce (3–37) to

$$T^* = \tfrac{1}{2}Mv_c^2 + \tfrac{1}{2}\boldsymbol{\omega} \cdot \mathbf{H}_c. \tag{3–38}$$

This expansion exhibits the total kinetic coenergy as the sum of a translational coenergy (which has the same form as the kinetic coenergy of a particle of mass M and velocity $\mathbf{v}_c$) plus a rotational coenergy [which is the kinetic coenergy the body would have according to (3–35) if the angular velocity of the body were to remain unchanged and the linear velocity of the centroid were to be reduced to zero].

The expressions (3–35) and (3–38) for kinetic coenergy may be recast in matrix form by introducing the matrix representation for angular momentum (3–26). When the rigid body has a point O which is fixed in inertial space, the kinetic coenergy is

$$T^* = \tfrac{1}{2}\{\omega\}^t[I]_o\{\omega\}, \tag{3–39}$$

where $[I]_o$ is the inertia matrix of the rigid body with respect to a coordinate system originating from the motionless point O, $\{\omega\}$ is a column matrix of the components of the angular velocity (with respect to the same coordinate system), and $\{\omega\}^t$ is a row matrix which is the transpose of the column $\{\omega\}$. Similarly, (3–38) can be written

$$T^* = \tfrac{1}{2}Mv_c^2 + \tfrac{1}{2}\{\omega\}^t[I]_c\{\omega\}, \tag{3–40}$$

where $[I]_c$ is the inertia matrix with respect to a coordinate system originating from the centroid C, and $\{\omega\}$ and $\{\omega\}^t$ are column and row matrices containing the components of the angular velocity (with respect to the same coordinate system).

Example 3–2. Let us reconsider the rigid cube of Example 3–1. Suppose it is rotating about the fixed corner O, with a general angular velocity $\boldsymbol{\omega}$ having components ω_x, ω_y, and ω_z. What is the kinetic coenergy of the cube under these circumstances?

Here the formula (3–39) applies, and the inertia matrix $[I]_o$ is given by (3–31). The kinetic coenergy is thus

$$T^* = \tfrac{1}{2}ML^2\{\omega_x\ \omega_y\ \omega_z\} \begin{bmatrix} \tfrac{2}{3} & -\tfrac{1}{4} & -\tfrac{1}{4} \\ -\tfrac{1}{4} & \tfrac{2}{3} & -\tfrac{1}{4} \\ -\tfrac{1}{4} & -\tfrac{1}{4} & \tfrac{2}{3} \end{bmatrix} \begin{Bmatrix} \omega_x \\ \omega_y \\ \omega_z \end{Bmatrix}$$

$$= \tfrac{1}{2}ML^2[\tfrac{2}{3}(\omega_x^2 + \omega_y^2 + \omega_z^2) - \tfrac{1}{2}(\omega_x\omega_y + \omega_y\omega_z + \omega_z\omega_x)]. \tag{3–41}$$

3–5 Dynamic principles for a rigid body

The basic dynamic principles for a single particle are statements which relate velocity and momentum and which relate force and rate of change of momentum. For a rigid body considered as a single inertia element there are corresponding dynamic principles.

For *linear* momentum the dynamic principles for a rigid body are similar to those for a particle. The velocity-momentum relation is (3–7),

$$\mathbf{P} = M\mathbf{v}_c, \tag{3–42}$$

and the force-momentum relation is (2–260),

$$\mathbf{F} = \frac{d\mathbf{P}}{dt}. \tag{3–43}$$

For *angular* momentum the situation is more complex.

A complicating factor is the choice of the moment center B. As indicated in Sec. 3–2, the angular momentum of a rigid body can be computed with respect to *any* moment center, although the computation is simpler for certain classes of centers. There is also the relation (2–270) between the torque of the external forces acting on a rigid body and the rate of change of the angular momentum which is valid for *any* moment center, but the simple relation (2–271) is only valid for points B such that $\mathbf{v}_B \times \mathbf{P}$ vanishes. For the sake of simplicity we shall limit further discussion to the two special moment centers[1] of page 174, i.e., (1) a point O which is *fixed* in an inertial reference frame, or (2) the *centroid* C of the rigid body. For the fixed center O the angular-velocity–angular-momentum relation can be given in vector form (3–16) or in matrix form (3–26):

$$\mathbf{H}_o = \sum_{i=1}^{N} m_i \mathbf{r}_i \times (\boldsymbol{\omega} \times \mathbf{r}_i) \tag{3–44}$$

$$\{H\}_o = [I]_o\{\omega\}.$$

The torque–angular-momentum relation is (2–271),

$$\boldsymbol{\tau}_o = \frac{d\mathbf{H}_o}{dt}. \tag{3–45}$$

When the centroid C is taken as the moment center, the corresponding angular-velocity–angular-momentum relation is (3–11) in vector form and (3–26) in matrix form,

$$\mathbf{H}_c = \sum_{i=1}^{N} m_i \mathbf{c}_i \times (\boldsymbol{\omega} \times \mathbf{c}_i) \tag{3–46}$$

$$\{H\}_c = [I]_c\{\omega\},$$

and the torque–angular-momentum relation is

$$\boldsymbol{\tau}_c = \frac{d\mathbf{H}_c}{dt}. \tag{3–47}$$

[1] For the general case, see Prob. 3–9.

It is interesting to make a comparison between the dynamic principles for a particle and the dynamic principles for a rigid body. The major difference between the two inertia elements is that a rigid body translates *and* rotates (e.g., its velocity state is described by v_c and ω), whereas the motion of a particle is completely described by its velocity v. The *linear* dynamic principles for a rigid body are (3–42) and (3–43), which are completely parallel to the dynamic principles (2–86) and (2–88) for a particle. There is also a formal similarity between the *angular* dynamic principles for a rigid body and the dynamic principles for a particle. This formal similarity is displayed in Table 3–1.

Table 3–1
Comparison between the dynamic principles governing the motion of a rigid body about a fixed point O and the dynamic principles for a particle

Rigid body	*Particle*
$\{H\}_o = [I]_o\{\omega\}$	$p = mv$
$\tau_o = \dfrac{d\mathbf{H}_o}{dt}$	$f = \dfrac{dp}{dt}$
$T^* = \frac{1}{2}\{\omega\}'[I]_o\{\omega\}$	$T^* = \frac{1}{2}mv^2 = \frac{1}{2}vmv$
$T = \frac{1}{2}\{H\}_o^t[I]_o^{-1}\{H\}_o$	$T = \dfrac{1}{2}\dfrac{p^2}{m} = \frac{1}{2}pm^{-1}p$

Despite the similarity in the form of the dynamic principles, the angular motion of a rigid body is intrinsically more complex than the motion of a particle. The source of the divergence lies in the first pair of entries in Table 3–1, the velocity-momentum relations. The similarity here is limited because an inertia *tensor* differs essentially from a *scalar* mass.

When the momentum of a particle is known, the velocity of the particle is completely determined: the vector direction is the same as that of the momentum, and the magnitude is obtained by dividing the magnitude of momentum by the mass. If the momentum of a particle does not change with time, its velocity remains constant and the particle travels in a straight line.

The corresponding behavior of a rotating rigid body is quite different. We have already seen in Example 3–1 that angular momentum and angular velocity do not in general have the same direction. In Chap. 4 we shall also see that a constant angular momentum does not necessarily imply a constant angular velocity. A rigid body with constant angular momentum can sometimes gyrate in a fairly complex manner, with an angular velocity which is continually changing in magnitude and direction.

The remainder of this chapter is given over to developing properties of inertia matrices of rigid bodies.   These properties provide useful tools for the study of rigid-body dynamics in Chap. 4.

3–6   Parallel-axes theorem

In Sec. 3–3 it was pointed out that there is an inertia tensor associated with every point fixed in the frame of a rigid body.   At each point the inertia tensor has nine components (which form a symmetric matrix) with respect to any coordinate system originating from that point.   In this section we discuss the relations between the components of inertia tensors corresponding to *different points* in the rigid body but referred to *parallel* coordinate systems.   In Sec. 3–7 we discuss the relations between the components of an inertia tensor *at a single point* that are referred to different coordinate systems which are *rotated* with respect to one another.

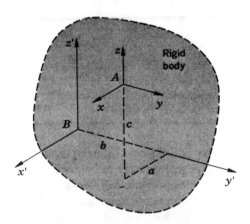

Fig. 3–8.   Parallel axes with origins at *A* and
B attached to the same rigid body.

The formulas for obtaining the components of the inertia tensor at one point in terms of the components of the inertia tensor at another point when the components are with respect to parallel coordinate systems are contained in the parallel-axes theorem, which is developed below.   This theorem is useful whenever the inertia properties of a rigid body are known with respect to one point but are required with respect to a different point. The theorem is helpful for determining the overall inertia properties of a composite rigid body which is made up of a number of smaller rigid bodies whose individual inertia properties are known.   The parallel-axes theorem deals with the situation shown in Fig. 3–8.   In a single rigid body there are two sets of parallel axes with origins at *A* and *B*.   The theorem provides a formula for obtaining the inertia matrix $[I]_B$ when the inertia matrix $[I]_A$ is known.

Let the origins be displaced by a vector whose components are a, b, and c, so that the relation between coordinates is

$$x' = x + a \qquad y' = y + b \qquad z' = z + c. \qquad (3\text{–}48)$$

This implies that in the primed frame the coordinates of point A are (a,b,c). We now compute the elements of $[I]_B$ according to (3–21) and (3–22) and insert (3–48). Thus, using (3–21), we have

$$
\begin{aligned}
I_{x'x'} &= \sum_{i=1}^{N} m_i(y_i'^2 + z_i'^2) \\
&= \sum_{i=1}^{N} m_i(y_i^2 + z_i^2 + 2by_i + 2cz_i + b^2 + c^2) \\
&= I_{xx} + 2M(by_c + cz_c) + M(b^2 + c^2), \qquad (3\text{–}49)
\end{aligned}
$$

where M is the total mass (3–3), and we have used the mass center relation (3–5) to write

$$\sum_{i=1}^{N} m_i x_i = M x_c \qquad \sum_{i=1}^{N} m_i y_i = M y_c \qquad \sum_{i=1}^{N} m_i z_i = M z_c. \qquad (3\text{–}50)$$

Similarly, using (3–22), we find

$$
\begin{aligned}
I_{x'y'} &= -\sum_{i=1}^{N} m_i x_i' y_i' \\
&= -\sum_{i=1}^{N} m_i(x_i y_i + bx_i + ay_i + ab) \\
&= I_{xy} - M(bx_c + ay_c) - Mab. \qquad (3\text{–}51)
\end{aligned}
$$

The other elements of the matrix $[I]_B$ are obtained in the same fashion. Assembling the elements, we obtain a general formula relating $[I]_B$ to $[I]_A$.

$$
[I]_B = [I]_A + M
\begin{bmatrix}
2(by_c + cz_c) & -(bx_c + ay_c) & -(cx_c + az_c) \\
-(bx_c + ay_c) & 2(cz_c + ax_c) & -(cy_c + bz_c) \\
-(cx_c + az_c) & -(cy_c + bz_c) & 2(ax_c + by_c)
\end{bmatrix}
$$

$$
+ M
\begin{bmatrix}
b^2 + c^2 & -ab & -ac \\
-ab & c^2 + a^2 & -bc \\
-ac & -bc & a^2 + b^2
\end{bmatrix}
\qquad (3\text{–}52)
$$

This result applies to *any* pair of origins A and B. A simpler result is obtained if the origin A coincides with the center of mass C because then $x_c = y_c = z_c = 0$ and the middle term on the right of (3–52) vanishes.

In this case (3–52) reduces to

$$[I]_B = [I]_c + M \begin{bmatrix} b^2 + c^2 & -ab & -ac \\ -ab & c^2 + a^2 & -bc \\ -ac & -bc & a^2 + b^2 \end{bmatrix}, \qquad (3\text{--}53)$$

which is known as the *parallel-axes theorem.* Note that (3–53) gives the inertia matrix $[I]_B$ when the centroidal inertia matrix $[I]_c$ for parallel axes is known and the coordinates of the mass center, in the frame with origin at B, are

$$x'_c = a \qquad y'_c = b \qquad z'_c = c. \qquad (3\text{--}54)$$

Example 3–3. Let us find the centroidal inertia matrix for the uniform cube of Fig. 3–9. Here we already know the inertia matrix $[I]_o$

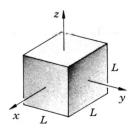

Fig. 3–9. Example 3–3. Cube with centroidal axes.

corresponding to an origin at the corner of the cube. See (3–31) and Fig. 3–6. Accordingly, we shall use (3–53) in reverse to obtain $[I]_c$. Considering the axes with origin at O in Fig. 3–6 to be the primed system, the mass-center coordinates (3–54) are $a = b = c = L/2$. Substituting in (3–53) gives

$$ML^2 \begin{bmatrix} \frac{2}{3} & -\frac{1}{4} & -\frac{1}{4} \\ -\frac{1}{4} & \frac{2}{3} & -\frac{1}{4} \\ -\frac{1}{4} & -\frac{1}{4} & \frac{2}{3} \end{bmatrix} = [I]_c + ML^2 \begin{bmatrix} \frac{1}{2} & -\frac{1}{4} & -\frac{1}{4} \\ -\frac{1}{4} & \frac{1}{2} & -\frac{1}{4} \\ -\frac{1}{4} & -\frac{1}{4} & \frac{1}{2} \end{bmatrix} \qquad (3\text{--}55)$$

or

$$[I]_c = ML^2 \begin{bmatrix} \frac{1}{6} & 0 & 0 \\ 0 & \frac{1}{6} & 0 \\ 0 & 0 & \frac{1}{6} \end{bmatrix} \qquad (3\text{--}56)$$

for the desired centroidal inertia matrix.

APPLICATION TO COMPOSITE BODIES

The individual moments and products of inertia (3–21) and (3–22) which make up the inertia matrix involve summations over all the mass particles in the rigid body. In some cases it may be convenient to subdivide the complete rigid body into n composite parts. The summations in (3–21) and (3–22) could correspondingly be subdivided, and we could write

$$[I]_o = \sum_{j=1}^{n} [I_j]_o, \qquad (3\text{–}57)$$

where $[I_j]_o$ represents the inertia matrix of the jth part. In (3–57) all matrices contain components referred to the *same* coordinate system with origin at O. The parallel-axes theorem can be used to evaluate each $[I_j]_o$ when inertia information for the jth part, expressed in terms of its own centroid, is known. Thus, if C_j, the centroid of the jth part, has

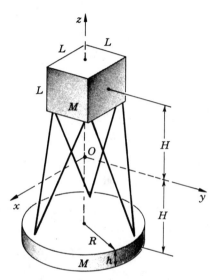

FIG. 3–10.   Example   3–4.   Satellite
package as a composite rigid body.

coordinates a_j, b_j, c_j in the frame with origin at O, and if the inertia matrix of the jth part (referred to parallel axes originating at C_j) is $[I_j]_{C_j}$, then the parallel-axes theorem yields

$$[I_j]_o = [I_j]_{C_j} + M_j \begin{bmatrix} b_j^2 + c_j^2 & -a_j b_j & -a_j c_j \\ -a_j b_j & c_j^2 + a_j^2 & -b_j c_j \\ -a_j c_j & -b_j c_j & a_j^2 + b_j^2 \end{bmatrix} \qquad (3\text{–}58)$$

where M_j is the mass of the jth part. The inertia matrix for the composite body can then be obtained by using (3–58) for each part and summing the results according to (3–57).

Example 3–4.   Figure 3–10 shows an idealized satellite configuration consisting of a cube and a disk rigidly connected by a truss of negligible mass. The cube and the disk have uniform mass densities, and they each have total mass M. It is desired to obtain the components of the inertia tensor with respect to the x, y, z axes whose origin O coincides with the center of mass of the composite body.

Since the two parts of the system have equal mass, the center of mass of the assembly is midway between the centroids of the individual parts, as indicated in Fig. 3–10. For the cube alone we already know the inertia matrix corresponding to its own centroid (see Example 3–3). We therefore turn to the disk to obtain the corresponding information.

In Fig. 3–11 the disk is shown with axes through its centroid which are

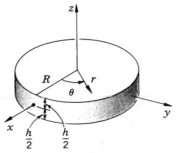

FIG. 3–11. Disk with its own centroidal axes.

parallel to those in Fig. 3–10. In cylindrical coordinates the mass in a differential volume $r \, d\theta \, dr \, dz$ is

$$dm = \frac{M}{\pi R^2 h} r \, d\theta \, dr \, dz, \tag{3–59}$$

and the coordinates x and y are represented in the form

$$x = r \cos \theta \qquad y = r \sin \theta. \tag{3–60}$$

The moments of inertia are obtained by replacing the summations of (3–21) with integrations; e.g.,

$$I_{xx} = \int_{-h/2}^{h/2} dz \int_0^{2\pi} d\theta \int_0^R (r^2 \sin^2 \theta + z^2) \frac{M}{\pi R^2 h} r \, dr$$

$$= M \left(\frac{R^2}{4} + \frac{h^2}{12} \right). \tag{3–61}$$

Similarly, we find

$$I_{yy} = M \left(\frac{R^2}{4} + \frac{h^2}{12} \right)$$

$$I_{zz} = M \frac{R^2}{2}. \tag{3–62}$$

The products of inertia are obtained by replacing the summations of

(3–22) with integrations; e.g.,

$$I_{xy} = \int_{-h/2}^{h/2} dz \int_0^{2\pi} d\theta \int_0^R r^2 \sin\theta \cos\theta \frac{M}{\pi R^2 h} r \, dr = 0. \qquad (3\text{–}63)$$

The other products of inertia also vanish, so that the inertia matrix of the disk with respect to the axes shown in Fig. 3–11 is

$$[I_{\text{disk}}]_{C_{\text{disk}}} = M \begin{bmatrix} \dfrac{R^2}{4} + \dfrac{h^2}{12} & 0 & 0 \\ 0 & \dfrac{R^2}{4} + \dfrac{h^2}{12} & 0 \\ 0 & 0 & \dfrac{R^2}{2} \end{bmatrix}. \qquad (3\text{–}64)$$

We are now ready to apply (3–57) and (3–58) to obtain the inertia matrix of the composite rigid body for the axes shown in Fig. 3–10. Applying the parallel-axes theorem (3–58) to the disk, we have

$$[I_{\text{disk}}]_o = M \begin{bmatrix} \dfrac{R^2}{4} + \dfrac{h^2}{12} & 0 & 0 \\ 0 & \dfrac{R^2}{4} + \dfrac{h^2}{12} & 0 \\ 0 & 0 & \dfrac{R^2}{2} \end{bmatrix} + M \begin{bmatrix} H^2 & 0 & 0 \\ 0 & H^2 & 0 \\ 0 & 0 & 0 \end{bmatrix}, \qquad (3\text{–}65)$$

since in the coordinate system of Fig. 3–10 the centroid of the disk is at $(0,0,-H)$. The corresponding parallel-axes statement for the cube is

$$[I_{\text{cube}}]_o = M \begin{bmatrix} \dfrac{L^2}{6} & 0 & 0 \\ 0 & \dfrac{L^2}{6} & 0 \\ 0 & 0 & \dfrac{L^2}{6} \end{bmatrix} + M \begin{bmatrix} H^2 & 0 & 0 \\ 0 & H^2 & 0 \\ 0 & 0 & 0 \end{bmatrix}, \qquad (3\text{–}66)$$

on using (3–56) and the fact that in the coordinate system of Fig. 3–10 the centroid of the cube is at $(0,0,H)$. Finally, the inertia matrix for the composite package is given by (3–57) as the sum of (3–65) and (3–66).

$$[I]_o = M \begin{bmatrix} \dfrac{R^2}{4} + \dfrac{h^2}{12} + \dfrac{L^2}{6} + 2H^2 & 0 & 0 \\ 0 & \dfrac{R^2}{4} + \dfrac{h^2}{12} + \dfrac{L^2}{6} + 2H^2 & 0 \\ 0 & 0 & \dfrac{R^2}{2} + \dfrac{L^2}{6} \end{bmatrix}.$$

$$(3\text{–}67)$$

3–7   Rotational transformation of inertia-tensor components

At any fixed point O in a rigid body there is associated an inertia tensor which characterizes the second moments of mass of the body with respect to O.   Given any rectangular coordinate system $Oxyz$ originating at O, the inertia tensor has nine components, which are evaluated according to (3–21) and (3–22).   The same formulas can be used to evaluate a different set of nine components with respect to a second coordinate system $Ox'y'z'$, which is rotated with respect to the first.   In this section we discuss the relations between these sets of components corresponding to different coordinate systems originating from the same point O.

We begin by reviewing the basic rotational-transformation law for the

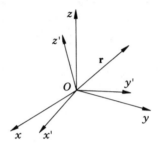

Fɪɢ. 3–12.   Rectangular coordinate systems $Oxyz$ and $Ox'y'z'$ and position vector **r**.

components of a vector.   Consider the rotated coordinate systems $Oxyz$ and $Ox'y'z'$ shown in Fig. 3–12, and the position vector **r**.   Let $\{r\}$ and $\{r'\}$ be column vectors of the components of **r** in the two coordinate systems; i.e.,

$$\{r\} = \begin{Bmatrix} x \\ y \\ z \end{Bmatrix} \qquad \{r'\} = \begin{Bmatrix} x' \\ y' \\ z' \end{Bmatrix}. \tag{3–68}$$

The relation between these components, according to (2–24), is

$$\{r'\} = [C]\{r\}, \tag{3–69}$$

where $[C]$ is the rotation matrix (2–17) made up of the directional cosines of the primed axes with respect to the unprimed axes.

We next give an explicit representation for the matrix (3–25) of inertia-tensor components.   The separate defining relations (3–21) and (3–22) for the tensor components can be cast into a common form by using the column $\{r\}$ of (3-68) and its transpose.   We have

$$[I]_o = \sum_{i=1}^{N} m_i(r_i^2[1] - \{r_i\}\{r_i\}^t),   \tag{3-70}$$

where r_i^2 is the scalar $x_i^2 + y_i^2 + z_i^2$, and [1] is the *unity*, or identity, matrix of (2-22).   The sum (3-70) is appropriate when the rigid body is modeled by a large number of mass particles rigidly attached to a frame, as indicated in Fig. 3-1*b*.   When the rigid body is modeled by a continuous distribution of mass, the corresponding representation of the inertia matrix is

$$[I]_o = \int dm(r^2[1] - \{r\}\{r\}^t),   \tag{3-71}$$

where the integral is extended over all the mass in the body.   For simplicity of notation we shall use the continuous model and the representation (3-71) in the following development.   Furthermore, since we shall always consider the same origin O in this section and the next, we shall temporarily omit the subscript o on the inertia matrix (and the associated angular-momentum column).

Now let us use the representation (3-71) to indicate the matrix of inertia-tensor components with respect to the rotated coordinate system $Ox'y'z'$ of Fig. 3-12.

$$[I'] = \int dm(r'^2[1] - \{r'\}\{r'\}^t).   \tag{3-72}$$

We can derive the transformation law for inertia-tensor components by relating (3-72) to (3-71).   To do this we introduce into (3-72) the vector-transformation law (3-69) and its transpose

$$\{r'\}^t = \{r\}^t[C]^t.   \tag{3-73}$$

First, we note that the scalar r'^2 (the square of the length of the vector **r**) is equal to the scalar r^2.   Then (3-72) becomes

$$\begin{aligned}
[I'] &= \int dm \ (r^2[1] - [C]\{r\}\{r\}^t[C]^t) \\
&= \int dm \ ([C]r^2[1][C]^t - [C]\{r\}\{r\}^t[C]^t) \\
&= [C]\int dm \ (r^2[1] - \{r\}\{r\}^t)[C]^t \\
&= [C][I][C]^t,
\end{aligned}   \tag{3-74}$$

on using the inverse relation of (2-22),

$$[C][C]^t = [1],   \tag{3-75}$$

and on interchanging the operation of matrix multiplication by the rotation matrix with the operation of integration over the rigid body.

The result of (3-74) is the transformation law for inertia-tensor components under rotation of the coordinate axes.   In compact form it indicates the operations which must be performed on the moments and products of inertia with respect to the xyz coordinate system in order to obtain the corresponding moments and products of inertia with respect to the $x'y'z'$ coordinate system. The formula (3–74) is convenient for programming digital-computer calculations of inertia properties. The calculations involved in even a single application of (3–74) generally represent a tedious amount of hand computation.

Example 3–5. As a simple illustration of (3–74) we shall obtain the inertia matrices $[I]$ and $[I']$ for the idealized rigid body shown in Fig. 3–13.

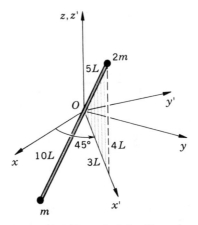

FIG. 3–13.   Example 3–5.   Two particles and massless connecting link.

The body consists of two mass particles connected by a rigid massless link. The origin O of both coordinate systems coincides with the center of mass of the rigid body. In the xyz system the positions of the particles $m_1 = 2m$ and $m_2 = m$ are

$$\{r_1\} = \begin{Bmatrix} \dfrac{3\sqrt{2}}{2}L \\ \dfrac{3\sqrt{2}}{2}L \\ 4L \end{Bmatrix} \qquad \{r_2\} = \begin{Bmatrix} -3\sqrt{2}\,L \\ -3\sqrt{2}\,L \\ -8L \end{Bmatrix}, \qquad (3\text{–}76)$$

and the inertia matrix obtained by evaluating (3–70) or (3–21) and (3–22) is

$$I = mL^2 \begin{bmatrix} 123 & -27 & -36\sqrt{2} \\ -27 & 123 & -36\sqrt{2} \\ -36\sqrt{2} & -36\sqrt{2} & 54 \end{bmatrix}. \qquad (3\text{–}77)$$

The rotation matrix $[C]$ between the unprimed and primed coordinates in Fig. 3–13 is

$$[C] = \begin{bmatrix} \dfrac{\sqrt{2}}{2} & \dfrac{\sqrt{2}}{2} & 0 \\[2ex] -\dfrac{\sqrt{2}}{2} & \dfrac{\sqrt{2}}{2} & 0 \\[2ex] 0 & 0 & 1 \end{bmatrix}. \tag{3-78}$$

The inertia matrix $[I']$ corresponding to the primed coordinate system in Fig. 3–13 is then obtained by substituting in the transformation law (3–74).

$$[I'] = mL^2 \begin{bmatrix} \dfrac{\sqrt{2}}{2} & \dfrac{\sqrt{2}}{2} & 0 \\[2ex] -\dfrac{\sqrt{2}}{2} & \dfrac{\sqrt{2}}{2} & 0 \\[2ex] 0 & 0 & 1 \end{bmatrix} \begin{bmatrix} 123 & -27 & -36\sqrt{2} \\[1ex] -27 & 123 & -36\sqrt{2} \\[1ex] -36\sqrt{2} & -36\sqrt{2} & 54 \end{bmatrix} \begin{bmatrix} \dfrac{\sqrt{2}}{2} & -\dfrac{\sqrt{2}}{2} & 0 \\[2ex] \dfrac{\sqrt{2}}{2} & \dfrac{\sqrt{2}}{2} & 0 \\[2ex] 0 & 0 & 1 \end{bmatrix}$$

$$= mL^2 \begin{bmatrix} 96 & 0 & -72 \\ 0 & 150 & 0 \\ -72 & 0 & 54 \end{bmatrix}. \tag{3-79}$$

This result can be verified by a direct computation of $[I']$, using the summation version of (3–72).

Transformation of the Angular-Velocity–Angular-Momentum Relation

The transformation law (3–74) for inertia-tensor components and the transformation law (3–69) for vector components can be applied to the relation (3–26) between angular velocity and angular momentum. In the primed coordinate system of Fig. 3–12 let us construct the difference,

$$\{H'\} - [I']\{\omega'\}. \tag{3-80}$$

Now applying (3–69) to the columns in (3–80) and (3–74) to the square matrix, we obtain

$$\{H'\} - [I']\{\omega'\} = [C]\{H\} - [C][I][C]^t[C]\{\omega\}$$
$$= [C](\{H\} - [I]\{\omega\}), \tag{3-81}$$

on using the identity

$$[C]^t[C] = [1]. \tag{3-82}$$

Equation (3–81) demonstrates that if the angular-velocity–angular-momentum relation (3–26) is satisfied in the unprimed coordinate system, it is also satisfied in the primed system. This result is a confirmation of the fact that the basic relation (3–16) between angular velocity and angular momentum is *independent* of the choice of coordinate system. To preserve this independence it is necessary for vector components to transform according to (3–69), and for tensor components to transform according to (3–74), whenever the coordinate axes are rotated.

3–8   Principal directions

The inertia-tensor transformation law (3–74) provides a means of evaluating the elements of the inertia matrix (for a given rigid body, with respect to a given origin) for any orientation of a rectangular coordinate system. If we imagine a continuous change in orientation of the coordinates, the corresponding elements of the inertia matrix will change continuously. In every orientation the inertia matrix is symmetric according to (3–27). In most orientations all nine elements of the matrix are nonzero. In this section we shall show that there always exists a special orientation of the coordinate system for which the inertia matrix has a *diagonal* form,

$$[I] = \begin{bmatrix} I_1 & 0 & 0 \\ 0 & I_2 & 0 \\ 0 & 0 & I_3 \end{bmatrix}. \tag{3-83}$$

The directions of the three coordinate axes in this special orientation are called *principal directions*, and the diagonal elements of (3–83) are called *principal moments of inertia*. In many cases it is more convenient to describe an inertia tensor by giving *three* mutually perpendicular principal directions and the *three* associated principal moments of inertia instead of giving the full inertia matrix (with *six* independent components) corresponding to some nonspecial orientation of the coordinate system.

Before entering into the demonstration that principal directions can always be found, let us pause to note an important implication of the diagonal form of the inertia matrix (3–83). Suppose that we do have a coordinate system with unit vectors $\mathbf{u}_1$, $\mathbf{u}_2$, and $\mathbf{u}_3$ along the coordinate axes for which the inertia matrix takes the form of (3–83). If the body were to have an angular velocity $\boldsymbol{\omega} = \omega_1\mathbf{u}_1$, the angular-momentum-vector components would be

$$\{H\} = \begin{bmatrix} I_1 & 0 & 0 \\ 0 & I_2 & 0 \\ 0 & 0 & I_3 \end{bmatrix} \begin{Bmatrix} \omega_1 \\ 0 \\ 0 \end{Bmatrix} = \begin{Bmatrix} I_1\omega_1 \\ 0 \\ 0 \end{Bmatrix}, \tag{3-84}$$

which implies that $\mathbf{H} = I_1\omega_1\mathbf{u}_1$. Similarly, we can see that if $\boldsymbol{\omega} = \omega_2\mathbf{u}_2$, then $\mathbf{H} = I_2\omega_2\mathbf{u}_2$, and if $\boldsymbol{\omega} = \omega_3\mathbf{u}_3$, then $\mathbf{H} = I_3\omega_3\mathbf{u}_3$. This means that when the body rotates so that its axis of rotation is parallel to a *principal axis*, the *angular-momentum* vector is *parallel* to the *angular-velocity* vector. We noted previously that generally angular-velocity and angular-momentum vectors for a rigid body are *not* parallel. Principal directions, then, are those exceptional directions such that, when the body rotates about an axis parallel to a principal direction, the angular-momentum vector also has this same direction.

We now return to the demonstration that a set of principal directions and a diagonal inertia matrix can always be found for an arbitrary inertia tensor. We start from a nonprincipal coordinate system with a full inertia matrix $[I]$. Let us try to find the orientation of a unit vector $\mathbf{u}$ such that, when the angular velocity is $\omega\mathbf{u}$, the angular momentum is $H\mathbf{u}$, where ω and H are scalar magnitudes. If the (unknown) components of $\mathbf{u}$, in the initial coordinate system, are placed in the column $\{u\}$, the relation (3-26) between the angular velocity $\omega\mathbf{u}$ and the angular momentum $H\mathbf{u}$ takes the form

$$H\{u\} = \omega[I]\{u\}. \tag{3-85}$$

Let λ denote the ratio of the scalar magnitudes of the angular momentum and angular velocity; i.e.,

$$\lambda = \frac{H}{\omega}. \tag{3-86}$$

Then (3-85) becomes

$$[I]\{u\} = \lambda\{u\}, \tag{3-87}$$

which is a matrix equation for determining the orientation of $\{u\}$ and the magnitude of the ratio λ from the given components of the inertia tensor in the matrix $[I]$. If we find a unit vector $\{u\}$ and a value of λ which satisfy (3-87), we shall have determined a principal direction along which angular velocity $\omega\mathbf{u}$ produces parallel angular momentum $H\mathbf{u}$ with magnitude fixed by (3-86).

The eigenvalue problem

The problem of solving (3-87) for $\{u\}$ and λ is a classic algebraic problem, known as the *eigenvalue problem*. The value of λ in a solution is called an *eigenvalue*, and the corresponding vector $\{u\}$ is called an *eigenvector*. Eigenvalue problems of this type occur frequently in engineering analysis.[1] Systematic procedures for finding numerical solutions are known and are well suited to automated computation. Programs for this purpose are available in most computer libraries.

Here we give an outline of the classical analysis of (3-87) which is able to develop the existence, and qualitative properties, of solutions without actually finding an explicit solution. In a numerical case the entire solution is determined by the components in the inertia matrix $[I]$. In the

[1] See, for example, S. H. Crandall, "Engineering Analysis," McGraw-Hill Book Company, New York, 1956, chaps. 2 and 5.

qualitative analysis the properties of $[I]$ that are required are

1. According to (3–27), $[I]$ is a *symmetric* matrix.
2. Since $\frac{1}{2}\{\omega\}^t[I]\{\omega\}$ is the kinetic coenergy of a rigid body (which is basically a sum of positive contributions $\frac{1}{2}m_i v_i^2$), the product $\{u\}^t[I]\{u\}$ must be *nonnegative* for any unit vector $\{u\}$.

To begin the solution of (3–87), we bring all terms to the left side and write out the matrix equation in full.

$$\begin{bmatrix} I_{xx} - \lambda & I_{xy} & I_{xz} \\ I_{xy} & I_{yy} - \lambda & I_{yz} \\ I_{xz} & I_{yz} & I_{zz} - \lambda \end{bmatrix} \begin{Bmatrix} u_x \\ u_y \\ u_z \end{Bmatrix} = 0. \qquad (3\text{–}88)$$

This is a system of three simultaneous homogeneous linear equations for the components of **u**. The trivial solution $u_x = u_y = u_z = 0$ must be ruled out because it does not represent a unit vector. Nontrivial solutions to (3–88) can exist only if the determinant of the coefficient matrix vanishes. If we expand the determinant and set it equal to zero, we obtain a cubic equation in λ. Only when λ is a root of this cubic equation can a unit vector **u** be found whose components satisfy (3–88).

It is a tour de force of classical analysis that the properties of the roots for λ and the corresponding unit vectors can be deduced without expanding the determinant of the coefficients in (3–88). The first step consists[1] in demonstrating that the roots for λ must be *real*. The proof depends on the *reality* and *symmetry* of the inertia matrix $[I]$. The next step consists[2] in demonstrating that the roots for λ must be *nonnegative*. This depends on the nonnegativeness of the triple product $\{u\}^t[I]\{u\}$ for arbitrary $\{u\}$. Therefore there are three real nonnegative roots, $\lambda = I_1$, $\lambda = I_2$, and $\lambda = I_3$, for which the determinant of the coefficient matrix in (3–88) vanishes. For each root I_j the system (3–88) can be solved to find a unit vector $\{u_j\}$. The complete solution to (3–88) could thus be displayed as follows:

$$\begin{array}{cccc} \text{Eigenvalues:} & I_1 & I_2 & I_3 \\ \text{Eigenvectors:} & \mathbf{u_1} & \mathbf{u_2} & \mathbf{u_3} \end{array} \qquad (3\text{–}89)$$

Next we shall show that the directions of the eigenvectors are mutually perpendicular in the usual case where the eigenvalues are unequal. If i and j represent particular values of the integers 1, 2, or 3, the fact that (3–89) is a solution to (3–87) means that

$$\begin{aligned} [I]\{u_i\} &= I_i\{u_i\} \\ [I]\{u_j\} &= I_j\{u_j\}. \end{aligned} \qquad (3\text{–}90)$$

[1] See Prob. 3–16.

[2] See Prob. 3–17.

We premultiply the first of (3–90) by $\{u_j\}^t$, premultiply the second of (3–90) by $\{u_i\}^t$, and then subtract the resulting equations to obtain

$$\{u_j\}^t[I]\{u_i\} - \{u_i\}^t[I]\{u_j\} = I_i\{u_j\}^t\{u_i\} - I_j\{u_i\}^t\{u_j\}$$
$$= (I_i - I_j)\{u_i\}^t\{u_j\}$$
$$= (I_i - I_j)\mathbf{u}_i \cdot \mathbf{u}_j. \qquad (3\text{–}91)$$

Now since $[I]$ is a symmetric matrix, the left side of (3–91) vanishes,[1] which implies that, when $I_i \neq I_j$, the unit vectors $\mathbf{u}_i$ and $\mathbf{u}_j$ must be *orthogonal.* Thus, when we have a rigid body whose inertia tensor at a point is represented by a matrix of components, $[I]$, with respect to a given coordinate system, the solution (3–89) of the eigenvalue problem (3–87) generally provides a unique set of three mutually orthogonal directions along which an angular velocity will develop a parallel angular momentum. These principal directions are described by the component representations $\{u_1\}$, $\{u_2\}$, and $\{u_3\}$, referred to the given coordinate system.

In the special case where the roots of the cubic are not all unequal, the principal directions are not uniquely determined. Thus, if $I_1 = I_2 \neq I_3$, the direction for $\mathbf{u}_3$ is unique, but *any* direction perpendicular to $\mathbf{u}_3$ is a principal direction. In this case we say the $\mathbf{u}_3$ vector defines an axis of *inertial symmetry.* We can, if we wish, select a particular pair of mutually perpendicular vectors $\mathbf{u}_1$ and $\mathbf{u}_2$ in the plane normal to $\mathbf{u}_3$ and place them in the display (3–89), but any other pair would do equally well. In the case of three equal roots (i.e., $I_1 = I_2 = I_3$), any direction at all is a principal direction, and any three mutually perpendicular unit vectors can be placed in the display (3–89).

Once the solution (3–89) to the eigenvalue problem is available, we can construct the rotation matrix $[C]$, which represents the rotation from the given coordinate system to a coordinate system aligned with the principal directions. According to (2–14) to (2–17), the elements of $[C]$ and $[C]^t$ are just the components of the principal-direction unit vectors arranged as follows:

$$[C] = \begin{bmatrix} \leftarrow\{u_1\}^t\rightarrow \\ \leftarrow\{u_2\}^t\rightarrow \\ \leftarrow\{u_3\}^t\rightarrow \end{bmatrix} \qquad [C]^t = \begin{bmatrix} \uparrow & \uparrow & \uparrow \\ \{u_1\} & \{u_2\} & \{u_3\} \\ \downarrow & \downarrow & \downarrow \end{bmatrix}. \qquad (3\text{–}92)$$

In the principal-direction coordinate system, the unit vectors in the principal directions are

$$[C]\{u_1\} = \begin{Bmatrix} 1 \\ 0 \\ 0 \end{Bmatrix} \qquad [C]\{u_2\} = \begin{Bmatrix} 0 \\ 1 \\ 0 \end{Bmatrix} \qquad [C]\{u_3\} = \begin{Bmatrix} 0 \\ 0 \\ 1 \end{Bmatrix}, \qquad (3\text{–}93)$$

[1] See Eq. (B–26).

according to (3–69), and the inertia matrix according to (3–74) is

$$[C][I][C]^t = [C]\begin{bmatrix} \uparrow & \uparrow & \uparrow \\ I_1\{u_1\} & I_2\{u_2\} & I_3\{u_3\} \\ \downarrow & \downarrow & \downarrow \end{bmatrix} = \begin{bmatrix} I_1 & 0 & 0 \\ 0 & I_2 & 0 \\ 0 & 0 & I_3 \end{bmatrix}, \quad (3\text{–}94)$$

on using (3–90).

The foregoing demonstration of the existence of principal directions with an associated diagonal matrix of inertia-tensor components has outlined a computational path which can actually be followed in a particular case to obtain quantitative results. The procedure is, however, quite tedious for hand computation, and is not very efficient for machine computation.[1]

Example 3–6.   To provide an illustration, we consider the inertia matrix (3–77) corresponding to the simple two-particle rigid body of Fig. 3–13.   For this matrix the eigenvalue problem in the form of (3–88) is

$$\begin{bmatrix} 123 - \dfrac{\lambda}{mL^2} & -27 & -36\sqrt{2} \\[2mm] -27 & 123 - \dfrac{\lambda}{mL^2} & -36\sqrt{2} \\[2mm] -36\sqrt{2} & -36\sqrt{2} & 54 - \dfrac{\lambda}{mL^2} \end{bmatrix} \begin{Bmatrix} u_x \\ u_y \\ u_z \end{Bmatrix} = 0. \quad (3\text{–}95)$$

Expanding the determinant of the coefficient matrix yields the following cubic equation for λ:

$$\left(\frac{\lambda}{mL^2}\right)^3 - 300\left(\frac{\lambda}{mL^2}\right)^2 + 22{,}500\left(\frac{\lambda}{mL^2}\right) = 0. \quad (3\text{–}96)$$

Calling the roots I_1, I_2 and I_3, we find the eigenvalues

$$I_1 = 0 \qquad I_2 = I_3 = 150mL^2. \quad (3\text{–}97)$$

The eigenvector **u** corresponding to I_1 is obtained by setting $\lambda = 0$ in (3–95) and solving for u_x and u_y in terms of u_z from any two rows.   For example, using the first and second rows, we have

$$\begin{bmatrix} 123 & -27 \\ -27 & 123 \end{bmatrix} \begin{Bmatrix} u_x \\ u_y \end{Bmatrix} = u_z \begin{Bmatrix} 36\sqrt{2} \\ 36\sqrt{2} \end{Bmatrix}, \quad (3\text{–}98)$$

from which we find

$$u_x = u_y = \frac{3\sqrt{2}}{8}u_z. \quad (3\text{–}99)$$

Since these values should also satisfy the unused third row of (3–95), a

[1] For an authoritative treatment of machine-oriented procedures, see J. H. Wilkinson, "The Algebraic Eigenvalue Problem," Oxford University Press, New York, 1965.

useful check on the computation is provided. The requirement that **u** be a *unit* vector, i.e.,

$$u_x^2 + u_y^2 + u_z^2 = 1, \tag{3-100}$$

permits us to solve for $u_z = 0.800$, and thus to obtain

$$\{u_1\} = \begin{Bmatrix} 0.424 \\ 0.424 \\ 0.800 \end{Bmatrix}. \tag{3-101}$$

Next we try to repeat this procedure by setting $\lambda = 150mL^2$ in (3-95). Using the first two rows, we have

$$\begin{bmatrix} -27 & -27 \\ -27 & -27 \end{bmatrix} \begin{Bmatrix} u_x \\ u_y \end{Bmatrix} = u_z \begin{Bmatrix} 36\sqrt{2} \\ 36\sqrt{2} \end{Bmatrix}. \tag{3-102}$$

In this case, because of the repeated root, we do not get unique values of u_x and u_y in terms of u_z. All three rows of (3-95) are equivalent to a single restriction on the components,

$$u_x + u_y + \frac{4\sqrt{2}}{3} u_z = 0. \tag{3-103}$$

There are infinitely many unit vectors which satisfy (3-103). They are all perpendicular to (3-101). A possible pair of unit vectors which satisfy (3-103) and which are mutually orthogonal are

$$\{u_2\} = \begin{Bmatrix} 0.707 \\ -0.707 \\ 0 \end{Bmatrix} \qquad \{u_3\} = \begin{Bmatrix} 0.566 \\ 0.566 \\ -0.600 \end{Bmatrix}. \tag{3-104}$$

Using these unit vectors, the rotational transformation to principal directions is described by the rotation matrix

$$[C] = \begin{bmatrix} 0.424 & 0.424 & 0.800 \\ 0.707 & -0.707 & 0 \\ 0.566 & 0.566 & -0.600 \end{bmatrix}. \tag{3-105}$$

The orientation of the principal directions is sketched in Fig. 3-14. The vector $\mathbf{u}_1$ is *along* the link joining the particles, and $\mathbf{u}_2$ and $\mathbf{u}_3$ are *at right angles* to the link. With respect to the principal axes, the inertia matrix takes the diagonal form

$$[C][I][C]^t = \begin{bmatrix} 0 & 0 & 0 \\ 0 & 150mL^2 & 0 \\ 0 & 0 & 150mL^2 \end{bmatrix}. \tag{3-106}$$

USE OF SYMMETRY

The principal directions associated with an inertia tensor can always be found by constructing the inertia matrix for an arbitrary coordinate system and then solving the eigenvalue problem (3–87). Although this procedure is possible by hand calculation, it is quite tedious. Fortunately, in many technical applications the symmetry of the rigid body

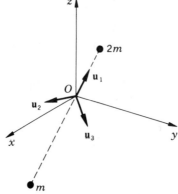

FIG. 3–14. Example 3–6. Principal directions for the rigid body of Example 3–5.

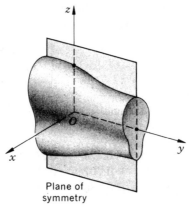

FIG. 3–15. A rigid body with a plane of mass symmetry.

under consideration can be utilized to determine (or partially determine) the principal directions by inspection.

Consider in Fig. 3–15 a rigid body with a single plane of mass symmetry, and let the y and z axes of a body-coordinate system lie in the plane of symmetry. The existence of a plane of mass symmetry implies that, for every mass particle with coordinates (x,y,z), there is an equal mass particle at $(-x,y,z)$. If this is the case, then the summations

$$I_{xy} = -\sum_{i=1}^{N} m_i x_i y_i \quad \text{and} \quad I_{xz} = -\sum_{i=1}^{N} m_i x_i z_i \quad (3\text{--}107)$$

of (3–22) must vanish, since the particles can be grouped into canceling pairs. The inertia matrix corresponding to Fig. 3–15 then has the form

$$[I] = \begin{bmatrix} I_{xx} & 0 & 0 \\ 0 & I_{yy} & I_{yz} \\ 0 & I_{yz} & I_{zz} \end{bmatrix}. \quad (3\text{--}108)$$

Since angular velocity about the axis Ox results in parallel angular momentum, the x axis is a principal direction. Thus, for an origin anywhere in a

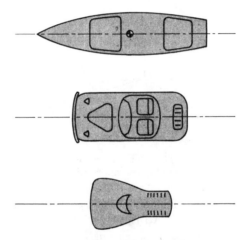

FIG. 3–16. Vehicles with planes of mass symmetry. An axis perpendicular to a plane of mass symmetry is a principal axis.

plane of symmetry, the direction perpendicular to that plane is a principal direction for the inertia tensor at that point.

Mass symmetry of the sort indicated in Fig. 3–15 occurs frequently in vehicles such as ships, automobiles, and reentry vehicles, as shown in Fig. 3–16. In aircraft and spacecraft design, particular care is taken in the location of electronic equipment, pumps, etc., so that the principal axes have locations which provide desirable dynamic behavior of the vehicle. An example of a vehicle which does not possess mass symmetry is a motorcycle with attached sidecar. Due to the asymmetry the vehicle tends to turn whenever it accelerates or decelerates. High speed maneuvers in these vehicles are difficult, not only for this reason, but also because, in comparison with an ordinary motorcycle, the range of corrective action available to the driver is limited.   In sidecar races, the passenger must compensate for this by executing violent contortions to counteract the frequent tendencies to spin and tip.

Further examples of mass symmetry are provided by Figs. 3–9 and 3–11.   In the disk of Fig. 3–11 the body axes shown do take advantage of symmetry.   All three coordinate planes are planes of mass symmetry, so that these axes are principal axes.   This is verified by the diagonal form of the matrix (3–64).   Note that there are two equal principal moments of inertia corresponding to the x and y axes which are diameters of the disk.   This means that any other pair of orthogonal diameters could have been used to construct a set of principal axes.   In the cube of Fig. 3–9 the body axes shown are also principal axes because the coordinate planes are

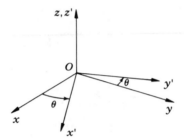

Fig. 3–17.   Rotated systems of body coordinates in a rigid body.

planes of mass symmetry.   This is verified by the diagonal form of (3–56). Here we can say even more.   All three principal moments of inertia are equal, which means that *any* axis through the center of the cube is a principal axis.   This last result is not obvious from symmetry considerations alone, as it would be in the case of a sphere.

MOHR'S CIRCLE

When the direction of one of the principal axes at a point is known, the remaining two-dimensional problem of locating the other two principal directions is much simpler than the general three-dimensional problem. Suppose that in a body-coordinate system $Oxyz$ the z axis is *known to be a principal direction*.   The inertia matrix in this case would have the form

$$I = \begin{bmatrix} I_{xx} & I_{xy} & 0 \\ I_{xy} & I_{yy} & 0 \\ 0 & 0 & I_{zz} \end{bmatrix}. \tag{3–109}$$

When $I_{xy} \neq 0$, the x and y axes are not principal axes.   Let us study transformations of the inertia-tensor components under rotations about the z axis, as indicated in Fig. 3–17. The rotation is described by the matrix $[C]$ of (2–17), where

$$[C] = \begin{bmatrix} \cos\theta & \sin\theta & 0 \\ -\sin\theta & \cos\theta & 0 \\ 0 & 0 & 1 \end{bmatrix}. \qquad (3\text{--}110)$$

According to (3–74), the inertia matrix $[I']$ corresponding to the primed coordinate system is given by

$$[I'] = [C][I][C]^t. \qquad (3\text{--}111)$$

When (3–111) is evaluated for (3–109) and (3–110), we find that the third row and column of the result are the same as the third row and column in

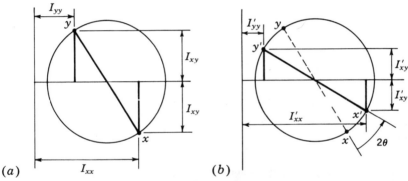

(a) (b)

FIG. 3–18. Mohr's circle is constructed from given data (a) and then used to read off transformed inertia components (b).

(3–109), but that the remaining components are transformed as follows:

$$I'_{xx} = I_{xx}\cos^2\theta + I_{yy}\sin^2\theta + 2I_{xy}\sin\theta\cos\theta$$
$$I'_{xy} = I_{xy}\cos^2\theta - I_{xy}\sin^2\theta - (I_{xx} - I_{yy})\sin\theta\cos\theta \qquad (3\text{--}112)$$
$$I'_{yy} = I_{yy}\cos^2\theta + I_{xx}\sin^2\theta - 2I_{xy}\sin\theta\cos\theta.$$

Introduction of the double-angle trigonometric identities yields

$$I'_{xx} = \tfrac{1}{2}(I_{xx} + I_{yy}) + \tfrac{1}{2}(I_{xx} - I_{yy})\cos 2\theta + I_{xy}\sin 2\theta$$
$$I'_{xy} = \qquad\qquad I_{xy}\cos 2\theta - \tfrac{1}{2}(I_{xx} - I_{yy})\sin 2\theta \qquad (3\text{--}113)$$
$$I'_{yy} = \tfrac{1}{2}(I_{xx} + I_{yy}) - \tfrac{1}{2}(I_{xx} - I_{yy})\cos 2\theta - I_{xy}\sin 2\theta.$$

A simple graphical interpretation of these equations was given by O. Mohr in 1882. When the given inertia components I_{xx}, I_{xy}, and I_{yy} are used to construct the diameter xy in Fig. 3–18a, it can be verified that transformed components I'_{xx}, I'_{xy}, and I'_{yy} which satisfy (3–113) can be read off from the diameter $x'y'$ in Fig. 3–18b, where the angle between diameters is *twice* the angle of rotation in Fig. 3–17.   Mohr's circle can be drawn carefully to scale to obtain a graphical solution, or it can be sketched to serve as a mnemonic device for reconstructing the analytical content of (3–113).

In Fig. 3–18 the moments of inertia are plotted horizontally and the products of inertia are plotted vertically.   The sign convention for products of inertia is indicated in Fig. 3–19.   For rotations about the z axis in Fig. 3–17 we identify the quadrant xOy as the *first* quadrant in the xy plane for the *right*-handed coordinate system used.   The quadrant xOy in Fig. 3–17 can be considered to be mapped into the diameter xy in

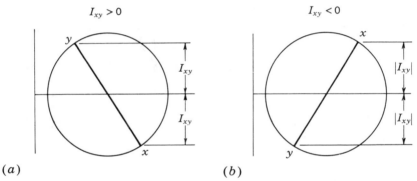

$I_{xy} > 0$ $I_{xy} < 0$

(*a*)   (*b*)

FIG. 3–19.   Sign convention for products of inertia.   (*a*) For positive I_{xy} the x end of the diameter is down and the y end up.   (*b*) For negative I_{xy} the combination is reversed.

the Mohr diagram of Fig. 3–19 by the double-angle transformation (3–113).   When I_{xy} is positive, the x end of the diameter should be plotted downward and the y end plotted upward, as indicated in Fig. 3–19.   If this convention is adopted, the diameter in Fig. 3–19 will rotate in the same sense as the first quadrant in Fig. 3–17.   In case the rotation is about the x axis, the rotating first quadrant is yOz, and for positive I_{yz} the y end of the Mohr diameter should be plotted downward and the z end plotted upward.   Similarly, when the rotation is about the y axis, the z end of the Mohr diameter goes down and the x end goes up for positive I_{zx}.

The Mohr diagram in Fig. 3–18 provides a comprehensive insight into the manner in which I'_{xx}, I'_{xy}, and I'_{yy} change as the angle θ in Fig. 3–17 is changed.   The principal directions correspond to a horizontal position of the $x'y'$ diameter; in this orientation $I'_{xy} = 0$ and I'_{xx} and I'_{xy} take on extreme values (maximum or minimum).

Example 3–7.   As an illustration, we consider the problem of locating the principal directions and principal moments of inertia of the inertia tensor at the origin for the rigid body shown in Fig. 3–20.   The yz plane is a plane of mass symmetry, and thus the x axis is a principal axis.

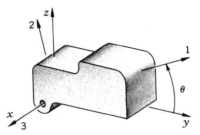

FIG. 3–20.   Example 3–7.   Location of principal directions is to be determined.

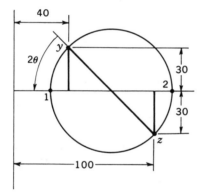

FIG. 3–21.   Mohr's-circle diagram corresponding to Fig. 3–20.

The inertia matrix with respect to the xyz coordinate system is given as

$$[I] = \begin{bmatrix} 120 & 0 & 0 \\ 0 & 40 & -30 \\ 0 & -30 & 100 \end{bmatrix} \text{lbm-ft}^2. \qquad (3\text{–}114)$$

Mohr's circle for representing transformations of the inertia components under rotation of coordinates about the x axis is constructed, as indicated in Fig. 3–21, by laying off $I_{yy} = 40$ lbm-ft^2 and $I_{zz} = 100$ lbm-ft^2 horizontally, and then using $I_{yz} = -30$ lbm-ft^2 to go up on y and down on z, thereby locating the diameter yz.   Once the diameter yz has been located in Fig. 3–21, the complete circle can be drawn.

The principal directions in the Mohr diagram are represented by the horizontal diameter $\widehat{12}$.   On the rigid body in Fig. 3–20 the corresponding principal directions are indicated by the axes 1 and 2.   The orientation angle θ is obtained quantitatively by noting from the geometry of Fig. 3–21 that

$$\tan 2\theta = \frac{|I_{yz}|}{\frac{1}{2}(I_{zz} - I_{yy})} = \frac{30}{\frac{1}{2}(100 - 40)} = 1, \qquad (3\text{–}115)$$

or that $2\theta = 45°$.   Therefore $\theta = 22.5°$.

The magnitudes of the principal moments of inertia are obtained by noting, in Fig. 3–21, that the abscissa of the center of the circle is 70 lbm-ft² and that the radius of the circle is $30 \sqrt{2} = 42.4$ lbm-ft².   Therefore $I_1 = 70 - 42.4 = 27.6$ lbm-ft², and $I_2 = 70 + 42.4 = 112.4$ lbm-ft². The inertia matrix corresponding to the principal directions is

$$\begin{bmatrix} I_1 & 0 & 0 \\ 0 & I_2 & 0 \\ 0 & 0 & I_3 \end{bmatrix} = \begin{bmatrix} 27.6 & 0 & 0 \\ 0 & 112.4 & 0 \\ 0 & 0 & 120 \end{bmatrix} \text{lbm-ft².} \qquad (3\text{–}116)$$

PROBLEMS

3–1. A uniform rigid sphere has mass M and radius a.   Evaluate the components of the inertia tensor at the center of the sphere.   (*Suggestion:* Show that, because of the symmetry of the sphere, $I_{xx} = I_{yy} = I_{zz}$ and

$$I_{xx} + I_{yy} + I_{zz} = 2 \int r^2 \, dm,$$

where $r^2 = x^2 + y^2 + z^2$.)

3–2. A uniform rectangular block has total mass M and dimensions L, h, and t. Evaluate the components of the centroidal inertia tensor with respect to the coordinate system shown.

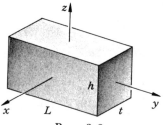

PROB. 3–2

3–3. Let **P** be the linear momentum of a rigid body, and let A and B be arbitrary points.   Show that the angular momenta of the rigid body with respect to A and B satisfy the relation

$$\mathbf{H}_A = \mathbf{H}_B + \overrightarrow{AB} \times \mathbf{P},$$

where $\overrightarrow{AB}$ is the vector distance from A to B.

3–4. Consider that the origin in Fig. 3–5 is an arbitrary point in the rigid body, with velocity $\mathbf{v}_o$. Show that the angular momentum of the rigid body with respect to O is

$$\mathbf{H}_o = M\mathbf{r}_c \times \mathbf{v}_o + \sum_{i=1}^{N} m_i \mathbf{r}_i \times (\boldsymbol{\omega} \times \mathbf{r}_i),$$

and hence that (3–26) gives the complete angular momentum with respect to O when $\mathbf{v}_o$ is either zero or directed toward the centroid, or when O coincides with the centroid.

3–5. Reconsider (3–33) and Fig. 3–7 for the case where the point O is an *arbitrary* point, fixed in the rigid body, with velocity $\mathbf{v}_o$. Use the result of Prob. 3–4 to show that

$$T^* = \tfrac{1}{2}\mathbf{v}_o \cdot \mathbf{P} + \tfrac{1}{2}\boldsymbol{\omega} \cdot \mathbf{H}_o.$$

Verify that this reduces to (3–35) and (3–38) under the appropriate restrictions.

3–6. The result of Prob. 3–5 divides the kinetic coenergy of a rigid body into translational and rotational contributions with respect to an arbitrary point O fixed in the body. These contributions are said to be *uncoupled* if the translational contribution depends on the linear velocity $\mathbf{v}_o$ but is independent of the angular velocity and if the rotational contribution depends on the angular velocity but is independent of the linear velocity $\mathbf{v}_o$. Determine the restrictions on the motion and on the choice of the point O within a rigid body such that these contributions will be uncoupled.

3–7. A rigid body whose mass is 10 lbm has the following inertia matrix with respect to an xyz body-coordinate system with its origin at the center of mass:

$$[I]_c = \begin{bmatrix} 0.0300 & 0.00707 & -0.00707 \\ 0.00707 & 0.04500 & 0.01500 \\ -0.00707 & 0.01500 & 0.04500 \end{bmatrix} \text{ ft-lbf-sec}^2.$$

At a certain instant, the body has an angular velocity $\boldsymbol{\omega}$ with components

$$\omega_x = 100 \text{ rad/sec}$$
$$\omega_y = 100 \text{ rad/sec}$$
$$\omega_z = 200 \text{ rad/sec},$$

and the center of mass has a speed of 100 ft/sec. Find

(*a*) The angular momentum relative to the center of mass
(*b*) The kinetic coenergy

3–8. Given the inertia matrix

$$[I]_o = \begin{bmatrix} 261 & -49.5 & -112 \\ -49.5 & 261 & -112 \\ -112 & -112 & 99 \end{bmatrix} \text{ lbm-ft}^2$$

and the angular-velocity components

$$\{\omega\} = \begin{Bmatrix} 62.8 \\ 62.8 \\ 0 \end{Bmatrix} \text{ rad/sec},$$

for a rigid body with respect to body axes originating at a point of zero velocity, find:

(a) The angular momentum $\mathbf{H}_o$
(b) The kinetic coenergy T^*

3-9. Consider a nonrigid system of particles under general motion, and consider an arbitrary moment center B. The angular momentum $\mathbf{H}_B$ is given by (3–14), and the general relation between torque and angular momentum is given by (2–270). Show that

$$\boldsymbol{\tau}_B = \frac{d\mathbf{H}_c}{dt} + \mathbf{b}_c \times \frac{d\mathbf{P}}{dt}.$$

Verify that this reduces to (3–45) and (3–47) under the appropriate restrictions.

3-10. The result of Prob. 3–9 can be used to justify *D'Alembert's principle* for a rigid body. The principle states that valid equations of motion are obtained if the actual external forces acting on the body are required to be in equilibrium with an *inertia force* $-d\mathbf{P}/dt$ acting through the centroid and an *inertia couple* $-d\mathbf{H}_c/dt$. Verify that the requirement of equilibrium with the inertia force and inertia couple is equivalent to the requirement that *both* the linear-momentum principle *and* the angular-momentum principle are to be satisfied.

3-11. A rigid body has angular momentum $\mathbf{H}_o$ with respect to a point O which is fixed in the body and which has zero velocity at the instant under consideration. Let $(d\mathbf{H}_o/dt)_0$ be the time rate of change of $\mathbf{H}_o$ at this instant under the condition that the acceleration of point O is zero, and let $d\mathbf{H}_o/dt$ be the corresponding rate under the condition that point O has arbitrary acceleration $\mathbf{a}_o$. Show that

$$\frac{d\mathbf{H}_o}{dt} = \left(\frac{d\mathbf{H}_o}{dt}\right)_0 + M\mathbf{r}_c \times \mathbf{a}_o,$$

where M is the mass of the rigid body, and $\mathbf{r}_c$ is the position vector of the centroid with respect to O.

3-12. Evaluate the matrix of components of the inertia tensor at O for the disk with respect to the xyz coordinate system shown. The disk has uniform density and a total mass of M.

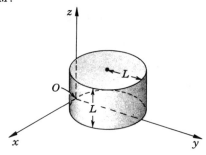

PROB. 3–12

3-13. The rigid body shown has uniform mass density and a total mass of $8m$. Evaluate the matrix of components of the inertia tensor at O with respect to the axes indicated.

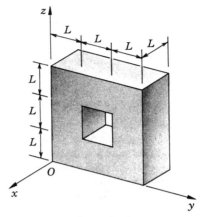

PROB. 3–13

3-14. The disk (with uniform mass density and total mass M) rolls without slip on a horizontal circle of radius L so that the arm OA rotates about the vertical with angular speed ω_o.

(a) Evaluate the components of the inertia tensor at O of the disk with respect to the xyz coordinate system.
(b) Determine the angular velocity *of the disk* and give its components with respect to this same coordinate system.
(c) Evaluate the angular momentum of the disk with respect to O.
(d) Evaluate the kinetic coenergy of the disk.
(e) Show the angular-velocity vector and the angular-momentum vector in a sketch.

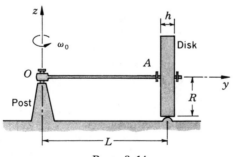

PROB. 3–14

3-15. A rigid body has a point O in it which is fixed in inertial space. With respect to a given $Oxyz$ coordinate system, the components of the inertia tensor and the angular velocity are

$$[I]_o = \begin{bmatrix} 3 & 2 & 1 \\ 2 & 4 & 1 \\ 1 & 1 & 5 \end{bmatrix} \text{lbm-ft}^2 \qquad \{\omega\} = \begin{bmatrix} 2 \\ 0 \\ -3 \end{bmatrix} \text{rad/sec.}$$

Find the corresponding components $\{H\}_o$ of the angular momentum $\mathbf{H}_o$. Now consider a coordinate system $Ox'y'z'$ where the rotation matrix from the unprimed to the primed axes is

$$[C] = \begin{bmatrix} \frac{3}{5} & \frac{4}{5} & 0 \\ -\frac{4}{5} & \frac{3}{5} & 0 \\ 0 & 0 & 1 \end{bmatrix}.$$

(a) Evaluate the components $\{\omega'\}$ of the angular velocity with respect to the primed coordinate system.

(b) Obtain $[I']_o$.

(c) Evaluate $\{H'\}_o = [I']_o\{\omega'\}$, and verify that $\{H'\}_o = [C]\{H\}_o$.

3-16. (a) Show that the roots of a cubic with real coefficients must occur as three real roots or as one real root and a pair of conjugate complex roots.

(b) Suppose that λ is a complex scalar and that $\{u\}$ is a column of complex components but that they satisfy (3-87). Show that the complex conjugates λ^* and $\{u^*\}$ must then satisfy

$$[I]\{u^*\} = \lambda^*\{u^*\}. \tag{i}$$

(c) Premultiply (3-87) by $\{u^*\}^t$ and premultiply (i) above by $\{u\}^t$. Subtract the two resulting equations, and use the symmetry of $[I]$ to show that λ cannot, in fact, have an imaginary part; i.e., the roots of the determinant of the coefficient matrix of (3-88) must be real.

3-17. Let λ and $\{u\}$ be a real scalar and the real components of a unit vector which together satisfy (3-87). Show that λ must be nonnegative.

3-18. The matrix of components of the inertia tensor at a point O in a rigid body with respect to a given coordinate system is

$$[I]_o = \begin{bmatrix} 30,000 & 7,071 & -7,071 \\ 7,071 & 45,000 & 15,000 \\ -7,071 & 15,000 & 45,000 \end{bmatrix} \text{lbm-ft}^2.$$

(a) Verify that the principal moments of inertia are

$$I_1 = 20,000 \text{ lbm-ft}^2$$
$$I_2 = 40,000 \text{ lbm-ft}^2$$
$$I_3 = 60,000 \text{ lbm-ft}^2.$$

(b) Determine the components, in the given coordinate system, of the unit vectors $\mathbf{u}_1$, $\mathbf{u}_2$, and $\mathbf{u}_3$ in the principal directions.

3-19. Consider the cube of Example 3-1 with the inertia matrix of (3-31). Use symmetry to locate the principal directions for the inertia tensor at O. Evaluate the principal moments of inertia.

3-20. Given the inertia matrix (3-109), where I_{zz} is a principal moment of inertia.

(a) Show that the remaining principal moments of inertia are given by

$$I_p, I_m = \tfrac{1}{2}(I_{xx} + I_{yy}) \pm \sqrt{\tfrac{1}{4}(I_{xx} - I_{yy})^2 + I_{xy}^2},$$

where the plus sign corresponds to I_p and the minus sign corresponds to I_m.

(b) Show that the corresponding principal directions lie in the xy plane and that the unit vector in the direction corresponding to I_p has components

$$u_x = \frac{I_{zz} - I_m}{\sqrt{(I_{zz} - I_m)^2 + I_{zy}^2}}$$

$$u_y = \frac{I_{zy}}{\sqrt{(I_{zz} - I_m)^2 + I_{zy}^2}},$$

while the unit vector in the direction corresponding to I_m has components

$$u_x = \frac{I_{zz} - I_p}{\sqrt{(I_{zz} - I_p)^2 + I_{zy}^2}}$$

$$u_y = \frac{I_{zy}}{\sqrt{(I_{zz} - I_p)^2 + I_{zy}^2}}.$$

3-21. A rigid body has the following components of the inertia tensor at O with respect to a coordinate system $Oxyz$:

$$[I]_o = \begin{bmatrix} 1 & 1 & 0 \\ 1 & 1 & 0 \\ 0 & 0 & 2 \end{bmatrix} mL^2.$$

(a) Locate the principal directions.
(b) Determine the principal moments of inertia.
(c) Describe a rigid body which could exhibit these inertia properties.

3-22. Apply Mohr's circle to the inertia matrix of (3–79). Locate the principal directions, and evaluate the corresponding principal moments of inertia. Verify the results by comparing them with Fig. 3–13.

3-23. The components of the inertia tensor of the centroid of the rigid airplane, with respect to the coordinates shown, are

$$[I]_c = \begin{bmatrix} 150,000 & 0 & 0 \\ 0 & 250,000 & 20,000 \\ 0 & 20,000 & 100,000 \end{bmatrix} \text{lbm-ft}^2.$$

Locate the principal directions and evaluate the principal moments of inertia.

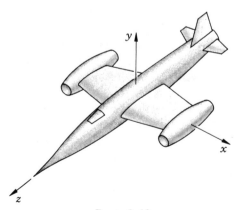

PROB. 3-23

CHAPTER FOUR

Dynamics of systems containing rigid bodies

In Chap. 2 the study of dynamics of mechanical systems is limited to systems in which the inertia elements are treated as mass particles. In Chap. 3 the properties of rigid bodies are discussed and the dynamic principles available are listed. In this chapter we return to the study of dynamics of mechanical systems. The emphasis here is on systems in which the rotation of a rigid body plays a dominant role. The early sections of the chapter deal with problems in which the rotation axis is always parallel to a fixed direction. Many problems of technical importance fall into this category. The later sections deal with more general problems, in which the orientation of the axis of rotation varies with time.

4–1 Plane motion of rigid bodies

A rigid body is said to be undergoing plane motion if the velocities of all its points remain parallel to a fixed plane in inertial space. Examples of plane motion are sketched in Fig. 4–1. The body (a) is undergoing *fixed-axis rotation* about the axis OZ, and the body (b) has a *general plane motion* in which the centroid C has an arbitrary linear velocity $\mathbf{v}_c$ (parallel to the XY plane) and the body has an arbitrary angular velocity $\boldsymbol{\omega}$ (parallel to OZ).

Plane motion is an especially simple type of rigid-body motion. Because all rotations are about parallel axes, the relation between angular displacement and angular velocity is elementary. Figure 4–2 shows the trace on a fixed XY plane of a rigid body with a body-coordinate system $Cxyz$. The angular displacement of the body is indicated by the angle θ between the fixed axes and the body axes, and the angular velocity of the rigid body is $\omega\mathbf{u}_z$, where $\omega = \dot\theta$.

208

The dynamic principles available for analyzing plane motion are special cases of the principles listed in Sec. 3–5.   Because there is no velocity normal to the XY plane in Fig. 4–1, there is no linear momentum in this direction.   Furthermore, there is no change in this component of linear momentum, so that for a body to remain in a state of plane motion, it is

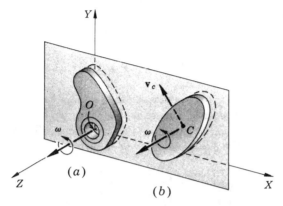

Fig. 4–1.   Rigid bodies moving parallel to XY plane. (a) Body rotates about fixed point O.   (b) Body with centroid C has general plane motion.

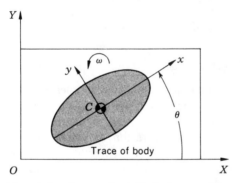

Fig. 4–2.   Trace of rigid body on plane through centroid parallel to motion.   Magnitude of angular velocity is $\omega = \dot{\theta}$.

necessary that there be no unbalanced external force acting on the body normal to the plane of motion.   For the forces *in the plane of motion* the linear-momentum principles (3–42) and (3–43) still apply:

$$\mathbf{P} = M\mathbf{v}_c \quad \text{and} \quad \mathbf{F} = \frac{d\mathbf{P}}{dt}. \tag{4-1}$$

The angular momentum $\mathbf{H}_c$ of (3–46) has the following components in the body-coordinate system $Cxyz$ of Fig. 4–2:

$$\{H\}_c = [I]_c\{\omega\} = \begin{bmatrix} I_{xx} & I_{xy} & I_{xz} \\ I_{xy} & I_{yy} & I_{yz} \\ I_{xz} & I_{yz} & I_{zz} \end{bmatrix}_c \begin{bmatrix} 0 \\ 0 \\ \omega \end{bmatrix} = \omega \begin{bmatrix} I_{xz} \\ I_{yz} \\ I_{zz} \end{bmatrix}_c. \qquad (4\text{–}2)$$

The relation (3–47) between the external torque and the rate of change of angular momentum with respect to an inertial reference frame can be evaluated in terms of changes observed from the body-coordinate system by applying (2–53).

$$\boldsymbol{\tau}_c = \left(\frac{\partial \mathbf{H}_c}{\partial t}\right)_{\text{rel}} + \boldsymbol{\omega} \times \mathbf{H}_c. \qquad (4\text{–}3)$$

Since the inertia-tensor components in (4–2) remain constant in the body-coordinate system, the only change in $\mathbf{H}_c$ observed in the body-coordinate system is due to the change in the magnitude ω. Thus, in the body-coordinate system, the components of (4–3) are

$$\begin{Bmatrix} \tau_x \\ \tau_y \\ \tau_z \end{Bmatrix}_c = \dot{\omega} \begin{Bmatrix} I_{xz} \\ I_{yz} \\ I_{zz} \end{Bmatrix}_c + \omega^2 \begin{Bmatrix} -I_{yz} \\ I_{xz} \\ 0 \end{Bmatrix}_c, \qquad (4\text{–}4)$$

which implies that in general a three-dimensional torque is required to maintain a plane motion of a rigid body. If, however, the rigid body is so aligned that its z axis is a *principal direction*, then $I_{xz} = I_{yz} = 0$, and I_{zz} is a principal moment of inertia, which we shall call I_c. In this case (4–2) and (4–4) reduce to

$$\mathbf{H}_c = I_c \omega \mathbf{u}_z \quad \text{and} \quad \boldsymbol{\tau}_c = I_c \dot{\omega} \mathbf{u}_z. \qquad (4\text{–}5)$$

The linear- and angular-momentum principles of (4–1) and (4–5) are sufficient for the direct analysis of general plane motion of any rigid body which has a principal direction normal to the plane of motion.

For the case of fixed-axis rotation it usually is more convenient to dispense with the linear-momentum principles (4–1) and consider only the angular-momentum relations with respect to the fixed point O. Thus, instead of (4–2) and (4–4), we have

$$\{H\}_o = \omega \begin{Bmatrix} I_{xz} \\ I_{yz} \\ I_{zz} \end{Bmatrix}_o$$

$$\begin{Bmatrix} \tau_x \\ \tau_y \\ \tau_z \end{Bmatrix}_o = \dot{\omega} \begin{Bmatrix} I_{xz} \\ I_{yz} \\ I_{zz} \end{Bmatrix}_o + \omega^2 \begin{Bmatrix} -I_{yz} \\ I_{xz} \\ 0 \end{Bmatrix}_o, \qquad (4\text{–}6)$$

where the inertia-tensor components are with respect to a body-coordinate system $Oxyz$ which has Oz normal to the plane of motion. In case

Oz is also a *principal direction* for the rigid body, then $I_{zz} = I_{yz} = 0$ and I_{zz} is a principal moment of inertia, which we shall call I_o. Then (4–6) reduces to

$$\mathbf{H}_o = I_o\omega\mathbf{u}_z \quad \text{and} \quad \boldsymbol{\tau}_o = I_o\dot{\omega}\mathbf{u}_z. \tag{4–7}$$

When the indirect methods of analysis based on Hamilton's principle or Lagrange's equations are applied to a rigid body, it is necessary to evaluate the kinetic coenergy T^*. The general formulas (3–39) and (3–40) reduce, in the plane motions of Fig. 4–1, to

$$T^* = \tfrac{1}{2}I_o\omega^2, \tag{4–8}$$

for the case (*a*), where the rigid body rotates with angular speed ω about a fixed axis OZ, and to

$$T^* = \tfrac{1}{2}Mv_c^2 + \tfrac{1}{2}I_c\omega^2, \tag{4–9}$$

for the case (*b*), where there is general plane motion.

Example 4–1. A uniform disk of mass M, radius r, and thickness h rolls without slip on a cylindrical surface of radius R, as shown in Fig.

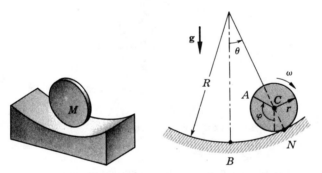

Fig. 4–3.   Problem 4–1.   Rocking disk.

4–3. It is desired to obtain equations of motion for the disk under the influence of the uniform gravitational field.

In Fig. 4–3 it is assumed that when the disk passes through the central position, the point A on the disk is in contact with point B on the surface. The no-slip requirement implies that $BN = AN$, or that

$$R\theta = r(\varphi + \theta). \tag{4–10}$$

Since the angle φ measures the displacement between a fixed direction (the vertical) and a line imbedded in the disk, the magnitude of the angular velocity of the disk is

$$\omega = \dot{\varphi} = \frac{R - r}{r}\,\dot{\theta}. \tag{4–11}$$

The principal moment of inertia of the disk with respect to a centroidal axis normal to the plane of motion is

$$I_c = \tfrac{1}{2}Mr^2, \qquad (4\text{--}12)$$

according to the second of (3–62).

We shall now formulate the equations of motion in two ways: First, we apply the momentum principles directly, and then we illustrate the use of Lagrange's equations to implement the indirect variational approach.

To apply the momentum principles it is necessary to identify the total external force **F** and the total external torque $\boldsymbol{\tau}_c$ acting on the disk. In Fig. 4–4 the resultant gravity force is shown acting through the centroid

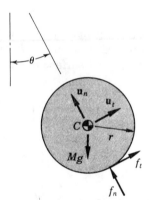

Fig. 4–4.   External forces
acting on disk.

(see Prob. 2–54), and the contact force is shown decomposed into a normal component f_n and a tangential (frictional) component f_t. The total force and moment are

$$\mathbf{F} = (f_t - Mg \sin \theta)\mathbf{u}_t + (f_n - Mg \cos \theta)\mathbf{u}_n$$
$$\boldsymbol{\tau}_c = -rf_t\mathbf{u}_z, \qquad (4\text{--}13)$$

where $\mathbf{u}_z$ is a unit vector normal into the plane of motion.

The linear momentum (4–1) of the disk is

$$\mathbf{P} = M\mathbf{v}_c = M(R - r)\dot{\theta}\mathbf{u}_t. \qquad (4\text{--}14)$$

To obtain the time rate of change of this, we note that $\mathbf{u}_t$ turns with θ, so that, according to (2–9),

$$\frac{d\mathbf{u}_t}{dt} = \dot{\theta}\mathbf{u}_n. \qquad (4\text{--}15)$$

Using (4-15), we calculate

$$\frac{d\mathbf{P}}{dt} = M(R - r)\ddot{\theta}\mathbf{u}_t + M(R - r)\dot{\theta}^2\mathbf{u}_n. \tag{4-16}$$

The angular momentum (4-5) of the disk with respect to its centroid is

$$\mathbf{H}_c = I_c\omega\mathbf{u}_z = \tfrac{1}{2}Mr^2\,\frac{R - r}{r}\,\dot{\theta}\mathbf{u}_z, \tag{4-17}$$

on using (4-11) and (4-12). The time rate of change of angular momentum follows on differentiation:

$$\frac{d\mathbf{H}_c}{dt} = \frac{r}{2}\,M(R - r)\ddot{\theta}\mathbf{u}_z. \tag{4-18}$$

Finally, on equating the external forces and torques of (4-13) with the time rates of change of the linear and angular momenta in (4-16) and (4-18), respectively, we obtain three simultaneous equations for f_t, f_n, and θ,

$$\begin{aligned} f_t - Mg\sin\theta &= M(R - r)\ddot{\theta} \\ f_n - Mg\cos\theta &= M(R - r)\dot{\theta}^2 \end{aligned} \tag{4-19}$$

$$-rf_t = \frac{r}{2}M(R - r)\ddot{\theta}.$$

A single equation for the motion, in terms of θ, is obtained by eliminating the force f_t between the first and third of (4-19).

$$\tfrac{3}{2}(R - r)\ddot{\theta} + g\sin\theta = 0. \tag{4-20}$$

An alternative route to this equation of motion is furnished by the indirect approach. Here we begin by forming the state functions T^* and V (and noting that there are no nonconservative forces acting). The holonomic constraint (4-10) implies that the system has a single degree of freedom and that only the single generalized coordinate θ is necessary. According to (4-9), the kinetic coenergy is

$$T^* = \tfrac{1}{2}M(R - r)^2\dot{\theta}^2 + \tfrac{1}{2}\left[\tfrac{1}{2}Mr^2\left(\frac{R - r}{r}\right)^2\dot{\theta}^2\right] = \tfrac{3}{4}M(R - r)^2\dot{\theta}^2, \tag{4-21}$$

on using (4-11) and (4-12). The potential energy of the disk (see Prob. 2-54) is

$$V = Mg(R - r)(1 - \cos\theta). \tag{4-22}$$

Setting $\mathcal{L} = T^* - V$ and applying Lagrange's equation,

$$\frac{d}{dt}\left(\frac{\partial\mathcal{L}}{\partial\dot{\theta}}\right) - \frac{\partial\mathcal{L}}{\partial\theta} = 0, \tag{4-23}$$

leads immediately to

$$\tfrac{3}{2}(R - r)\ddot{\theta} + g \sin \theta = 0, \qquad (4\text{-}24)$$

which is identical with (4–20).

A quantitative description of the motion of the disk can be obtained by integrating (4–24) subject to initial conditions.   A qualitative description is obtained by noting the analogy between (4–24) and the equation of motion for a simple pendulum.   For the simple pendulum of Fig. 4–5, the equation of motion is easily found to be

$$L\ddot{\theta} + g \sin \theta = 0, \qquad (4\text{-}25)$$

and thus the motions of the disk and the pendulum would be "equivalent" if $L = \tfrac{3}{2}(R - r)$.   This equivalence gives considerable insight into the

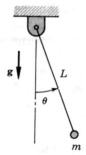

FIG. 4–5. Simple pendulum is "equivalent" to rocking disk of Fig. 4–3 if $L = \tfrac{3}{2}(R - r)$.

nature of the motion of the rocking disk.   For small motions the pendulum oscillates with a period very nearly equal to $2\pi \sqrt{L/g}$.   The rocking disk then has a long period when the curved surface is nearly flat (R, large) and a short period when the curved surface is nearly as curved as the periphery of the disk ($R - r$, small).

In comparing the application of the direct and indirect methods of analysis in this problem, we note that the primary objective of obtaining the equation of motion is most easily attained by the latter approach. The former approach has the advantage, however, in providing ready access to a number of subsidiary quantities, such as the forces f_n and f_t, should they be required.   For example, our problem has been posed on the assumption that the disk rolled without slip.   It might be necessary to ascertain how much friction[1] is necessary to ensure this assumption.

[1] See Prob. 4–7.

To investigate this question it would be necessary to study the contact forces which are given in the direct-momentum method by (4–19) but which are not available at all in the indirect procedure leading to (4–24).

Example 4–2. In Fig. 4–6 we consider an idealized torsional-vibration system. The torsional vibrations of systems involving rotating machinery such as turbines, motors, engines, pumps, generators, or propellers are usually analyzed[1] by idealized models of the sort indicated in Fig. 4–6. The rigid rotors represent the rotating parts of three machines. It is assumed that the axis of rotation of the system is a principal axis of inertia for each of the rotors and that the corresponding principal moments of inertia are I_1, I_2, and I_3, respectively. The rotors are connected by linearly elastic shafts which are assumed to be massless

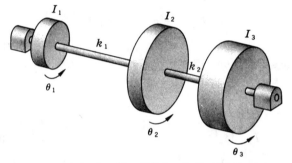

Fig. 4–6. Example 4–2. Torsional vibration system.

(in comparison with the rotors). The torsional spring constants of the lengths of shafting are indicated by k_1 and k_2. It is assumed that the only motions permitted are fixed-axis rotations about the system axis defined by the bearing centers. In normal operation all three rotors rotate at the same fixed speed. Under certain circumstances, however, there may be a torsional oscillation superposed on the steady rotation. When this occurs the twisting angles of the shafts and the relative positions of the disks fluctuate with time. We shall obtain the equations of motions for these torsional fluctuations.

For linearly elastic shafts, as we have assumed here, the character of the torsional vibration is independent[2] of the steady rotation of the system, and hence can be analyzed most simply by taking the steady speed to be zero. We consider in Fig. 4–6 that θ_1, θ_2, and θ_3 represent angular displacements of the rotors from a static-equilibrium configuration of the

[1] See, for example, S. H. Crandall, Vibrations of Rotating and Reciprocating Machines, chap. 58 in W. Flügge (ed.), "Handbook of Engineering Mechanics," McGraw-Hill Book Company, New York, 1962.

[2] See Prob. 4–8.

system.   The system is holonomic, and these three angles are a complete set of independent generalized coordinates.   When the system is in a configuration represented by θ_1, θ_2, and θ_3, the twist angles in the two lengths of shafting are $\theta_2 - \theta_1$ and $\theta_3 - \theta_2$.

We shall use Lagrange's equations to derive equations of motion under the assumption that no unbalanced torques, other than those of the elastic shafting, act on the rotors during the vibration.   The kinetic coenergy of each rotor in fixed-axis rotation is given by (4–8).   The total kinetic coenergy of the system is the sum

$$T^* = \tfrac{1}{2}I_1\dot{\theta}_1^2 + \tfrac{1}{2}I_2\dot{\theta}_2^2 + \tfrac{1}{2}I_3\dot{\theta}_3^2. \tag{4–26}$$

The total potential energy of the shafting is

$$V = \tfrac{1}{2}k_1(\theta_2 - \theta_1)^2 + \tfrac{1}{2}k_2(\theta_3 - \theta_2)^2. \tag{4–27}$$

When the lagrangian $\mathcal{L} = T^* - V$, based on (4–26) and (4–27), is inserted in Lagrange's equations (2–228), we obtain the following equations of motion:

$$\begin{bmatrix} I_1 & 0 & 0 \\ 0 & I_2 & 0 \\ 0 & 0 & I_3 \end{bmatrix} \begin{Bmatrix} \ddot{\theta}_1 \\ \ddot{\theta}_2 \\ \ddot{\theta}_3 \end{Bmatrix} + \begin{bmatrix} k_1 & -k_1 & 0 \\ -k_1 & k_1 + k_2 & -k_2 \\ 0 & -k_2 & k_3 \end{bmatrix} \begin{Bmatrix} \theta_1 \\ \theta_2 \\ \theta_3 \end{Bmatrix} = \begin{Bmatrix} 0 \\ 0 \\ 0 \end{Bmatrix}. \tag{4–28}$$

Note in Fig. 4–6 that the elastic torque acting on the rotor I_1, in the direction of increasing θ_1, is $k_1(\theta_2 - \theta_1)$.   The first of (4–28) is thus a statement of (4–7), the relation between the torque and time rate of change of angular momentum, applied to the first rotor.   The second and third of (4–28) are corresponding statements for the second and third rotors; i.e., (4–28) could be written immediately from Fig. 4–6 by direct application of the angular-momentum principle.   Careful attention to detail is necessary, however, when the direct method is used in order to obtain the correct algebraic signs for the shaft-torque terms.   In the indirect method the correct signs are supplied automatically by the formalism, provided that the geometric admissibility requirements have been met.

Example 4–3.   A rigid block of mass M is suspended by a massless cantilever beam and vibrates in the plane indicated in Fig. 4–7.   The direction normal to the plane of motion is a principal direction for the centroidal inertia tensor of the block, and the principal moment of inertia associated with this direction is I_c.   The cantilever beam has length L and bending modulus EI.   Of special interest in this system is the role played by the beam.   Essentially, the beam is a massless "spring" which supports the block, but it is a spring with two independent displacements (one linear and one angular) and two corresponding *force quantities* (one a force and one a torque).   We shall obtain equations of small motion for the system by the direct application of the momentum principles and by the application of Lagrange's equations.

In Fig. 4–8 the generalized coordinates y and θ are indicated. We assume that y and θ are sufficiently small so that the engineering theory of beam bending can be used. Furthermore, we assume that the transverse

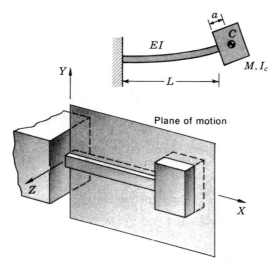

FIG. 4–7.   Example 4–3.   Cantilever vibration.

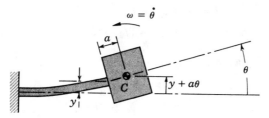

FIG. 4–8.   Generalized coordinates for small motions
are y and θ.

displacement of C is adequately represented by $y + a\theta$ and that the longitudinal displacement of C can be neglected. Under these assumptions the linear momentum (4–1) of the block is $M(\dot{y} + a\dot{\theta})\mathbf{u}_Y$ and its time rate of change is

$$\frac{d\mathbf{P}}{dt} = M(\ddot{y} + a\ddot{\theta})\mathbf{u}_Y. \tag{4–29}$$

The angular momentum (4–5) of the block with respect to the centroid is

$I_c\dot\theta \mathbf{u}_z$, and its time rate of change is

$$\frac{d\mathbf{H}_c}{dt} = I_c\ddot\theta \mathbf{u}_z. \qquad (4\text{--}30)$$

In order to obtain the force and torque acting on the block, we make the force analysis shown in Fig. 4–9. The block and beam are separated, and the contact interaction is indicated in terms of a force f and a torque τ.

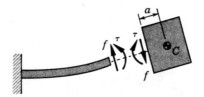

FIG. 4–9.   Force and torque at juncture of beam and block.

For small θ, the linear- and angular-momentum principles applied to the block yield

$$f = -M(\ddot y + a\ddot\theta)$$
$$af - \tau = I_c\ddot\theta. \qquad (4\text{--}31)$$

These, when solved for f and τ, become

$$f = -M(\ddot y + a\ddot\theta)$$
$$\tau = -Ma\ddot y - (I_c + Ma^2)\ddot\theta, \qquad (4\text{--}32)$$

which can be displayed as follows in matrix notation:

$$\begin{Bmatrix} f \\ \tau \end{Bmatrix} = -\begin{bmatrix} M & Ma \\ Ma & I_c + Ma^2 \end{bmatrix} \begin{Bmatrix} \ddot y \\ \ddot\theta \end{Bmatrix}. \qquad (4\text{--}33)$$

Next we turn to the cantilever beam to evaluate f and τ in terms of the deflections y and θ of the beam. By superposing standard solutions of engineering beam theory,[1] we find

$$y = \frac{L^3}{3EI}f + \frac{L^2}{2EI}\tau$$
$$\theta = \frac{L^2}{2EI}f + \frac{L}{EI}\tau. \qquad (4\text{--}34)$$

---

[1] See, for example, S. H. Crandall and N. C. Dahl (eds.), "An Introduction to Mechanics of Solids," McGraw-Hill Book Company, New York, 1959, p. 378.

This result may be abbreviated by introducing the *compliance matrix*

$$
\begin{bmatrix} \alpha_{11} & \alpha_{12} \\ \alpha_{21} & \alpha_{22} \end{bmatrix} = \begin{bmatrix} \dfrac{L^3}{3EI} & \dfrac{L^2}{2EI} \\[2ex] \dfrac{L^2}{2EI} & \dfrac{L}{EI} \end{bmatrix}.
\tag{4-35}
$$

We then can write (4–34) in matrix form as

$$
\begin{Bmatrix} y \\ \theta \end{Bmatrix} = \begin{bmatrix} \alpha_{11} & \alpha_{12} \\ \alpha_{21} & \alpha_{22} \end{bmatrix} \begin{Bmatrix} f \\ \tau \end{Bmatrix}.
\tag{4-36}
$$

At this stage the momentum principles have been used to relate the accelerations to the force and torque in (4–33), and elastic-beam theory has been used to relate the deflections to the force and torque in (4–36). To obtain equations of motion for the deflections, it is only necessary to eliminate the force and torque between (4–33) and (4–36). The simplest way is to substitute (4–33) into (4–36) to get

$$
\begin{Bmatrix} y \\ \theta \end{Bmatrix} = - \begin{bmatrix} \alpha_{11} & \alpha_{12} \\ \alpha_{21} & \alpha_{22} \end{bmatrix} \begin{bmatrix} M & Ma \\ Ma & I_c + Ma^2 \end{bmatrix} \begin{Bmatrix} \ddot{y} \\ \ddot{\theta} \end{Bmatrix},
\tag{4-37}
$$

which represents a pair of coupled linear second-order differential equations with constant coefficients.

An alternative procedure requires that we calculate the *inverse* of (4–35),

$$
\begin{bmatrix} k_{11} & k_{12} \\ k_{21} & k_{22} \end{bmatrix} = \begin{bmatrix} \alpha_{11} & \alpha_{12} \\ \alpha_{21} & \alpha_{22} \end{bmatrix}^{-1} = \begin{bmatrix} \dfrac{12EI}{L^3} & -\dfrac{6EI}{L^2} \\[2ex] -\dfrac{6EI}{L^2} & \dfrac{4EI}{L} \end{bmatrix},
\tag{4-38}
$$

which is called the *stiffness matrix*. Multiplying both sides of (4–36) by the stiffness matrix yields an alternative relation between the deflections and the force and torque,

$$
\begin{bmatrix} k_{11} & k_{12} \\ k_{21} & k_{22} \end{bmatrix} \begin{Bmatrix} y \\ \theta \end{Bmatrix} = \begin{Bmatrix} f \\ \tau \end{Bmatrix}.
\tag{4-39}
$$

If we now eliminate the force and torque between (4–33) and (4–39), we obtain, instead of (4–37),

$$
\begin{bmatrix} k_{11} & k_{12} \\ k_{21} & k_{22} \end{bmatrix} \begin{Bmatrix} y \\ \theta \end{Bmatrix} = - \begin{bmatrix} M & Ma \\ Ma & I_c + Ma^2 \end{bmatrix} \begin{Bmatrix} \ddot{y} \\ \ddot{\theta} \end{Bmatrix},
\tag{4-40}
$$

as equations of motion for the system.

We may note in passing that compliance matrices and stiffness matrices are widely used to characterize linearly elastic structures of arbitrary complexity. In general, the *compliance coefficient* α_{ij} is the deflection at station i due to a unit load at j (and no additional forces permitted at the other stations).   Similarly, the *stiffness coefficient* k_{ij} is the force reaction at i due to a unit displacement at j (and no additional displacements permitted at the other stations).

POTENTIAL ENERGY

The application of Lagrange's equations to the cantilever-vibration problem is straightforward, except for the problem of representing the potential energy of the bent beam.   Careful study of this problem is illuminating in its own right, and also serves as a useful introduction to the theory of energy-storing transducers in Chap. 6.

The geometric state of the beam in Figs. 4–8 and 4–9 is defined by the tip deflections y and θ.   The potential energy $V(y,\theta)$ is the work done *on* the beam by the force f and the torque τ when the beam passes through

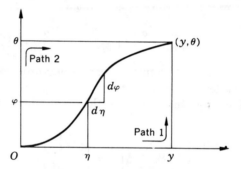

FIG. 4–10.   Integration paths for potential
energy $V(y,\theta)$.

any succession of equilibrium states leading from a datum state to the state under consideration.   If we take $y = 0$ and $\theta = 0$ as the datum configuration, the potential energy is the path integral

$$V(y,\theta) = \int_{(0,0)}^{(y,\theta)} (f \, d\eta + \tau \, d\varphi), \qquad (4\text{–}41)$$

where η and φ are used to represent intermediate values of y and θ, as shown in Fig. 4–10.   The integral (4–41) must be independent of the path followed from $(0,0)$ to (y,θ).   In order to evaluate (4–41), we use (4–39) to substitute for the intermediate values of f and τ.

$$V(y,\theta) = \int_{(0,0)}^{(y,\theta)} (k_{11}\eta + k_{12}\varphi) \, d\eta + (k_{21}\eta + k_{22}\varphi) \, d\varphi. \qquad (4\text{–}42)$$

Integration of (4–42) is especially simple along the paths 1 and 2 indicated in Fig. 4–10. Along path 1 we have

$$V(y,\theta) = \int_0^y k_{11}\, d\eta + \int_0^\theta (k_{21}y + k_{22}\varphi)\, d\varphi = \tfrac{1}{2}k_{11}y^2 + k_{21}y\theta + \tfrac{1}{2}k_{22}\theta^2,$$
$$\text{(4–43)}$$

while along path 2 we have

$$V(y,\theta) = \int_0^\theta k_{22}\varphi\, d\varphi + \int_0^y (k_{11}\eta + k_{12}\theta)\, d\eta = \tfrac{1}{2}k_{22}\theta^2 + k_{12}\theta y + \tfrac{1}{2}k_{11}y^2.$$
$$\text{(4–44)}$$

Comparing (4–43) and (4–44), we note that if V is to be independent of path, it is necessary[1] for k_{12} to equal k_{21}; i.e., the stiffness matrix must be symmetrical [note that (4–38) is in fact symmetrical]. In terms of the stiffness matrix, the potential energy $V(y,\theta)$ can be written

$$V(y,\theta) = \tfrac{1}{2}\{y,\theta\} \begin{bmatrix} k_{11} & k_{12} \\ k_{21} & k_{22} \end{bmatrix} \begin{Bmatrix} y \\ \theta \end{Bmatrix}. \qquad \text{(4–45)}$$

Once the potential energy is available, the formulation of the equations of motion, using Lagrange's equations, is straightforward. The kinetic coenergy for the block of Fig. 4–8 follows from (4–9):

$$T^* = \tfrac{1}{2}M(\dot{y} + a\dot{\theta})^2 + \tfrac{1}{2}I_c\dot{\theta}^2$$
$$= \tfrac{1}{2}\{\dot{y}\ \dot{\theta}\} \begin{bmatrix} M & Ma \\ Ma & I_c + Ma^2 \end{bmatrix} \begin{Bmatrix} \dot{y} \\ \dot{\theta} \end{Bmatrix}; \qquad \text{(4–46)}$$

and since there are no nonconservative forces acting, the lagrangian $\mathcal{L} = T^* - V$ contains all the required dynamic information. Routine application of Lagrange's equations (2–228) yields the equation of motion in the form (4–40).

4–2 Balancing of rotors

When a rotor must rotate at high speed in fixed bearings, it is desirable to minimize any fluctuating reaction forces at the bearings. If the rotor can be considered to be a rigid body, the origin of these fluctuating forces can be readily explained. First of all, if the center of mass of the rotor does not lie on the axis of rotation, the rotating rigid body will possess constantly changing linear momentum. The external force required by the linear-momentum principle to change the momentum must be delivered to the rotor through the bearings. Second, even if the centroid lies on the axis of rotation, there will still be fluctuating bearing reactions if the

---

[1] This is an instance of a general *reciprocity* property of linear-elastic structures due to J. C. Maxwell: for any such structure the compliance and stiffness matrices must be symmetrical.

axis of rotation is not a principal axis of inertia for the rotor.   In this case the angular-momentum vector will not be aligned with the axis of rotation, but will itself rotate with the body, as indicated in Fig. 4–11.

If its center of mass does not lie on the bearing axis, a nonspinning rotor will tend to hang in a gravity field "heavy side down."   In this case the rotor is said to be *statically unbalanced*.   Automobile wheels often develop static unbalance.   They can be restored to a state of static balance by adding small lead weights to the rim in such a way as to bring the centroid back onto the axis of rotation.

When a rotor is statically balanced but the axis of rotation is not a principal axis, the rotor is said to be *dynamically unbalanced*.   This type of unbalance cannot be detected unless the rotor is spinning.   When the rotor spins steadily, the angular-momentum vector is continually forced to change its orientation, and the external torque required is supplied by

Fig. 4–11.   Rotor is unbalanced if centroid is not on axis or if angular momentum is not aligned with axis.

equal and opposite forces at the two bearings.   Measurement of dynamic unbalance requires instrumentation capable of indicating the magnitude and orientation of these rotating bearing reactions.   Correction for dynamic unbalance is achieved by changing the mass distribution (by adding or removing mass) in such a way as to bring the angular-momentum vector back to the axis of rotation (without introducing any static unbalance in the process).   Over the years techniques for dynamic balancing have improved from trial-and-error methods to completely automated crankshaft production systems in which bearing-force information is processed by a special-purpose computer that activates a cutter to remove excess material.

Example 4–4.   To illustrate the development of rotating bearing reactions due to dynamic unbalance we consider the rotor of Fig. 4–12. The symmetrical rigid rotor has principal centroidal moments of inertia I_1, $I_2 = I_1$, and I_3.   The shaft is assumed to be rigid but massless.   The rotor is attached to the shaft so that the center of mass of the rotor lies on the axis of rotation, but there is a misalignment θ between the axis of rotation and the principal axis 3 of the rotor. We shall determine the bearing reactions when the rotor spins at the steady speed ω_0.

Let the xyz frame shown in Fig. 4–12 be a body-coordinate system with origin at the centroid of the rotor. Then, since $\dot{\omega} = 0$, the external torques which must act on the rotor are given by the relation (4–4)

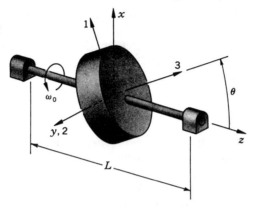

FIG. 4–12. Example 4–4. Dynamically unbalanced rotor.

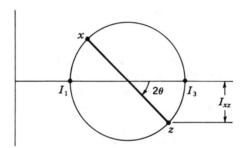

FIG. 4–13. Determination of I_{xz}.

between the torque and time rate of change of angular momentum

$$\begin{Bmatrix} \tau_x \\ \tau_y \\ \tau_z \end{Bmatrix}_c = \omega_0^2 \begin{Bmatrix} -I_{yz} \\ I_{xz} \\ 0 \end{Bmatrix}_c . \qquad (4\text{–}47)$$

In order to evaluate (4–47), it is necessary to determine the inertia-tensor components I_{yz} and I_{xz} from the given principal moments of inertia. The fact that the y axis coincides with the principal axis 2 implies that $I_{yz} = 0$. To obtain I_{xz}, we construct the Mohr circle shown in Fig. 4–13, from which we read off $I_{xz} = \frac{1}{2}(I_3 - I_1) \sin 2\theta$. Therefore the torque acting on the rotor is

$$\tau_y = \tfrac{1}{2}(I_3 - I_1)\omega_0^2 \sin 2\theta. \qquad (4\text{--}48)$$

This torque is supplied by a pair of forces τ_y/L in magnitude at the bearings. At the right-hand bearing the force acting on the shaft is in the positive x direction, and at the left-hand bearing the force on the shaft is in the negative x direction. These forces retain a fixed orientation in the body-coordinate system and thus rotate with the shaft at speed ω_0. Note that the magnitude of (4–48) depends on the *square* of the rotational speed. This means, in practice, that massive high-speed rotors require careful alignment and balancing if the rotating bearing reactions are to be kept below a tolerable level.

4–3 Euler's angles

In preparation for analyzing three-dimensional rotations of rigid bodies, we introduce a convenient set of generalized coordinates for the purpose of describing the orientation of a rectangular frame or body-coordinate system. In Secs. 2–3 and 3–7, the rotation matrix $[C]$ is used for this purpose. This matrix has nine elements, but there are[1] six equations interrelating these nine elements. The matrix representation lends formal simplicity to many operations with rotations, but it does not lead directly to a complete and independent set of three generalized coordinates. Such a set is, however, provided by the three angles θ, φ, and ψ shown in Fig. 4–14.

In Fig. 4–14 the origin of all the frames is taken to be a point fixed in the rigid body under consideration. If the origin is also fixed in inertial space, the XYZ frame is an *inertial* reference frame. If, however, the origin is a moving point, the XYZ frame is a moving frame, but we suppose that it always remains *parallel* to some *inertial* reference frame.

The rectangular frame 123 is taken to be a *body*-coordinate system; i.e., the rigid body is rigidly attached to the 123 frame. In addition, we shall usually assume that the 123 frame is aligned with the *principal directions* of the rigid body's inertia tensor at the origin. The problem, then, of describing the orientation of the rigid body reduces to the problem of describing the location of the 123 frame with respect to the XYZ frame.

This is done in Fig. 4–14 with the aid of the auxiliary xyz frame. A sequence of rotations which carries a moving rectangular frame from coincidence with the XYZ frame to coincidence with the 123 frame can be described as follows: Consider first a rotation φ about the Z axis. This brings the axis which started along the X axis into coincidence with the x axis. Next, consider a rotation θ about the x axis. This brings the

[1] See Prob. 2–9.

moving frame into coincidence with the xyz frame.   Finally, consider a rotation ψ about the z axis which brings the moving frame into coincidence with the body frame 123.   The three individual rotation angles φ, θ, and ψ are called *Euler's angles.*

The sequence described above indicates how the body can be oriented if the three Euler angles are given.   Conversely, if the body frame is in a given orientation, the corresponding Euler angles are determined as follows: The angle θ is measured directly between the Z axis and the 3 axis.   The x axis, or the *nodal line,* is located as the line perpendicular to the plane determined by the Z axis and the 3 axis.   Then the angles

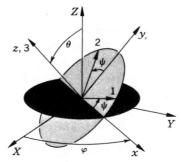

FIG. 4–14.   The Euler angles θ, φ, and ψ.

φ and ψ are measured from the nodal line to the X axis and to the 1 axis, respectively.

When the body changes its orientation, the Euler angles change.   The time rates of change $\dot{\varphi}$, $\dot{\theta}$, and $\dot{\psi}$ are angular-velocity components directed along the Z, x, and 3 axes, respectively.   The resultant angular velocity of the body axes with respect to the XYZ reference frame is

$$\boldsymbol{\omega} = \dot{\varphi}\mathbf{u}_Z + \dot{\theta}\mathbf{u}_x + \dot{\psi}\mathbf{u}_3. \tag{4–49}$$

The vector (4–49) may be decomposed into components with respect to any convenient frame.   For this purpose it is useful to note the following relations between the unit vectors directed along the axes in Fig. 4–14:

$$\begin{aligned}
\mathbf{u}_Z &= \mathbf{u}_3 \cos\theta + \mathbf{u}_y \sin\theta \\
\mathbf{u}_x &= \mathbf{u}_1 \cos\psi - \mathbf{u}_2 \sin\psi \\
\mathbf{u}_x &= \mathbf{u}_X \cos\varphi + \mathbf{u}_Y \sin\varphi \\
\mathbf{u}_y &= \mathbf{u}_1 \sin\psi + \mathbf{u}_2 \cos\psi.
\end{aligned} \tag{4–50}$$

By use of (4–50) we can represent the angular velocity (4–49) in terms of components along the body axes,

$$\boldsymbol{\omega} = (\dot{\varphi} \sin \psi \sin \theta + \dot{\theta} \cos \psi)\mathbf{u}_1$$
$$+ (\dot{\varphi} \cos \psi \sin \theta - \dot{\theta} \sin \psi)\mathbf{u}_2 + (\dot{\varphi} \cos \theta + \dot{\psi})\mathbf{u}_3, \quad (4\text{-}51)$$

or along the intermediate xyz axes,

$$\boldsymbol{\omega} = \dot{\theta}\mathbf{u}_x + \dot{\varphi} \sin \theta \mathbf{u}_y + (\dot{\varphi} \cos \theta + \dot{\psi})\mathbf{u}_z, \quad (4\text{-}52)$$

or along the XYZ reference axes,

$$\boldsymbol{\omega} = (\dot{\theta} \cos \varphi + \dot{\psi} \sin \theta \sin \varphi)\mathbf{u}_X$$
$$+ (\dot{\theta} \sin \varphi - \dot{\psi} \sin \theta \cos \varphi)\mathbf{u}_Y + (\dot{\psi} \cos \theta + \dot{\varphi})\mathbf{u}_Z. \quad (4\text{-}53)$$

These representations are useful for calculating the angular momentum or the kinetic coenergy of the rigid body.

For example, if the rigid body has inertial *symmetry* with respect to the 3 axis, so that the principal moments of inertia corresponding to the 1, 2, and 3 axis are I_1, $I_2 = I_1$, and I_3, then the x and y axes are also principal directions with principal moment of inertia equal to I_1. In this case we may use (4-52) and (3-39) to evaluate the kinetic coenergy when the origin of Fig. 4-14 is fixed in inertial space.

$$T^* = \tfrac{1}{2}I_1(\dot{\varphi}^2 \sin^2 \theta + \dot{\theta}^2) + \tfrac{1}{2}I_3(\dot{\varphi} \cos \theta + \dot{\psi})^2. \quad (4\text{-}54)$$

4-4   Euler's equations

The dynamic principles available for the analysis of three-dimensional rotations of rigid bodies are the linear- and angular-momentum principles listed in Sec. 3-5. Explicit expressions are given in (3-44) and (3-46) for angular momenta, but the time rates of change of angular momenta are not given explicitly in (3-45) and (3-47). In this section we show how these rates can be evaluated. We consider rotation about a point O fixed in the body and in inertial space. The angular-momentum principle (3-45) states

$$\boldsymbol{\tau}_o = \frac{d\mathbf{H}_o}{dt}, \quad (4\text{-}55)$$

where the derivative is a rate of change with respect to an inertial reference frame. If we consider an observer within a body-coordinate system, we can use (2-53) to write

$$\boldsymbol{\tau}_o = \left(\frac{\partial \mathbf{H}_o}{\partial t}\right)_{\text{rel}} + \boldsymbol{\omega} \times \mathbf{H}_o, \quad (4\text{-}56)$$

where $\boldsymbol{\omega}$ is the angular velocity of the body. When the vector equation (4-56) is resolved into components, the three resulting scalar equations are called *Euler's equations*. This resolution may be performed with respect to the reference-frame coordinate system, or with respect to a body-coordinate system, or with respect to any[1] convenient coordinate system.   Most commonly, the resolution is made with respect to *principal-direction body coordinates*.   In such a coordinate system the components of the angular momentum are

$$\begin{Bmatrix} H_1 \\ H_2 \\ H_3 \end{Bmatrix}_o = \begin{Bmatrix} I_1\omega_1 \\ I_2\omega_2 \\ I_3\omega_3 \end{Bmatrix}, \tag{4-57}$$

and the components of $(\partial \mathbf{H}_o/\partial t)_{\text{rel}}$ are

$$\begin{Bmatrix} I_1\dot{\omega}_1 \\ I_2\dot{\omega}_2 \\ I_3\dot{\omega}_3 \end{Bmatrix}. \tag{4-58}$$

The three component equations corresponding to (4–56) may then be written out as follows:

$$\begin{aligned} \tau_1 &= I_1\dot{\omega}_1 + (I_3 - I_2)\omega_2\omega_3 \\ \tau_2 &= I_2\dot{\omega}_2 + (I_1 - I_3)\omega_3\omega_1 \\ \tau_3 &= I_3\dot{\omega}_3 + (I_2 - I_1)\omega_1\omega_2. \end{aligned} \tag{4-59}$$

These are Euler's equations referred to principal axes.   They constitute a useful restatement of the basic angular-momentum principle (4–55). In (4–55) the moment center is taken to be a *fixed point O*.   An identical development can be applied to (3–47), in which the *centroid* is taken as the moment center; i.e., the Euler equations (4–59) can be interpreted to apply when the moment center is either (1) a fixed point O or (2) the center of mass C of the rigid body.   In either case the moment center must also be the point for which the inertia-tensor components are evaluated.

ALTERNATIVE DERIVATION OF EULER'S EQUATIONS

An interesting alternative development of (4–59) uses Lagrange's equations and the representation (4–51) of angular velocity in terms of Euler's angles.   We assume, as above, that a rigid body has a fixed point O and is subjected to an external torque τ_o having principal-direction components τ_1, τ_2, and τ_3.   If the principal moments of inertia at O are I_1, I_2, and I_3, the kinetic coenergy is

$$\begin{aligned} T^* &= \tfrac{1}{2}I_1\omega_1^2 + \tfrac{1}{2}I_2\omega_2^2 + \tfrac{1}{2}I_3\omega_3^2 \\ &= \tfrac{1}{2}I_1(\dot{\varphi}\sin\psi\sin\theta + \dot{\theta}\cos\psi)^2 + \tfrac{1}{2}I_2(\dot{\varphi}\cos\psi\sin\theta - \dot{\theta}\sin\psi)^2 \\ &\qquad\qquad\qquad\qquad + \tfrac{1}{2}I_3(\dot{\varphi}\cos\theta + \dot{\psi})^2, \quad (4-60) \end{aligned}$$

[1] See Prob. 4–13.

on using (4–51). Since there is no potential energy, the Lagrange equation for the generalized coordinate ψ is

$$\frac{d}{dt}\left(\frac{\partial T^*}{\partial \dot\psi}\right) - \frac{\partial T^*}{\partial \psi} = \Xi_\psi, \qquad (4\text{–}61)$$

where the generalized force here is the torque τ_3 about the axis for ψ. Substitution of (4–60) into (4–61) yields

$$I_3 \frac{d}{dt}(\dot\varphi \cos\theta + \dot\psi) - I_1(\dot\varphi \sin\psi \sin\theta + \dot\theta \cos\psi)(\dot\varphi \cos\psi \sin\theta - \dot\theta \sin\psi)$$

$$+ I_2(\dot\varphi \cos\psi \sin\theta - \dot\theta \sin\psi)(\dot\varphi \sin\psi \sin\theta + \dot\theta \cos\psi) = \tau_3. \qquad (4\text{–}62)$$

By introducing the angular-velocity components ω_1, ω_2, and ω_3 from (4–51) we obtain

$$I_3\dot\omega_3 + (I_2 - I_1)\omega_1\omega_2 = \tau_3, \qquad (4\text{–}63)$$

which is the third equation of (4–59). The other two Euler equations can be obtained from (4–63) by systematic permutation of the subscripts.

4–5 Torque-free motions

If a rigid body has a fixed point O and there is no external torque about O applied to the body, the angular momentum $\mathbf{H}_o$ will remain constant. Similarly, if an unconstrained rigid body experiences no external torque about its centroid, the angular momentum $\mathbf{H}_c$ will remain constant. If the body rotates about a principal axis, the angular velocity and the angular momentum are parallel, and both remain constant as long as no torque acts. If, however, the initial motion of the body is such that the angular-velocity direction does not coincide with that of the angular momentum, we shall see that, despite the constancy of the angular momentum, the subsequent motion is surprisingly complex. In this section we study the nature of motions in which the angular momentum remains constant. The analyses of this section are applicable to many important problems. A dramatic example is provided by the problem of tumbling of a space vehicle. When no jets are acting, the total torque on the vehicle is generally so small that, at least for short time intervals, it is permissible to consider the vehicle to be in a torque-free environment. A more homely example of nearly torque-free motion occurs whenever a rigid object is tossed up in the air. For compact heavy objects the aerodynamic torques are so small that there is very little change in angular momentum during the time the object is in the air.

We consider first the case of a rigid body with an axis of inertial *symmetry*. This case is examined in detail by both the direct method, using Euler's equations, and the indirect method, using Lagrange's equations. This is followed by a qualitative discussion of torque-free motion of a *general* rigid body. In order to be specific, we consider the rigid bodies in question to be rotating about a fixed point O, although the analyses apply equally well to the case of an unconstrained rigid body when we take the centroid C as the moment center.

APPLICATION OF EULER'S EQUATIONS

We consider a rigid body with a fixed point O having principal directions at O as sketched in Fig. 4–15 and principal moments of inertia $I_1, I_2 = I_1$, and I_3; i.e., the 3 axis is an axis of *inertial symmetry*. There is also an inertial reference frame $OXYZ$ originating at O, from which we shall eventually describe the motion, but for the present we concentrate our

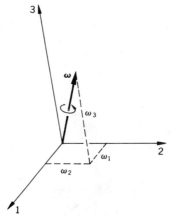

FIG. 4–15. Principal-axis body-coordinate system rotates with body.

attention on the body-coordinate system. Since no torque acts, Euler's equations (4–59) reduce to

$$I_1\dot{\omega}_1 = (I_1 - I_3)\omega_3\omega_2$$
$$I_1\dot{\omega}_2 = -(I_1 - I_3)\omega_3\omega_1 \qquad (4\text{–}64)$$
$$I_3\dot{\omega}_3 = 0.$$

From the third of (4–64) we learn that, because of the symmetry, the angular-velocity component ω_3 remains constant (its value in a particular case is determined by initial conditions). The first two of (4–64) are simplified by introducing the abbreviation s for the constant quantity,

$$s = \frac{I_1 - I_3}{I_1}\,\omega_3. \qquad (4\text{–}65)$$

Then the equations for ω_1 and ω_2 take the form

$$\dot{\omega}_1 = s\omega_2 \qquad \dot{\omega}_2 = -s\omega_1. \tag{4–66}$$

It is easily verified that the general solution of (4–66) is

$$\omega_1 = \omega_0 \cos (\gamma - st)$$
$$\omega_2 = \omega_0 \sin (\gamma - st), \tag{4–67}$$

where ω_0 and γ are constants of integration (which depend on the initial conditions). The nature of this solution is sketched in Fig. 4–16. As time goes on the two components (4–67) trace out in the $\widehat{12}$ plane a circle of radius ω_0 in a clockwise sense at the uniform angular rate s. When the constant ω_3 component is added, the resultant ω vector traces out a circular cone as viewed from the body.

At this stage we have integrated Euler's equations and, in a sense, we

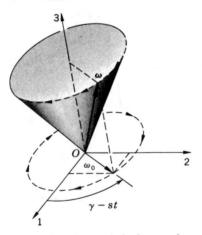

FIG. 4–16.   As seen in body, angular
velocity traces out cone.

have the solution. The initial conditions, through ω_3 and ω_0, fix the cone angle and the magnitude of ω; and through γ, they also fix the starting position of ω on the cone. In addition, through ω_3, the initial conditions also fix the rate (4–65) at which the cone is traversed. Now the angular velocity ω is *the* angular velocity of the body; i.e., the vector ω gives the direction of the instantaneous axis of rotation of the body-coordinate system and the magnitude of the angular speed at which the body system is rotating with respect to the reference frame. It is, however, difficult to use this information directly in visualizing how the motion appears to an observer in the reference frame.

INTERPRETATION OF THE SOLUTION

To facilitate the transfer from the body axes to the inertial reference frame, we construct the angular-momentum vector $\mathbf{H}_o$ with components $I_1\omega_1$, $I_2\omega_2$, and $I_3\omega_3$ in Fig. 4–17. Because of the inertial symmetry, $\mathbf{H}_o$ lies in the plane through the 3 axis, which contains ω. Thus, as ω traces out the cone indicated in Fig. 4–16, the vector $\mathbf{H}_o$ also traces out a cone centered about the 3 axis as viewed from the body. The bridge to the reference-coordinate system is now established, because we know that in the inertial frame the angular momentum remains permanently fixed in an *invariable* direction.

In the body-coordinate system $\mathbf{H}_o$ traces out a cone at a uniform rate

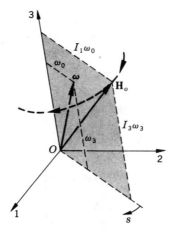

FIG. 4–17. Angular-momentum vector $\mathbf{H}_o$ lies in plane determined by ω and 3 axis.

about the 3 axis. In the reference frame the angular-momentum vector is fixed; so the 3 axis must appear to trace out a cone about the axis of $\mathbf{H}_o$. In Fig. 4–18 we have chosen the Z axis of the reference-coordinate system to coincide with the invariable direction of the angular momentum. Note that the plane carrying ω, $\mathbf{H}_o$, and the 3 axis, and which appeared to revolve about the 3 axis in Fig. 4–17, is still a plane in Fig. 4–18, but it now appears to revolve about the invariable direction of $\mathbf{H}_o$, This revolution of the symmetry axis 3 about the invariable direction is called *precession*. To determine the rate of precession, we draw the velocity diagram shown in Fig. 4–19. · The angular velocity ω, which in Fig. 4–16 is compounded from the components ω_o and ω_3, is here decomposed into the component Ω in the invariable direction and the component s parallel to the symmetry axis 3 of the body.

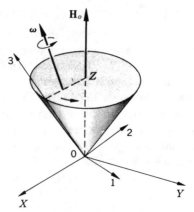

FIG. 4–18.   In inertial reference frame, 3 axis traces out cone centered on fixed direction of the angular momentum **H**$_o$.

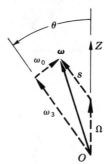

FIG. 4–19.   Velocity diagram in plane containing **ω**, **H**$_o$, and 3 axis.

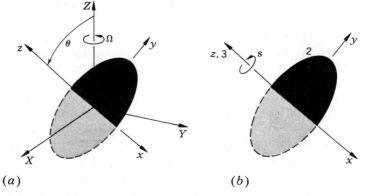

(a)   (b)

FIG. 4–20.   Decomposition of angular velocity **ω** into precession Ω and spin s.

The significance of this decomposition is indicated in Fig. 4–20. The component Ω about the invariable direction is called the *precessional velocity*. In Fig. 4–20a the effect of this component is pictured as carrying around an intermediate frame xyz. The z axis is tipped at the (constant) angle θ, and the x axis always lies in the XY plane (it is the nodal line of Fig. 4–14). To an observer in the xyz frame the direction of the angular momentum (i.e., the Z axis) remains stationary. In Fig. 4–20b the body-axis frame 123 is shown revolving about the z axis of the intermediate frame. The angular rate of this revolution is called the *spin*. We can now see that the magnitude of the spin is just the rate s pictured in Figs. 4–16 and 4–17. As the 123 frame spins counterclockwise at rate s in Fig. 4–20b, an observer in this frame would see the angular-momentum direction (i.e., the Z axis) trace out a cone, in a clockwise sense, at the rate s. This is in agreement with the description of Fig. 4–17.

Quantitative relations between the several angular-velocity components can be obtained from the geometry of Fig. 4–19; e.g.,

$$\omega_0 = \Omega \sin \theta$$
$$\omega_3 = \Omega \cos \theta + s. \tag{4-68}$$

A number of interesting relations can be developed by combining (4–68) and (4–65). To an observer in the inertial reference frame, the most easily observed characteristics of the motion are the precession velocity Ω and the tip angle θ between the symmetry axis of the body and the axis of the precession (i.e., the invariable direction of the angular momentum). In terms of these we find

$$s = \frac{I_1 - I_3}{I_3} \Omega \cos \theta$$

$$\omega_3 = \frac{I_1}{I_3} \Omega \cos \theta \tag{4-69}$$

$$\boldsymbol{\omega} \cdot \boldsymbol{\omega} = \Omega^2 \left[\sin^2 \theta + \left(\frac{I_1}{I_3} \right)^2 \cos^2 \theta \right]$$
$$\mathbf{H}_o \cdot \mathbf{H}_o = I_1^2 \Omega^2.$$

ROLLING-CONE REPRESENTATION

We have seen that the plane of Fig. 4–19 precesses about the invariable direction of $\mathbf{H}_o$ in Fig. 4–18 at the rate Ω. The angular-velocity vector $\boldsymbol{\omega}$ shown in Fig. 4–18 thus traces out a fixed cone in the inertial reference frame. Since the $\boldsymbol{\omega}$ vector passes through the fixed point 0 it also represents the instantaneous axis of rotation of the body as viewed from the reference frame. The fact that this axis is continually moving around on a fixed cone can be visualized by imagining that the body is rolling without slip on the fixed cone as shown in Fig. 4–21.   In order that the symmetry axis 3 of the body may also trace out a cone, it is necessary for the body to be itself a cone centered on the 3 axis.   It is readily verified that, when this is the case, all the relations we have derived are maintained.   Thus we have the interpretation that when the body rotates so as to preserve its angular momentum, it moves *as if* it were a cone rolling

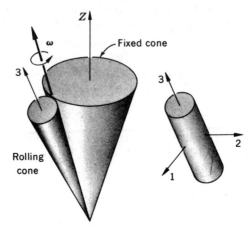

FIG. 4–21.   Rolling-cone interpretation for torque-free   motion   of   rigid   body   with $I_3 < I_1 = I_2$.

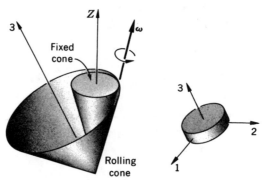

FIG. 4–22.   Rolling-cone interpretation for torque-free motion of rigid body with $I_1 = I_2 < I_3$.

on a fixed cone.   The sizes and orientations of the cones are determined by the initial conditions of the motion and by the moments of inertia I_1 and I_3.   All the sketches in Figs. 4–17 to 4–21 have been drawn for the case of $I_1 > I_3$, i.e., for a pencil-shaped body.   In the case of a disk-shaped body, $I_3 > I_1$, the spin direction is reversed. The corresponding rolling-cone representation is sketched in Fig. 4–22.

APPLICATION OF LAGRANGE'S EQUATIONS

We consider again an inertially symmetric rigid body with a fixed point O and principal moments of inertia I_1, $I_2 = I_1$, and I_3, at O. This time we use the Euler angles of Fig. 4–14 to define the position of the body with respect to the inertial reference frame $OXYZ$. Since no torques act on the body, there is no potential energy and there are no nonconservative forces. The system dynamics are completely determined by the kinetic

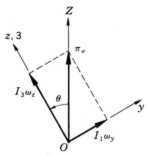

FIG. 4–23. Component of $\mathbf{H}_o$ along Z is vector sum of components along y and z.

coenergy (4–54). Lagrange's equations for θ, φ, and ψ, respectively, are

$$\frac{d}{dt}(I_1\dot{\theta}) - I_1\dot{\varphi}^2 \sin\theta\cos\theta + I_3(\dot{\varphi}\cos\theta + \dot{\psi})\dot{\varphi}\sin\theta = 0$$

$$\frac{d}{dt}[I_1\dot{\varphi}_1 \sin^2\theta + I_3(\dot{\varphi}\cos\theta + \dot{\psi})\cos\theta] = 0 \quad (4\text{–}70)$$

$$\frac{d}{dt}[I_3(\dot{\varphi}\cos\theta + \dot{\psi})] = 0.$$

Note that φ and ψ are cyclic coordinates in (4–54), and thus we have conservation of the corresponding generalized momenta, as indicated by the last two of (4–70). The third equation of (4–70) states that the component $I_3\omega_3$ of the angular momentum conserves its magnitude, and therefore

$$\omega_3 = \dot{\varphi}\cos\theta + \dot{\psi} \quad (4\text{–}71)$$

is a *constant* of the motion (to be determined by the initial conditions). The second equation of (4–70) states that the generalized momentum π_φ corresponding to φ is also a *constant* of the motion. The interpretation of π_φ is facilitated by introducing the velocity components ω_y and ω_z defined by (4–52),

$$\pi_\varphi = (I_1\omega_y) \sin \theta + (I_3\omega_z) \cos \theta. \tag{4-72}$$

The significance of (4–72) is indicated in the sketch of Fig. 4–23.

The general integration of (4–70) is nevertheless a formidable task. A particular solution is, however, readily obtained if we try $\theta \equiv 0$. Then θ is a *constant*, and (4–70) becomes a set of *algebraic* equations for $\dot{\varphi}$ and $\dot{\psi}$. This implies that $\dot{\varphi}$ and $\dot{\psi}$ remain constant during the motion. Insertion of (4–71) in the first of (4–70) yields

$$\dot{\varphi} \sin \theta(-I_1\dot{\varphi} \cos \theta + I_3\omega_3) = 0, \tag{4-73}$$

from which we conclude that either $\dot{\varphi} \sin \theta = 0$ (which can only be the case for exceptional initial conditions) or

$$\omega_3 = \frac{I_1}{I_3} \dot{\varphi} \cos \theta, \tag{4-74}$$

which must hold for all other initial conditions. When (4–74) is inserted in (4–71) we find

$$\dot{\psi} = \frac{I_1 - I_3}{I_3} \dot{\varphi} \cos \theta. \tag{4-75}$$

At this point we have completed the analysis, and it remains only to interpret the results. One solution to Lagrange's equations consists of a steady rotation $\dot{\varphi}$ of the line of nodes in Fig. 4–14, with a fixed tip-angle θ and with a steady rotation $\dot{\psi}$ given by (4–75).

This solution is the same as that obtained from Euler's equations. If we call $\dot{\varphi}$ the precessional velocity Ω, and $\dot{\psi}$ the spin s, the results (4–74) and (4–75) are *identical* with the first two of (4–69). In these solutions the invariable direction of $\mathbf{H}_o$ has been aligned with the Z axis. Equation (4–70) also possesses solutions of the same nature for *arbitrary* fixed orientations of $\mathbf{H}_o$ with respect to $OXYZ$.

General rigid body

The torque-free motion of a general unsymmetric rigid body (i.e., unequal principal moments of inertia) can be analyzed by an extension of the preceding methods. The analysis is simple in principle but difficult in detail. The resulting motion can still be viewed as the rolling of one cone on another, but now the cones are no longer circular, and moreover, the fixed cone does not generally close on itself. Here we shall only show how the conservation principles can be used to obtain a qualitative picture of the motion as viewed from a body-coordinate system.

We consider a principal-direction body-coordinate system with principal moments of inertia $I_1 < I_2 < I_3$. Since there is no torque, the angular momentum has a constant magnitude H_o. In terms of the principal-direction components of $\mathbf{H}_o$, we have

$$H_1^2 + H_2^2 + H_3^2 = H_o^2. \tag{4-76}$$

Furthermore, the kinetic energy T is also conserved. Using the expression for T given in Table 3–1 in Sec. 3–5, and introducing principal-direction components, we have

$$T = \tfrac{1}{2}\{H\}_o^t[I]_o^{-1}\{H\}_o$$

$$= \tfrac{1}{2}\{H_1, H_2, H_3\}
\begin{bmatrix}
I_1 & 0 & 0 \\
0 & I_2 & 0 \\
0 & 0 & I_3
\end{bmatrix}^{-1}
\begin{Bmatrix}
H_1 \\
H_2 \\
H_3
\end{Bmatrix}$$

$$= \frac{1}{2}\left(\frac{H_1^2}{I_1} + \frac{H_2^2}{I_2} + \frac{H_3^2}{I_3}\right). \tag{4–77}$$

In a torque-free motion the principal-axes components of $\mathbf{H}_o$ are always constrained by the two conservation requirements (4–76) and (4–77).

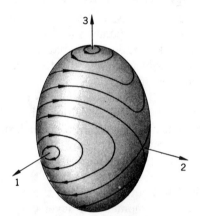

FIG. 4–24. Energy ellipsoid showing loci of constant magnitude of angular momentum.

If we take H_1, H_2, and H_3 as rectangular coordinates along the principal directions, the requirement (4–76) is represented by a *sphere* of radius H_o, and the requirement of (4–77) is represented by an *ellipsoid*. To satisfy both requirements simultaneously, the tip of the angular-momentum vector must remain on the intersection of the sphere and the ellipsoid. In Fig. 4–24 the ellipsoid (4–77) has been sketched for a fixed value, T_0, of kinetic energy. For any given magnitude of H_o, between the limits $H_{\min}$ and $H_{\max}$, where

$$H_{\min}^2 = 2T_0 I_1 \qquad H_{\max}^2 = 2T_0 I_3, \tag{4–78}$$

the sphere will intersect the ellipsoid in a pair of symmetrical closed curves. Several of these curves[1] are sketched on the surface of the ellipsoid in Fig. 4–24.

[1] See Prob. 4–23.

The initial conditions of the motion determine where the angular-momentum vector is initially in the body-coordinate system. They also determine the energy ellipsoid and the path on the ellipsoid on which the tip of the angular momentum must remain as time goes on. This representation should be compared with Fig. 4–17 for the symmetrical case. In both cases we view how the direction of the angular momentum (which is actually invariable in inertial space) appears to change when viewed from the body-coordinate system. For the inertially symmetric body, the paths of Fig. 4–24 reduce to circles and the angular-momentum vector appears to trace out a circular cone.

Stability of torque-free motion

The representation of Fig. 4–24 can be used to discuss the stability of torque-free rotation of a rigid body about its *principal* axes. If the body is started spinning so that the angular-momentum vector is aligned with one of the body's principal directions, the paths in Fig. 4–24 degenerate to points and the angular momentum remains fixed in the body-coordinate system as well as in inertial space.

Now suppose that a small disturbance causes the angular-momentum vector to deviate slightly from the principal axis. Here there is a distinctive difference in behavior, according to which one of the principal directions the angular momentum is near. In the neighborhood of the principal direction 1 (with smallest moment of inertia) or the principal direction 3 (with largest moment of inertia) the path of the angular momentum is a small closed curve encircling the principal direction; i.e., as time goes on, the angular momentum never departs very far from its original principal direction, when viewed from the body. In inertial space (where the angular momentum stands still) this implies that the principal axis of the body never tilts very far away from the original axis of rotation.

The situation is quite different for small disturbances in the neighborhood of rotation about the principal direction 2 (with intermediate moment of inertia). As indicated in Fig. 4–24, the path of the angular-momentum vector near the 2 axis is not a small closed path about the 2 axis but is rather a *large* closed path about either the 1 axis or the 3 axis. Thus, once the angular-momentum vector departs at all from the 2 axis, it embarks on a journey which takes it very far from its original orientation.

This qualitative[1] difference in behavior is described by saying that rotation about a principal axis having smallest or largest moment of inertia is *stable*, while rotation about the principal axis with intermediate

[1] See Prob. 4–24.

moment of inertia is *unstable*. This difference is readily demonstrated by tossing an object like a tennis racket in the air. It is an easy matter to throw the racket in the air so that it spins about either the 1 axis or the 3 axis, as indicated in Fig. 4–25. It is, however, very difficult to

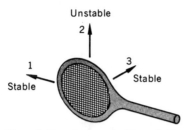

FIG. 4–25. Demonstration of stability and instability of rotation about principal axes.

produce a simple rotation about the 2 axis. Most attempts result in a complicated tumbling motion.

4–6 Gyroscopes

In the preceding section the torque-free motion of rigid bodies is examined in some detail. In this section we consider the effect of torques applied to rotating rigid bodies.

The word *gyroscope* can be applied to any rotating rigid body which is suspended so that the orientation of its axis of rotation can be changed. In technical applications the rigid body is usually inertially symmetric and has a large spin about the axis of symmetry. A simple gyroscope is indicated in Fig. 4–26. A massive disk is mounted in bearings in a light rigid box and given a high spin velocity (a small motor may be used to maintain the speed against the friction of the bearings). As a result the gyroscope has a large axial angular momentum with respect to its centroid. As long as the orientation of the box does not change with respect to inertial space, the angular momentum remains constant.

When the orientation of the box changes, the angular momentum also changes, and according to the angular-momentum principle, torque must be applied. What is unusual about a gyroscope is the direction in which the torque must be applied. The angular-momentum principle requires that the torque vector be directed parallel to the time rate of change of angular momentum. In a gyroscope the important changes in angular momentum are not changes in magnitude, but changes in orientation. For example, if the box in Fig. 4–26 turns slowly about the y axis (counterclockwise when viewed from above), the significant change in angular momentum involves a swiveling to the right of the tip of the vector $\mathbf{H}_c$ (which remains essentially constant in magnitude). The change in angular momentum is directed parallel to the x direction. Thus the major torque required to turn the box about the y axis is a torque about the x axis. Cause and effect are at right angles!

The box of Fig. 4–26 can be considered as a model of the primary element of most gyroscopic instruments and devices for inertial navigation. For example, suppose the box is mounted on a horizontal table fixed in a

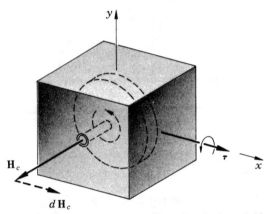

FIG. 4–26. Gyroscope consists of spinning rigid body with large axial angular momentum.

vehicle in such a way that a signal is produced which is proportional to the torque about the x axis (e.g., the box could be supported by a servo-driven torque motor). Then, whenever the vehicle changes its course (turns about the y axis), a signal proportional to the rate of turning is produced. In this application the system is called a *rate* gyro. Precision rate gyros can detect rotation rates which are small fractions of the earth's rotation rate of one revolution per day.

GENERALIZED COORDINATES FOR GYROSCOPES

We shall restrict our analysis of gyroscopes to the technically important case where the axial spin is large and the angular deviations in the orientation of the spin axis are small. In Fig. 4–27 the Euler angles of Fig. 4–14 are shown for a configuration in which $\varphi = \frac{1}{2}\pi$ and $\theta = \frac{1}{2}\pi$. This position is a convenient configuration to use as a center for small deviations in the orientation of the z, or 3, axis of the rigid body. Changes in the three Euler angles are about axes which are nearly perpendicular to one another. It may be noted that the neighborhood of $\theta = 0$ is *not* well suited to describing small deviations. The position $\theta = 0$ is, in fact, a singular position for the Euler angles. When $\theta = 0$, the angles φ and

ψ are measured about the same axis, and hence only their sum is uniquely determined by the location of the body. When θ is not zero, but is small, the angles φ and ψ are uniquely determined by the position of the body,

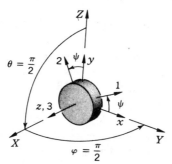

Fig. 4–27. Gyroscope in position defined by Euler angles $\varphi = \frac{1}{2}\pi$, $\theta = \frac{1}{2}\pi$, and ψ.

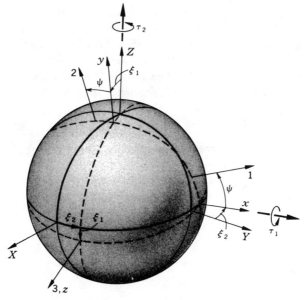

Fig. 4–28. Spherical representation of orientation defined by Euler angles $\varphi = \frac{1}{2}\pi + \xi_2$, $\theta = \frac{1}{2}\pi + \xi_1$, and ψ.

but a very small change in position can correspond to large changes in both φ and ψ.

We shall use the coordinates shown in Fig. 4–28 to describe the orientation of a gyroscope with principal directions 1, 2, and 3 which has

242 Dynamics of systems containing rigid bodies

inertial symmetry about the 3 axis; i.e., $I_1 = I_2$. We consider that the *spin* ψ is large and that the generalized coordinates ξ_1 and ξ_2, defined by

$$\theta = \frac{\pi}{2} + \xi_1 \qquad \varphi = \frac{\pi}{2} + \xi_2 \tag{4-79}$$

remain small. The z, or 3, axis is often called the *axle*, or *axial direction*, of the gyroscope. As in Fig. 4–14, the common origin of the frames in Fig. 4–28 may be either a fixed point O or the centroid C, depending on the application. Note that the intermediate frame xyz which contains the nodal line (the x axis) and the axle (the z axis) partakes in the small motion defined by ξ_1 and ξ_2. To an observer fixed in the xyz frame the gyroscope appears to have a single component of angular velocity, the spin ψ. It is useful to note that because of the inertial symmetry, even though the gyroscope spins with respect to the xyz frame, the components of the inertia tensor in the xyz coordinate system do *not* change with time;[1] i.e., $I_x = I_y = I_1$ and $I_z = I_3$.

We shall obtain linearized equations for *small* motions of the gyroscope axle under the influence of the torques τ_1 and τ_2, shown in Fig. 4–28. In this case Lagrange's equations depend only on the kinetic coenergy T^* and the generalized forces τ_1 and τ_2 corresponding to the generalized coordinates ξ_1 and ξ_2. In order to derive linearized equations, it is only necessary[2] to retain terms in T^* that are of second order in the small quantities ξ_1, ξ_2, $\dot{\xi}_1$, and $\dot{\xi}_2$. When (4–79) is introduced into the kinetic coenergy (4–54), and terms of higher order than second are omitted, the result is

$$T^* = \tfrac{1}{2}I_1(\dot{\xi}_1^2 + \dot{\xi}_2^2) + \tfrac{1}{2}I_3(\dot{\psi} - \xi_2\dot{\xi}_1)^2. \tag{4-80}$$

Lagrange's equations corresponding to ξ_1, ξ_2, and ψ are

$$\frac{d}{dt}(I_1\dot{\xi}_1) + I_3(\dot{\psi} - \xi_2\dot{\xi}_1)\dot{\xi}_2 = \tau_1$$

$$\frac{d}{dt}[I_1\dot{\xi}_2 - I_3(\dot{\psi} - \xi_2\dot{\xi}_1)\xi_1] = \tau_2 \tag{4-81}$$

$$\frac{d}{dt}[I_3(\dot{\psi} - \xi_2\dot{\xi}_1)] = 0.$$

The last of (4–81) implies that

$$\omega_3 = \dot{\psi} - \xi_2\dot{\xi}_1 \tag{4-82}$$

remains *constant*. When (4–82) is inserted into the first two of (4–81) and the differentiations are carried out, the resulting equations of motion

[1] See Prob. 4–27.

[2] See p. 134.

are

$$I_1\ddot{\xi}_1 + I_3\omega_3\dot{\xi}_2 = \tau_1$$
$$I_1\ddot{\xi}_2 - I_3\omega_3\dot{\xi}_1 = \tau_2. \qquad (4\text{–}83)$$

These equations are useful for analyzing gyroscopic instruments. In any particular case the torques τ_1 and τ_2 must be identified at this stage before further progress can be made. Note the manner in which the terms containing the axial angular-momentum component $I_3\omega_3$ enter (4–83). These terms are said to provide *gyroscopic coupling*. When the axle motions have sufficiently small accelerations so that the first terms in each of (4–83) can be neglected, we have an analytical description of the phenomenon discussed in connection with Fig. 4–26. Rotation $\dot{\xi}_2$ about the Z axis of Fig. 4–28 requires torque τ_1 about the x axis, and conversely, rotation $\dot{\xi}_1$ requires torque τ_2.

In deriving (4–83) we assumed that there was no net torque acting about the axle of the gyroscope. In some gyroscopic instruments the rotor is driven by a synchronous motor which maintains the spin $\dot{\psi}$ (very nearly) constant. For small motions there is no difference in behavior between an instrument with no axle torque ($\omega_3 = $ constant) and an instrument with $\dot{\psi} = $ constant, since according to (4–82) the difference between ω_3 and $\dot{\psi}$ is of second order.

Example 4–5. An inertially symmetric gyroscope supported in gimbal rings is shown in Fig. 4–29. The base C-C is fixed in inertial space. The gimbal bearings at A-A and B-B permit the gyro axle to turn through angles ξ_1 and ξ_2, respectively (note the correspondence between Figs. 4–28 and 4–29), while keeping the gyro center fixed in inertial space. We take the gyro to be spinning initially, with large spin about its (stationary) axle in the position $\xi_1 = 0$, $\xi_2 = 0$. Our problem is to analyze the subsequent motion of the system when at $t = 0$ an initial angular velocity $\dot{\xi}_2(0) = \omega_0$ is imparted to the outer gimbal ring about the axis B-B.

We assume that the mass of the gimbal rings can be neglected, but we take the gimbal bearings at A-A and B-B to have appreciable viscous friction. Because of the bearings A-A, we assume a retarding torque proportional to the relative angular velocity of the two gimbal rings

$$\tau_1 = -b\dot{\xi}_1, \qquad (4\text{–}84)$$

and because of the bearings B-B, we assume a retarding torque proportional to the absolute angular velocity of the outer gimbal ring,

$$\tau_2 = -b\dot{\xi}_2, \qquad (4\text{–}85)$$

where b is a constant damping coefficient. Finally, we assume that the motion of the axle is sufficiently small so that we can apply the linearized equations of motion of (4–83).

When (4–84) and (4–85) are substituted in (4–83) the result is

$$\begin{bmatrix} I_1 & 0 \\ 0 & I_2 \end{bmatrix} \begin{Bmatrix} \ddot{\xi}_1 \\ \ddot{\xi}_2 \end{Bmatrix} + \begin{bmatrix} b & I_3\omega_3 \\ -I_3\omega_3 & b \end{bmatrix} \begin{Bmatrix} \dot{\xi}_1 \\ \dot{\xi}_2 \end{Bmatrix} = \begin{Bmatrix} 0 \\ 0 \end{Bmatrix}, \qquad (4\text{--}86)$$

where, correct to first order, we can take ω_3 to be equal to the initial spin.

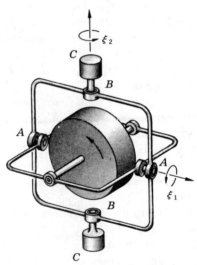

FIG. 4–29.   Example 4–5.   Gyroscope with viscous friction in the gimbal bearings.

The equations (4–86) are to be integrated subject to the initial conditions

$$\begin{aligned} \xi_1(0) &= 0 & \dot{\xi}_1(0) &= 0 \\ \xi_2(0) &= 0 & \dot{\xi}_2(0) &= \omega_0. \end{aligned} \qquad (4\text{--}87)$$

A possible procedure[1] for obtaining the solution is outlined below.

Since (4–86) is linear with constant coefficients, we assume a solution of the form

$$\begin{Bmatrix} \xi_1 \\ \xi_2 \end{Bmatrix} = \begin{Bmatrix} a_1 \\ a_2 \end{Bmatrix} e^{st}, \qquad (4\text{--}88)$$

where a_1, a_2, and s are undetermined parameters. Substitution in (4–86)

[1] The same general procedure can be applied to obtain solutions to other systems of linear differential equations with constant coefficients, such as (2–235), (4–28), and (4–40).

yields

$$\begin{bmatrix} I_1s^2 + bs & I_3\omega_3 s \\ -I_3\omega_3 s & I_1s^2 + bs \end{bmatrix} \begin{Bmatrix} a_1 \\ a_2 \end{Bmatrix} e^{st} = 0. \tag{4-89}$$

Nontrivial solutions occur only for those values of s which make the determinant of the coefficient matrix vanish. Setting the determinant equal to zero,

$$s^2[(I_1s + b)^2 + I_3^2\omega_3^2] = 0, \tag{4-90}$$

we find the four roots

$$s_1 = 0 \qquad s_2 = 0 \qquad s_3 = -\frac{b}{I_1} + j\frac{I_3\omega_3}{I_1} \qquad s_4 = -\frac{b}{I_1} - j\frac{I_3\omega_3}{I_1}, \tag{4-91}$$

where $j^2 = -1$. For the roots s_1 and s_2, Eq. (4–89) does not restrict the amplitudes a_1 and a_2 at all. For the root s_3, Eq. (4–89) requires that $a_2 = -ja_1$, and for the root s_4, Eq. (4–89) requires $a_2 = ja_1$. Taking these restrictions into account, the general solution to (4–86), in real form, is

$$\xi_1 = c_1 + e^{-bt/I_1}\left(c_3 \sin\frac{I_3}{I_1}\omega_3 t + c_4 \cos\frac{I_3}{I_1}\omega_3 t\right)$$

$$\xi_2 = c_2 + e^{-bt/I_1}\left(-c_3 \cos\frac{I_3}{I_1}\omega_3 t + c_4 \sin\frac{I_3}{I_1}\omega_3 t\right), \tag{4-92}$$

where c_1, c_2, c_3, and c_4 are arbitrary constants of integration. Finally, these constants are determined by imposing the initial conditions (4–87) on the solution (4–92). We find

$$-c_1 = c_4 = \frac{I_1 I_3\omega_3}{b^2 + I_3^2\omega_3^2}\omega_0$$

$$c_2 = c_3 = \frac{bI_1}{b^2 + I_3^2\omega_3^2}\omega_0. \tag{4-93}$$

The corresponding path of the tip of the axle in the $\xi_1\xi_2$ plane is indicated in Fig. 4–30. Note that when the damping is small ($b \ll I_3\omega_3$), the axle of the gyro comes to rest in a position which is nearly *at right angles* to the direction of the initial velocity $\dot{\xi}_2(0) = \omega_0$. Note also that the angular rate $I_3\omega_3/I_1$ at which the axle spirals around its ultimate rest position is the limiting precessional velocity Ω of a torque-free rigid body (4–69) when the cone angle θ of the precession approaches zero. If $I_3 > I_1$, this precession rate will be faster than the spin of the gyro.

The motion of the gyro may be described qualitatively as follows. The initial velocity suddenly imparted at $t = 0$ changes the gyro's total angular momentum. As a result it begins to precess (like a torque-free gyro) about the new orientation of its angular momentum. The precession is not, however, truly torque-free. The bearing-friction torques act to displace further the position of the total angular momentum and

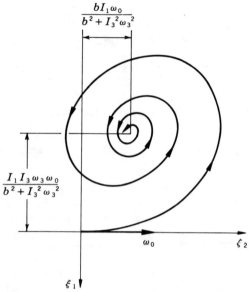

$$\frac{bI_1\omega_0}{b^2 + I_3^2\omega_3^2}$$

$$\frac{I_1 I_3 \omega_3 \omega_0}{b^2 + I_3^2 \omega_3^2}$$

ω_0 ζ_2

ξ_1

Fig. 4–30.   Response of gyro axle to initial velocity ω_o, under influence of viscous friction in gimbal bearings.

to cause the precession to spiral inward.   The rate at which the spiral decays is determined by the term e^{-bt/I_1} in the solution (4–92).   Note that this rate is independent of the spin.

PROBLEMS

4–1. A *simple* pendulum consists of a mass particle m suspended by a massless arm of length L.   The circular frequency of small oscillations is $\sqrt{g/L}$.   A *compound* pendulum consists of an extended rigid mass M suspended by a massless arm so that the distance from the point of suspension to the centroid of the rigid body is L.   If the centroidal moment of inertia of the rigid body about an axis perpendicular to the plane of the oscillation is I, find the circular frequency of small oscillations of the compound pendulum.

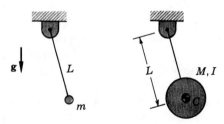

Prob. 4–1

4–2. A rigid body with mass M is suspended by two strings so that a principal direction at the centroid is vertical. The rigid body is given a *small* twist θ and released. If θ undergoes simple harmonic motion at circular frequency ω_0, what is the magnitude of the principal centroidal moment of inertia about the vertical?

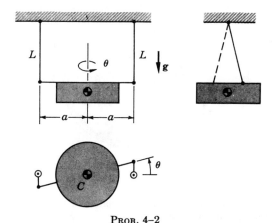

PROB. 4–2

4–3. A pendulum consists of a large ring of mass M suspended by a massless arm of length L. The pendulum can swing in the plane of the hoop about the frictionless pivot. A small ring of mass m slides without friction on the large ring. Assume that the small ring can be considered to be a mass particle.

(a) Select a set of generalized coordinates to describe the motions of the two rings.
(b) Formulate equations of motion by direct application of the momentum principles.
(c) Formulate equations of motion by constructing a lagrangian for the system.

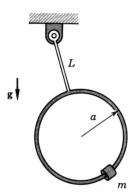

PROB. 4–3

4–4. A uniform cylinder of mass m and radius a rolls without slip inside a section of pipe of mass M, radius b, and wall thickness $h(h/b \ll 1)$. The pipe is supported on frictionless rollers.

(a) Select generalized coordinates to describe the motions of the cylinder and pipe.
(b) Formulate equations of motion.
(c) Obtain linearized equations of small motion in the neighborhood of the equilibrium position.

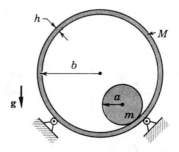

PROB. 4–4

4–5. A rigid rod of mass M, length L, and centroidal moment of inertia I, normal to the plane of motion, falls under the influence of gravity.

(a) Formulate equations of motion which apply when one end of the rod is in contact with a frictionless horizontal surface.
(b) Apply Hamilton's principle, using x, y, and θ as coordinates, and introduce the geometric constraint that the rod remain in contact with the surface by means of a Lagrange multiplier. Do this in such a way that the reaction force between the rod and the surface can be identified in terms of the Lagrange multiplier.

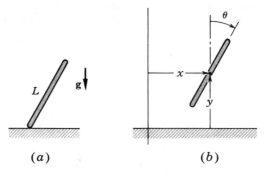

(a)   (b)

PROB. 4–5

4–6. A block of mass M slides without friction down an inclined plane. The block is connected to a cylinder by means of a massless connecting link. The cylinder has mass M, radius a, and centroidal moment of inertia I normal to the plane of motion and rolls without slip down the inclined plane.

(a) Find the acceleration of the block down the plane.
(b) Find the force in the connecting link. Is the link in tension or compression?

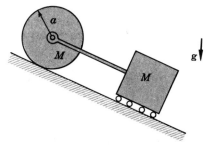

PROB. 4–6

4–7. Consider the rocking disk of Fig. 4–3 when it is oscillating back and forth between $-\theta_0$ and θ_0. What coefficient of friction is required to prevent slip when the disk is in the extreme position $\theta = \theta_0$?

4–8. Consider the torsional-vibration system of Fig. 4–6 under steady rotation at speed ω_0. Assume that a steady torque τ_1 is applied to I_1 and that I_2 and I_3 deliver steady torques of τ_2 and τ_3, respectively, with $\tau_1 = \tau_2 + \tau_3$.

(a) Construct the lagrangian for small torsional oscillations superposed on this steady state.
(b) Verify that the equations of motion for the vibration are independent of the steady-state speed and loading.

4–9. A rigid rod of mass M and length $4a$ is centered on a fixed cylinder of radius a. Consider small rocking motions of the rod in the plane of the sketch with no slip between the rod and the cylinder.

(a) Formulate the equation of motion.
(b) Find the natural frequency of small oscillations.

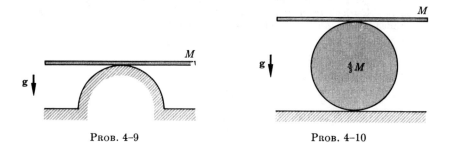

PROB. 4–9 PROB. 4–10

4–10. A rigid rod of mass M and length $4a$ is centered on a uniform rigid cylinder of mass $\frac{4}{3}M$ and radius a. The cylinder can roll without slip on the floor, and the rod can rock without slip on the cylinder.

(a) Formulate equations of small motion in the neighborhood of the equilibrium position shown.
(b) Investigate the stability of small motions by assuming that each generalized coordinate has a time history proportional to e^{st}.

250   *Dynamics of systems containing rigid bodies*

4-11. Consider a rigid body rotating with angular velocity ω with respect to a reference frame. Let a body-coordinate system have a triad of unit vectors **U**. Consider a vector **r** fixed in the body with body components $\{r\}$. Starting from

$$\mathbf{r} = \{r\}^t \mathbf{U}$$

and using (2–42), show that the body components of ω × **r** are

$$\left\{\frac{dr}{dt}\right\} = [\omega]^t\{r\},$$

where $[\omega]^t$ is the transpose of the skew-symmetric matrix of (2–42).

4-12. Consider a rigid body rotating with angular velocity ω with respect to a reference frame. Let **r** be a vector fixed in the body, and let its components with respect to a particular body-coordinate system be $\{r'\}$ and the components of the same vector with respect to the reference frame be $\{r\}$. Let $[C(t)]$ be the rotation matrix from the reference frame to the body-coordinate system. Show that

$$\frac{d}{dt}[C] = [\omega][C],$$

where $[\omega]$ is given by (2–42).

4-13. Consider a rigid body with a fixed point O rotating with angular velocity ω with respect to a reference frame. Let components with respect to a particular body-coordinate system be denoted by primes, and components with respect to the reference frame be unprimed. With the aid of Probs. 4–11 and 4–12, show that (4–56) may be written in body coordinates as

$$\{\tau'\} = [I']_o\{\dot{\omega}'\} + [\omega']^t[I']_o\{\omega'\},$$

and in the reference-frame coordinates as

$$\{\tau\}_o = [I]_o\{\dot{\omega}\} + [\omega]^t[I]_o\{\omega\},$$

where $[I]_o = [C(t)]^t[I']_o[C(t)]$.

4-14. The uniform rod of length L and mass M is pivoted, without friction, to the shaft OA, which revolves in fixed bearings at the steady rate ω_o. The rod is constrained to remain in a plane through OA which rotates with the shaft.

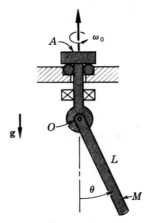

PROB. 4–14

(a) Formulate an equation of motion for $\theta(t)$.

(b) For each value of ω_0 there is at least one stationary angle θ_0 which the rod can maintain while steadily precessing at the rate ω_0. Find all stationary configurations as functions of ω_0.

(c) Examine the stability of the stationary configurations found in (b).

4-15. Two identical wheels of mass M and centroidal principal moments of inertia I_1, I_1, and I_3 are coupled by massless shafts and a frictionless joint at O. The top wheel undergoes fixed-axis rotation about the Z axis. The joint permits only a single degree of freedom, so that the top wheel must participate in the φ motion of the shaft.

(a) Find equations of motion when the lower wheel is fixed to the shaft ($\psi \equiv 0$).

(b) Find the relation between φ and θ for steady precession.

(c) Find equations of motion when the lower wheel is free to spin on the shaft without friction.

(d) Find the steady precession in case (c).

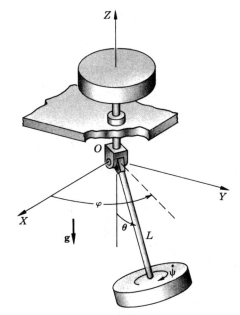

PROB. 4-15

4-16. A flywheel with axial moment of inertia I revolves in fixed bearings at a (variable) rate φ. Attached to the flywheel is a massless rigid wire frame. The upper part of the wire has a circular shape with radius a and carries a washer which can slide without friction on the wire. The washer has mass M, axial moment of inertia I_3, and centroidal moment of inertia I_1 about its diameters. The spin of the washer with respect to the wire is denoted by ψ.

(a) Select generalized coordinates to describe the motion of the flywheel and the washer.

(b) Formulate equations of motion.

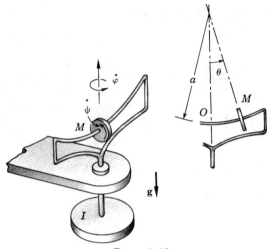

4-17. A table with axial moment of inertia I can turn in fixed bearings without friction. The table carries a massless rigid frame from which is suspended a massless pendulum arm carrying a spinning disk. The disk has mass M and principal moments of inertia I_1, I_1, and I_3 at the centroid.

(a) Select generalized coordinates to describe the motion of the table and the disk.
(b) Formulate equations of motion, including the effect of gravity but neglecting friction.
(c) Obtain integrals of the equations of motion which represent conservation laws.

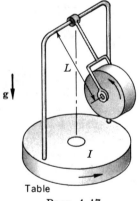

PROB. 4-17

4-18. A frictionless ball-and-socket joint suspends a rigid massless rod on which a disk can slide but not twist. The disk has mass M and centroidal moments of inertia I_1, I_1 and I_3, and is restrained by a linear spring of free length L and spring constant k.

(a) Select generalized coordinates to describe the motion of the disk.

(b) Formulate equations of motion.

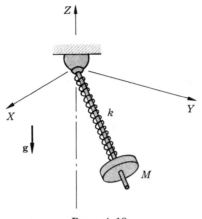

PROB. 4-18

4-19. A suspension system for an instrument to be installed in a vehicle consists of four equal leaf springs. In order to withstand the vibrational inputs, the system is designed to be "soft" in the Z direction. The suspension is, however, very "stiff" against motion in the XY plane. Take the instrument to be a rigid body of mass M, with centroid at the intersection of the center lines of the supporting leaf springs and with centroidal principal directions in the X, Y, and Z directions; i.e., I_{XX}, I_{YY}, and I_{ZZ} are principal moments of inertia. Take the springs to be elastic cantilevers which can bend and twist. Let the potential energy of a single cantilever be

$$V = \tfrac{1}{2}k_{11}y^2 + k_{12}y\theta + \tfrac{1}{2}k_{22}\theta^2 + \tfrac{1}{2}k_{33}\varphi^2,$$

where k_{11}, k_{12}, and k_{22} are the bending stiffnesses of (4-38), and k_{33} is a torsional stiffness.

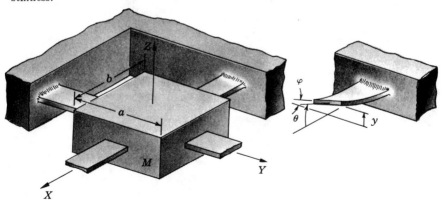

PROB. 4-19

(a)   Select generalized coordinates to describe small motions of the instrument.
(b)   Formulate the equations of motions.
(c)   Find the natural frequencies of simple harmonic oscillations.

   4-20. A rigid cone with apex half angle α rolls steadily without slip on a horizontal surface so that it precesses about the Z axis at a constant angular rate Ω.   The cone has mass M and principal moments of inertia I_1, I_1, and I_3 at the tip.

(a)   Determine the angular velocity of the cone.
(b)   Determine the angular momentum of the cone.
(c)   What forces and torques are required to maintain this motion?

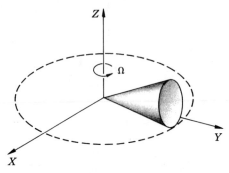

PROB. 4-20

   4-21. A top of mass M and principal moments of inertia I_1, I_1, and I_3 at the tip has large spin and is precessing and nutating (nodding), while its tip remains over a fixed point on the floor.   The centroid of the top is a distance h above the tip.

(a)   Select generalized coordinates to describe this motion.
(b)   Formulate equations of motion under the influence of gravity, neglecting frictional torques at the tip.
(c)   Obtain the conditions for steady precession without nutation.

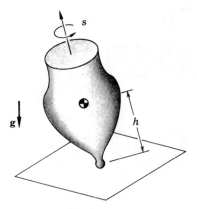

PROB. 4-21

4-22. A billiard ball rolls without slipping on a horizontal turntable which turns at constant angular rate Ω. Take the ball to be a rigid sphere of radius a, mass M, and centroidal moment of inertia $I = \frac{2}{5}Ma^2$ (in any direction).

(a) Formulate the nonholonomic constraints imposed by the requirement that the ball should not slip on the turning table.

(b) Apply the linear- and angular-momentum principles to the ball.

(c) Derive, from (a) and (b), equations of motion for the trajectory of the center of the ball, as seen from a nonrotating coordinate system.

(d) Show that every trajectory is a circle whose center and radius depend on the initial conditions.

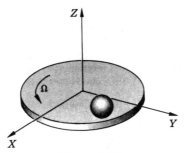

PROB. 4-22

4-23. Use Euler's equations to verify that the tip of the angular-momentum vector must trace out the paths shown in Fig. 4-24 in the sense indicated by the arrows.

4-24. Consider a torque-free rigid body with principal-direction angular-momentum components H_1, H_2, and H_3. Suppose that, initially, $H_3 = H_0$, a *large* quantity, while H_1 and H_2 are small quantities of first order in comparison with H_0. Use Euler's equations to obtain linearized differential equations describing the changes in the angular-momentum components in the neighborhood of the initial values. Investigate how the stability depends on the relative magnitudes of the principal moments of inertia I_1, I_2, and I_3.

4-25. A bicycle wheel has a short handle attached to the axle, and the end of the handle is supported from above by a length of string. The wheel has mass M and

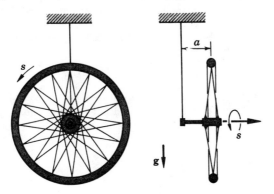

PROB. 4-25

centroidal moments of inertia I_1, I_1 and I_3.   The wheel is given a large spin s with respect to the handle, and then the handle is released from rest in the horizontal position shown.

(a)   Formulate equations of motion to describe the subsequent motion.
(b)   Estimate the average precessional rate.
(c)   Does the string continue to hang vertically?
(d)   When first released, in what direction does the axle move?

4–26. Consider the gyroscope of Fig. 4–29 for the case where there is no bearing friction but where there are linear torsional springs resisting the turning of the gimbal rings.   In place of (4–84) and (4–85) take

$$\tau_1 = -k\xi_1 \quad \text{and} \quad \tau_2 = -k\xi_2,$$

where k is the spring constant.   Formulate the equations of motion and find the natural frequencies of free oscillation.

4–27. Let the components of the rigid body's angular velocity in Fig. 4–14 or 4–28 be ω_x, ω_y, and ω_z with respect to the intermediate frame.   For a body with *inertial symmetry*, $I_1 = I_2 = I_x = I_y$, and the angular-momentum vector is

$$\mathbf{H} = I_1\omega_x\mathbf{u}_x + I_1\omega_y\mathbf{u}_y + I_3\omega_z\mathbf{u}_z.$$

Apply the differentiation formula

$$\frac{d}{dt} = \left(\frac{\partial}{\partial t}\right)_{\text{rel to } i} + \boldsymbol{\omega}_i \times \ ,$$

where ω_i is the angular velocity of the intermediate frame, to the angular-momentum principle, and thus obtain the *modified Euler equations*

$$\tau_x = I_1\dot{\omega}_x + I_3\omega_z\omega_y - I_1\omega_y\omega_{iz}$$
$$\tau_y = I_1\dot{\omega}_y - I_3\omega_z\omega_x + I_1\omega_x\omega_{iz}$$
$$\tau_z = I_3\dot{\omega}_z.$$

where $\omega_{iz} = \dot{\phi}\cos\theta$.   To what linear equations do these reduce when ω_x, ω_y, and ω_{iz} are first order small quantities, ω_z is large, and $\tau_z = 0$?   Compare with (4-83).

4–28. Consider Eq. (4–83) for the small motions of the axle of a symmetric gyro. Let $\omega_1 = \xi_1$ and $\omega_2 = \xi_2$ be the angular rates corresponding to the angular displacements ξ_1 and ξ_2.   Study steady-state simple harmonic motion of the system when τ_1, τ_2, ω_1, and ω_2 are all proportional to the real part of $e^{i\omega t}$.   Find the amplitudes of ω_1 and ω_2 in terms of the amplitudes of τ_1 and τ_2.   Show that, for low-frequency operation (i.e., $\omega I_1/\omega_3 I_3 \ll 1$), the relations between applied torques and response angular rates can be approximated by

$$\tau_1 = G\omega_2 \quad \text{and} \quad \tau_2 = -G\omega_1,$$

where $G = I_3\omega_3$.   A device with these constitutive relations is said to $\vdash$ $\!$n *ideal gyrator*.   Show that the power flowing *into* an ideal gyrator at 1 (due to th$\curvearrowright$orque τ_1 acting with angular velocity ω_1) is at every instant equal to the power flowing *out* at 2.

CHAPTER FIVE

Electrical networks

We consider in this chapter the dynamics of systems composed of lumped-parameter electrical elements.   The governing principles and the methods of analysis for electrical systems are, to a great extent, parallel to the principles and methods described in the preceding chapters for mechanical systems.   The chapter begins with a brief review of basic electrical concepts, a description of lumped-parameter circuit elements, and statements of the governing principles for the direct analysis of electric-circuit dynamics.   It is expected that the reader will have some familiarity with these topics.   The remainder of the chapter is mainly devoted to developing the indirect method of analyzing electric circuits, based on the variational approach embodied in Hamilton's principle and Lagrange's equations.   We are aware that this latter approach is seldom used for the routine solution of linear electrical networks.

Our motive here is to lay the groundwork for the discussion in Chap. 6 of electromechanical systems, where variational methods are of great utility and are widely employed.[1]

It is interesting to note that, historically, the formalism of the variational approach provided a suggestive outline for the development of electrical theory and that early workers[2] in the field leaned heavily on "energy methods" and Lagrange's equations.   These methods are still in use today for electrical applications which lie outside the standard areas of linear circuit analysis, e.g., nonlinear systems[3] and systems such as

---

[1] See, for example, D. C. White and H. H. Woodson, "Electromechanical Energy Conversion," John Wiley & Sons, Inc., New York, 1959.

[2] See, for example, J. C. Maxwell, "A Treatise on Electricity and Magnetism," 3d ed., 1891, reprinted by Dover Publications, Inc., 1954, vol. 2, pt. III, chaps. 6 and 7.

[3] E. C. Cherry, Some General Theorems for Non-linear Systems Possessing Reactance, *Phil. Mag.*, 42(7):1161–1177 (1951).

transducers, in which there is coupling between electrical and nonelectrical variables.

5-1   Basic concepts

We present a summary of the fundamental physical principles involved in the description of the behavior of lumped-parameter electric-circuit elements.   Our discussion is within the framework of quasi-static electromagnetic field theory,[1] which is limited to the case where the velocities of all charges can be neglected in comparison with the speed of light.

The system of units[2] used to describe electrical and mechanical quantities in this chapter is the Georgi, or rationalized *mks*, system.   In this system the units of mechanical quantities are as follows:

length—meter
mass —kilogram
time —second
force —newton
energy—newton-meter, or *joule*

The fundamental electrical quantity is the *electrical charge*, measured in *coulombs*.   We shall generally reserve the letter q to stand for a quantity of charge.   Charge can be carried from place to place on charged particles such as electrons or ions.   When a great many such charged particles flow along a path such as a metallic wire, one may speak of a continuous flow of charge past a reference point on the path.   The rate of flow of charge (or charged particles) is called *current*,

$$i = \frac{dq}{dt}. \tag{5-1}$$

The unit of current is the *ampere*, which is one coulomb per second.   A positive current can result from a flow of positive charge in a positive direction or of negative charge in a negative direction.

Stationary or slowly moving charges exert forces on one another according to Coulomb's law.   The force $\mathbf{f}_{21}$ acting on the particle with charge q_2 in Fig. 5–1 due to the charge q_1 is

$$\mathbf{f}_{21} = -\mathbf{f}_{12} = \frac{1}{4\pi\epsilon} \frac{q_1 q_2}{r^2} \mathbf{u}_r, \tag{5-2}$$

where the magnitude of the permittivity ϵ depends on the electrical prop-

[1] See, for example, R. M. Fano and L. J. Chu, "Fields, Energy, and Forces," John Wiley & Sons, Inc., New York, 1959.

[2] See, for example, D. C. Ipsen, "Units, Dimensions, and Dimensionless Numbers," McGraw-Hill Book Company, New York, 1960, sec. 7-7.

erties of the intervening medium.   For a vacuum,

$$\epsilon_0 = 8.854 \times 10^{-12} \text{ coul}^2/\text{newton-m}^2.$$

Coulomb's law is similar to the law of gravitational attraction, except that the electric force may be attractive or repulsive according to whether the charges involved have unlike or like signs.   Usually, the electric force between charged particles is very much stronger than their gravitational attraction.

The force on a particle with charge q due to a fixed constellation of other charged particles can be found by applying Coulomb's law to every possible interaction pair and adding the forces vectorially.   The resulting force is a function of the position of the particle in question, or probe, relative to the fixed constellation of charges.   We thus have a *field* of electric force $\mathbf{f(R)}$, where $\mathbf{f}$ is the force experienced by the probe when it

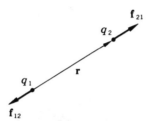

Fig. 5–1.   Coulomb's law for the force between two charged particles.

occupies the position $\mathbf{R}$.   This, like the gravitational field, is a *conservative* force field, and the potential energy *per unit charge* on the probe is

$$e(\mathbf{R}) = -\int_{\mathbf{R}_0}^{\mathbf{R}} \frac{1}{q} \mathbf{f} \cdot d\mathbf{R}, \tag{5–3}$$

where $\mathbf{R}_0$ is a convenient datum position.   The potential e defined by (5–3) is called the *electric potential*.   The unit of potential is the *volt*, which is one newton-meter per coulomb.   The change in potential energy V of any charge q moving from $\mathbf{R}_1$ to $\mathbf{R}_2$ with respect to the fixed constellation of charges can be evaluated in terms of $e(\mathbf{R})$.

$$\Delta V = q \, \Delta e = q[e(\mathbf{R}_2) - e(\mathbf{R}_1)]. \tag{5–4}$$

Note that the choice of the datum for e does not affect *changes* in potential energy.   In the following discussion we often choose the datum so that the potential difference of interest is simply e itself.

Suppose, now, that in a time interval dt a current i flows along a path from $\mathbf{R}_1$ to $\mathbf{R}_2$.   The amount of charge transferred from $\mathbf{R}_1$ to $\mathbf{R}_2$ will be

$dq = i \, dt$. The increase in potential energy of these charges will then be $i \, dt[e(\mathbf{R}_2) - e(\mathbf{R}_1)]$, according to (5–4). This increase must be supplied to the charges by some external agent. The increase in potential energy of the charges per unit time is the instantaneous *power P* which must be delivered to the charges by the external agent,

$$P = i \, \Delta e = i[e(\mathbf{R}_2) - e(\mathbf{R}_1)]. \qquad (5\text{–}5)$$

The unit of power is the *watt*, which is one volt-ampere. If Δe is negative, the charges deliver power to the external agent. The derivation of (5–5) is based on a static electric field, for which $e(\mathbf{R})$ is independent of time. The result, however, applies to slowly varying fields, provided all quantities are evaluated at the same instant.

5–2   Passive circuit elements

In this section we consider the capacitor, the resistor, and the inductor as lumped-parameter circuit elements. These three elements are passive

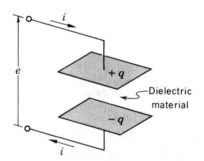

FIG. 5–2.   Parallel-plate capacitor.

in the sense of not incorporating an energy source. Two of the elements, the capacitor and the inductor, are capable of storing energy and yielding up previously stored energy. The resistor dissipates electrical power in the form of heat.

THE CAPACITOR

A capacitor, or *condenser*, is an electric-circuit element whose characteristics can be measured statically in much the same way as the characteristics of a mechanical spring can be measured. A parallel-plate capacitor is shown in Fig. 5–2. The capacitor is said to be uncharged when both plates contain the same quantity of electric charge. The capacitor is said to have acquired a *charge q* when q units of charge have been added to one plate *and* q units of charge have been taken away from the other;

i.e., the charge on a capacitor is one-half the difference between the amounts of charge carried by the two plates. Note that, according to (5–1), the current i flowing through the capacitor is the rate of change of the charge q on the capacitor,

$$i = \frac{dq}{dt}. \tag{5-6}$$

Now, it is found that the process of inducing a charge q on a condenser also creates a *potential difference* e between the plates. The constitutive relation between e and q can be measured statically by applying a known charge and measuring the corresponding voltage. A sketch of such a relation appears in Fig. 5–3. The constitutive relation can be given directly in the form of an equation, $e = e(q)$, or inversely, $q = q(e)$. An alternative description is in terms of a pair of complementary state functions, W_e and W_e^*.

The *electrical energy* $W_e(q)$ stored in a capacitor is the work done in

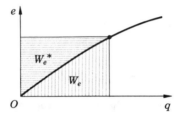

Fig. 5–3. Constitutive relation
for a capacitor.

charging the capacitor from a datum state of no charge to a state with charge q. This work can be computed as the time integral of the power delivered. Using (5–5) and (5–6), we have

$$W_e(q) = \int_0^t ei\,dt = \int_0^q e\,dq. \tag{5-7}$$

The unit of electrical energy is the *joule*, which is one watt-second (a joule is equivalent to 0.73756 ft-lbf). The integral (5–7) is indicated as the area under the curve in Fig. 5–3. When the electrical-energy function is known, the voltage across the capacitor can be recovered by differentiation,

$$\frac{dW_e}{dq} = e. \tag{5-8}$$

For real capacitors, the constitutive relation is often very nearly linear over the operating range of the variables. The linear constitutive relation is written

$$q = Ce, \tag{5-9}$$

where C is called the *capacitance*. The unit of capacitance is the *farad*, which is one coulomb per volt. For a linear capacitor, the electrical energy is

$$W_e(q) = \frac{q^2}{2C}. \tag{5-10}$$

Note that differentiation of (5-10) according to (5-8) leads back to (5-9).

The constitutive relation for a capacitor can also be described in terms of the complementary state function W_e^*, which is related to the electrical energy by a Legendre transformation,

$$W_e^*(e) = qe - W_e(q). \tag{5-11}$$

The function $W_e^*(e)$ is called the *electrical coenergy* of the capacitor. The electrical coenergy is indicated as the area to the left of the curve in Fig. 5-3, and can be represented by the integral,

$$W_e^*(e) = \int_0^e q\, de. \tag{5-12}$$

When the electrical-coenergy function is known, the charge on the capacitor can be obtained by differentiation,

$$\frac{dW_e^*}{de} = q. \tag{5-13}$$

For the *linear* capacitor with the constitutive relation (5-9), the electrical coenergy is

$$W_e^*(e) = \tfrac{1}{2}Ce^2, \tag{5-14}$$

and (5-13) becomes

$$\frac{dW_e^*}{de} = Ce = q. \tag{5-15}$$

THE RESISTOR

The resistor is a lumped-parameter element which resists the flow of current. In order to have a current flow through the element, a voltage must be impressed. The relation between current and voltage in passive resistors is such that the moving charges always lose potential energy. The electrical energy is not stored in a recoverable form as it is in the capacitor, but is transformed into thermal energy.

The constitutive relation for a resistor can be obtained by applying a voltage and measuring the corresponding current (see Fig. 5-4). A *linear* constitutive relation for a resistor is called *Ohm's law*,

$$e = Ri, \tag{5-16}$$

where R is the *resistance*. The unit of resistance is the *ohm*, which is one volt per ampere.

THE INDUCTOR

An inductor, or *coil*, consists of one or more loops in a conducting path. Every closed current-path is a coil of at least one loop and hence possesses in some degree the properties of a coil. The characteristics of an inductor are particularly evident if a coil of many close-packed turns is studied.

The properties of an inductor stem from magnetic-field effects, just as those of a capacitor stem from electric-field effects. When a current flows through a coil, a magnetic field is induced, and the current-path *links* the lines of magnetic flux. A complete description here requires careful investigation of the properties of magnetic fields. We give only

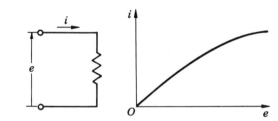

FIG. 5-4.   Constitutive relation for a resistor.

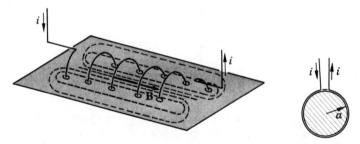

FIG. 5-5.   Iron filings indicate magnetic field **B**.   Cross-sectional area
of coil is πa^2.

a qualitative discussion sufficient to motivate the terminology. In Fig. 5-5 a classical experiment is sketched in which current flows through a coil wound through a sheet of paper. When iron filings are sprinkled on the paper, they line up with the magnetic field and form a pattern of closed curves called *flux lines*. The magnetic-flux density **B** is a vector directed along a flux line with a magnitude proportional to the number of lines which cross a unit area perpendicular to B. The unit of flux density is the *weber per square meter*, which is one volt-second per square meter. The amount of magnetic flux that is linked by a coil is called the *flux linkage* λ. For example, if in Fig. 5-5 the flux density **B** is uniform in the interior of the coil, the flux linked by a single turn is $\pi a^2 B$, and the total flux linkage for a coil of n turns, $n\pi a^2 B$.   The unit of flux linkage is the *weber*, which is one volt-second.

In an ideal inductor the flux linkage λ depends only on the instantaneous current i, as indicated in Fig. 5–6.   In real inductors, particularly those with iron cores, the flux linkage can depend on the history of i, as well as the instantaneous value of i.   When this occurs the coil is said to possess magnetic *hysteresis*.   The effect is analogous to plastic yielding of an elastic spring.

The flux linkage is made accessible in terms of lumped-parameter circuit variables through *Faraday's law*, which states that the voltage developed across an inductor is equal to the rate of change of flux linkage,

$$e = \frac{d\lambda}{dt}. \tag{5-17}$$

The *magnetic energy* $W_m(\lambda)$ stored in an ideal inductor is the work done

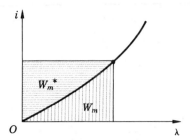

Fig. 5–6.   Constitutive relation for an inductor.

on the inductor when its magnetic state is changed from a condition of no flux linkage to a state with flux linkage λ.   This work can be evaluated as a time integral of the power delivered by using (5–5).   Inserting (5–17), we have

$$W_m(\lambda) = \int_0^t ei\,dt = \int_0^\lambda i\,d\lambda. \tag{5-18}$$

The magnetic energy W_m is indicated as the area below the curve in Fig. 5–6.   When the magnetic-energy function is known, the current in the coil can be recovered by differentiation,

$$\frac{dW_m}{d\lambda} = i. \tag{5-19}$$

If a coil exhibits a linear relation between λ and i,

$$\lambda = Li, \tag{5-20}$$

the coefficient L is called the *inductance*.   The unit of inductance is the *henry*, which is one weber per ampere, or one volt-second per ampere. For a linear inductor the magnetic energy is

$$W_m(\lambda) = \frac{\lambda^2}{2L}. \tag{5-21}$$

Note that differentiation of (5–21) according to (5–19) leads back to (5–20).

A *magnetic-coenergy* state-function W_m^* is defined by the Legendre transformation,

$$W_m^*(i) = \lambda i - W_m(\lambda). \tag{5-22}$$

The magnetic coenergy is represented by the area to the left of the curve in Fig. 5–6 and can be evaluated as follows:

$$W_m^*(i) = \int_0^i \lambda \, di. \tag{5-23}$$

The derivative of the magnetic coenergy yields the flux linkage,

$$\frac{dW_m^*}{di} = \lambda. \tag{5-24}$$

In the linear case, the magnetic coenergy is

$$W_m^*(i) = \tfrac{1}{2}Li^2, \tag{5-25}$$

and (5–24) becomes

$$\frac{dW_m^*}{di} = Li = \lambda. \tag{5-26}$$

MAGNETIC-FIELD RELATIONS

In our discussion of electric-circuit dynamics we use flux linkage λ as a convenient variable, but beyond this we make no direct use of magnetic-field properties. In Chap. 6, however, we study transducers which couple mechanical and electrical variables by way of magnetic-field interactions. In preparation for this we state two additional properties of a magnetic field.

Faraday's law states that the voltage developed across a coil equals the rate of change of flux linkage. As applied to an inductor, the coil stands still and the strength of the field changes with time. In a transducer the situation is often reversed: the field is stationary, and the coil moves so as to change the flux linkage. A basic relation which follows from Faraday's law is illustrated in Fig. 5–7. Here a small element of a coil moving in a magnetic field is pictured. The sense of the vector $d\mathbf{l}$ along the length of the element is determined by the direction of current flow. It is a consequence of Faraday's law that the increment in voltage, *de*, in the direction of the current flow induced by the motion is

$$de = \mathbf{v} \times \mathbf{B} \cdot d\mathbf{l}. \tag{5-27}$$

A second relation follows from the Lorentz-force law (2–87), which gives the force on a single charged particle due to motion in a magnetic field. If we consider that a current flow is made up of a very large number of charged particles moving along a conductor, we can deduce that the total force *of the field* acting on the conducting element of Fig. 5–7 is

$$d\mathbf{f} = i\, d\mathbf{l} \times \mathbf{B}. \tag{5-28}$$

The relations (5–27) and (5–28) represent the fundamental principles underlying many transducers, including electric generators and motors. Because of the magnetic field, mechanical motion induces voltage (5–27), and current flow exerts force (5–28).   If an external mechanical agent

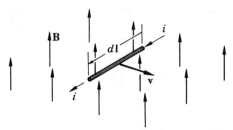

Fig. 5-7. Conducting element *dl* carries current *i* and moves with velocity **v** through magnetic field with flux density **B**.

maintains the velocity of the conductor and an external electric circuit maintains the current flow in the conductor, it can be shown[1] from (5–27) and (5–28) that the mechanical power delivered (or absorbed) by the external agent is equal at every instant to the electric power absorbed (or delivered) by the external electric circuit.   The moving conducting element in a magnetic field thus acts as an ideal energy-conversion device.

5-3   Electrical sources

The passive circuit elements store or dissipate energy.   Energy can be supplied to a circuit by a battery, by a generator, or by another electric circuit.   The characteristics of these energy sources can vary widely. For the purpose of analysis it is convenient to introduce two idealized sources, an ideal voltage source and an ideal current source.   It is then

[1] See Prob. 5–18.

usually possible to model the behavior of a real source by combining an ideal source with appropriately selected passive elements.

An *ideal voltage source* $E(t)$ produces a prescribed time history of voltage difference across its terminals independently of the current which flows through it. An *ideal current source* $I(t)$ produces a prescribed time history of current flow through it independently of the voltage difference across its terminals. These ideal sources can be approximated by real devices over limited ranges of voltage and current. An ideal source can-

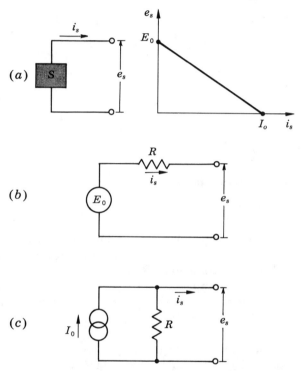

FIG. 5–8. (a) Real source with measured characteristic. (b) Ideal voltage-source model.   (c) Ideal current-source model.

not, however, be completely realized because infinite power capability is implied by the idealization. Thus, if a resistance R were connected across the terminals of an ideal voltage source and then the value of R were made to approach zero, the source would maintain its prescribed voltage, while the current supplied by the source would approach infinity. According to (5–5), the power delivered would also approach infinity.

In Fig. 5–8 we show how a real-source characteristic can be modeled by ideal sources combined with resistors. The real source has a linear characteristic. It has an open-circuit voltage E_0. When current flows, the voltage across the terminals decreases until, when the terminals are short-circuited, the voltage is zero and a maximum current I_0 is flowing. In Fig. 5–8b, an ideal voltage source with constant voltage E_0 is connected in series with a resistance R. Since the voltage e_s of the real source can be written

$$e_s = E_0 - \frac{E_0}{I_0} i_s, \tag{5–29}$$

we see that the voltage e_s in Fig. 5–8b will have this same dependence on i_s, provided that $R = E_0/I_0$. Similarly, in Fig. 5–8c, an ideal current source with constant current I_0 is connected in parallel with a resistance R. Since the current i_s of the real source can be written

$$i_s = I_0 - \frac{I_0}{E_0} e_s, \tag{5–30}$$

we see that the current i_s in Fig. 5–8c will have this same dependence on e_s, provided that $R = E_0/I_0$.

5–4 Interconnection laws

Electrical networks are constructed by connecting passive elements and sources. The electrical variables describing each element are constrained by the constitutive relations of the individual element. In addition, these variables must satisfy the fundamental interconnection relations for circuits embodied in Kirchhoff's laws.

Kirchhoff's *current law* applies to every junction, or *node*, of the circuit. It states that the sum of the currents flowing into any node must be zero. Kirchhoff's current law is essentially a statement of conservation of electric charge.

Kirchhoff's *voltage law* applies to every closed loop, or *mesh*, of the circuit. It states that the sum of the voltage drops across every element encountered in a complete trip around any mesh must be zero. Kirchhoff's voltage law assures that the voltage (i.e., potential) at any point in a circuit is independent of the path followed in reaching that point.

With the statement of Kirchhoff's laws, we have concluded an enumeration of the governing principles underlying the dynamic behavior of electric circuits. These requirements may be divided into categories, as follows:

1. *Requirements on currents.* These include Kirchhoff's current law and relations between current variables i_j and charge variables q_j of

the form

$$\frac{dq_j}{dt} = i_j.$$

2. *Requirements on voltages.* These include Kirchhoff's voltage law and relations between voltage variables e_j and flux-linkage variables λ_j of the form

$$\frac{d\lambda_j}{dt} = e_j.$$

3. *Constitutive relations.* These include the constitutive relations for the passive elements and the characteristics of the sources.

The direct method of formulating dynamic "equations of motion" for an electric circuit consists in stating each and every one of the above requirements in analytical form, and then combining the equations, eliminating

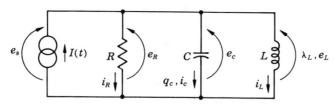

FIG. 5–9.   Example 5–1.   *RLC* circuit with ideal current source.

variables where possible.   Two possible systematic procedures for carrying this out are described below.

In the first procedure a set of independent charge variables q_j is selected, and the first-category requirements are used to express all other charges and all currents in terms of the q_j.   Next the constitutive relations are used to express voltages or flux linkages across all circuit elements in terms of the q_j.   Finally, when the second-category requirements are imposed on these voltages and flux linkages, the result is a complete set of differential equations for the q_j.

The second procedure is complementary to the first.   A set of independent flux-linkage variables λ_j is selected, and the second-category requirements are used to express all other flux linkages and all voltages in terms of the λ_j.   Next the constitutive relations are used to express the currents or charges in all circuit elements in terms of the λ_j.   Finally, when the first-category requirements are imposed on these currents and charges, the result is a complete set of differential equations for the λ_j.

Example 5–1.   Consider the circuit of Fig. 5–9, which contains an ideal current source $I(t)$, a linear resistor with resistance R, a linear capacitor with capacitance C, and a linear inductor with inductance L.

We shall formulate dynamic equations for this circuit by following the second of the two procedures described above.

We begin by examining the second-category requirements, i.e., Kirchhoff's voltage law and the voltage–flux-linkage relation for the inductor. When the circuit of Fig. 5–9 is sketched as in Fig. 5–10, it is clear that the circuit has only two nodes. Kirchhoff's voltage law will be satisfied if we assign the potential zero to one of these nodes, and an unknown potential e to the other node. Then the potential is everywhere single-valued, and the sum of the potential drops around any closed loop is zero.

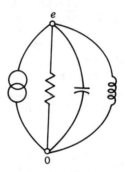

FIG. 5–10. Node potentials consistent with Kirchhoff's voltage law.

Faraday's law (5–17) for the inductor requires that

$$\frac{d\lambda_L}{dt} = e_L. \tag{5-31}$$

In this case only a single independent flux-linkage variable λ is required. All the second-category requirements will be met provided that we take

$$\lambda_L = \lambda$$
$$e = e_s = e_R = e_C = e_L = \frac{d\lambda}{dt}. \tag{5-32}$$

Next we use the constitutive relations for the passive elements to express currents and charges in terms of λ. For the resistor, Ohm's law (5–16) yields

$$i_R = \frac{1}{R} e_R = \frac{1}{R} \frac{d\lambda}{dt}. \tag{5-33}$$

For the capacitor, the linear constitutive relation (5–9) yields

$$q_C = Ce_C = C \frac{d\lambda}{dt}. \tag{5-34}$$

For the inductor, the linear constitutive relation (5–20) yields

$$i_L = \frac{1}{L}\lambda_L = \frac{1}{L}\lambda. \tag{5-35}$$

Finally, we impose the first-category requirements on the currents and charges. For the capacitor, the charge and current are related by (5–6). Thus

$$i_C = \frac{dq_C}{dt} = C\frac{d^2\lambda}{dt^2}. \tag{5-36}$$

Kirchhoff's current law requires that the sum of the currents flowing into any node be zero. In Fig. 5–10 there are only two nodes. Furthermore, if the net current entering the top node is zero, then the net current leaving the bottom node must also be zero. The first-category requirements will therefore be met if we ensure that the currents (5–33), (5–35), and (5–36), together with the prescribed current $I(t)$ from the current source, satisfy Kirchhoff's current law at the top node. Taking into account the directions indicated on Fig. 5–9, we write

$$C\frac{d^2\lambda}{dt^2} + \frac{1}{R}\frac{d\lambda}{dt} + \frac{1}{L}\lambda = I(t) \tag{5-37}$$

as the dynamic equation of the network in terms of the flux-linkage variable λ. A solution for $\lambda(t)$ can be obtained by integrating (5–37) for a given $I(t)$ subject to given initial conditions. Once $\lambda(t)$ is available, the time history of all other variables can be obtained by substitution in (5–32) to (5–36).

Example 5–2. It is instructive to rework Example 5–1 using the alternative procedure, in which independent charge variables are employed. In this procedure we begin by considering the charge and current requirements which are Kirchhoff's current law,

$$I(t) = i_R + i_C + i_L, \tag{5-38}$$

and the charge-current relation (5–6) for the capacitor,

$$\frac{dq_C}{dt} = i_C. \tag{5-39}$$

Equation (5–38) is one constraint on three currents. There are thus two independent currents. Suppose we select i_R and i_C to be independent and use (5–38) to determine i_L in terms of i_R and i_C. The capacitor current i_C is represented in (5–39) as the time derivatives of the capacitor charge q_C. Let us introduce an additional charge variable q_R, *defined* by

$$\frac{dq_R}{dt} = i_R. \tag{5-40}$$

The variable q_R can be considered to be a *generalized* charge variable. It does not represent an accumulation of charge stored anywhere in the circuit. It is simply a convenient variable from which the resistor current can be obtained by differentiation. The current and charge requirements will all be met if we take q_C and q_R as independent charge variables, provided we use (5–39) and (5–40) and

$$i_L = I(t) - \frac{dq_R}{dt} - \frac{dq_C}{dt}, \tag{5–41}$$

which follows from (5–38), as expressions for the currents in terms of these charge variables.

Next we employ the constitutive relations to express potentials and flux linkages in terms of the charge variables. For the resistor, Ohm's law (5–16) yields

$$e_R = Ri_R = R \frac{dq_R}{dt}. \tag{5–42}$$

For the capacitor, the linear constitutive relation (5–9) yields

$$e_C = \frac{1}{C} q_C. \tag{5–43}$$

For the inductor, the linear constitutive relation (5–20) yields

$$\lambda_L = Li_L = L \left[I(t) - \frac{dq_R}{dt} - \frac{dq_C}{dt} \right]. \tag{5–44}$$

It remains to impose the voltage and flux-linkage requirements on (5–42) to (5–44). Faraday's law (5–17) requires that

$$e_L = \frac{d\lambda_L}{dt} = L \left[\frac{dI}{dt} - \frac{d^2q_R}{dt^2} - \frac{d^2q_C}{dt^2} \right]. \tag{5–45}$$

Kirchhoff's voltage law requires that

$$e_L = e_C \quad \text{and} \quad e_L = e_R, \tag{5–46}$$

in order for the upper node in Fig. 5–10 to have a unique potential with respect to the lower node. Substitution from (5–42), (5–43), and (5–45) into (5–46) leads to the following pair of dynamic equations for the network of Fig. 5–9:

$$L \frac{d^2q_C}{dt^2} + \frac{1}{C} q_C + L \frac{d^2q_R}{dt^2} = L \frac{dI}{dt}$$
$$\tag{5–47}$$
$$L \frac{d^2q_C}{dt^2} + L \frac{d^2q_R}{dt^2} + R \frac{dq_R}{dt} = L \frac{dI}{dt}.$$

These are a pair of coupled differential equations for the two charge variables q_C and q_R.[1] Given a particular $I(t)$ and a particular set of initial conditions, it is possible to solve for $q_C(t)$ and $q_R(t)$ by integrating (5–47).   Once $q_C(t)$ and $q_R(t)$ are available, the time history of any other variable can be obtained by substitution in (5–39) to (5–45).

It is interesting to compare the complementary solution procedures of Examples 5–1 and 5–2.   The same requirements are met, but in opposite sequence.   Note that for this circuit the formulation in terms of a flux-linkage variable resulted in a single second-order differential equation, whereas the formulation in terms of charge variables resulted in a pair of coupled second-order differential equations.   The root of this discrepancy can be traced to the topology of Fig. 5–10.   There is only one independent node, and hence only one independent node voltage (and one corresponding flux-linkage variable).   There are, however, two independent loops, or meshes, and hence two independent branch currents (and two corresponding charge variables).   In a network of greater complexity than this, the topology of the network can also be used in this manner to indicate which analysis procedure will lead to the smaller number of final equations.

5–5 Electromechanical analogies

The *RLC* network of Examples 5–1 and 5–2 is analogous to the mass-spring-dashpot mechanical system of Example 2–9, in the sense that the dynamic behavior of the electrical system is described by a differential equation (5–37), which is completely analogous to the equation (2–136) describing the dynamic behavior of the mechanical system.   The analogous elements and analogous variables are displayed in corresponding positions in the symbolic diagrams of Fig. 5–11.   Note that each element in Fig. 5–11 connects points labeled with the variables which enter into the constitutive relation for that element.   For example, the constitutive relation for the capacitor is $q = Ce$, while the analogous mechanical element, the newtonian mass particle, has the velocity-momentum relation $p = mv$.

Based on Fig. 5–11, we can say that for any linear network there is an analogous linear mechanical system[2] in which all motion is along a single axis.   Ideal voltage and current sources in the electrical network are modeled by prescribed velocity sources and prescribed force loadings.   Not only will the analogous systems have analogous elements as indicated by Fig. 5–11, but the elements will be connected in such a way that the systems are topologically equivalent.   This follows from the fact

---

[1] See Prob. 5–8.

[2] See Prob. 5–11.

that the analogy also extends to the interconnection laws in electrical and mechanical systems.

In electric circuits Kirchhoff's current law requires that the sum of all the currents entering a node must be zero. In mechanical systems at every mechanical node (i.e., at every massless junction) there must be equilibrium of forces.   The net force acting on the junction must be zero.

Kirchhoff's voltage law requires that the sum of all the voltage drops around a closed circuit must be zero.   In mechanical systems geometric compatibility requires that the sum of the velocity differences across elements in any connected closed chain must be zero.

The analogy just described is not strictly limited to systems with linear

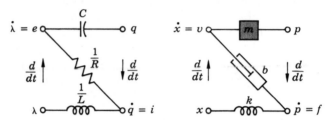

FIG. 5–11.   Analogy between electrical networks and translational mass-spring-dashpot systems.

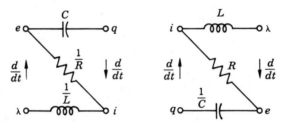

FIG. 5–12.   Dual analogy for electrical circuits.

electrical and mechanical elements.   In principle, there could be analogous nonlinear constitutive relations; e.g., a nonlinear resistor could be analogous to a nonlinear dashpot.   The analogous variables employed would not be affected, nor would the interconnection laws.

In addition to the analogy just described, there are a number of other possibilities.   For example, one electric circuit can be analogous to another (different) electric circuit in the sense that their dynamics are described by analogous differential equations.   The basis for this so-called *dual analogy* is indicated schematically in Fig. 5–12, where analogous elements and analogous variables occupy corresponding positions in the two diagrams.   In this analogy the roles of the two Kirchhoff laws are interchanged.   The analogous circuits have complementary connection topologies in that nodes in one circuit are analogous to meshes in the other. Because of this, a dual analogy exists for a given network only when the network topology is such that a network with complementary topology can be found.   Such a network cannot be found[1] if the original network is *nonplanar,* i.e., if the network diagram cannot be drawn on a plane without crossing connections.   There also exists a similar dual analogy for mechanical systems,[2] in which the roles of springs and masses are interchanged.

There is a well-known electromechanical analogy based on the analogy between the right-hand diagrams of Figs. 5–11 and 5–12.   In this analogy the "potential"-energy storage elements, the mechanical spring and the electrical capacitor, are analogous, and the "kinetic"-energy storage elements, the mass particle and the inductor, are analogous.   This analogy is particularly apt in describing the behavior of the individual elements. It is not, however, so well adapted to describing the behavior of circuits and systems as a whole, because of the interchange in connection topology that is involved.   The electrical *node* law is analogous to the mechanical *mesh* law, and vice versa.   Furthermore, the analogy breaks down for a circuit or system that is nonplanar.

5–6   Hamilton's principle for electrical networks

In Sec. 5–4 the basic requirements of network dynamics are divided into three categories.   Two direct procedures for obtaining dynamic equations consistent with these requirements are described.   We now discuss two indirect procedures for reaching the same goal.   These indirect procedures are variational methods based on electrical extensions of Hamilton's principle.

The manner in which the variational methods indirectly meet the three categories of requirements can be outlined as follows: Hamilton's principle provides a variational indicator.   The constitutive relations for the network are employed in constructing the indicator.   This takes care of the third category of requirements.   Admissible variations of the indicator are then considered.   The restrictions on admissible variations are equivalent to one of the first two categories of requirements.   Finally, the vanishing of the indicator for arbitrary variations is equivalent to meeting the requirements of the remaining category.

There are two forms of Hamilton's principle for electric circuits, and two corresponding variational indicators.   In one form the independent

[1] See, for example, E. C. Cherry, The Duality between Interlinked Electric and Magnetic Circuits and the Formation of Transformer Equivalent Circuits, *Proc. Phys. Soc. (London),* **B62**:101–111 (1949).

[2] See Prob. 5–12.

variables are *charges*, and in the other form the independent variables are
flux linkages. These two forms correspond with the two direct methods
for formulating dynamic equations for networks. The validity of Hamilton's principle in these forms can be proved by an all-electrical development paralleling Sec. 2–11, or more simply, by using the analogies of the previous section to pass directly from the results of Sec. 2–11 to the corresponding electrical counterparts.

When charge variables are employed, *Hamilton's principle* for an electrical network states that an admissible dynamic "motion" of the charges from a fixed configuration at $t = t_1$ to another fixed configuration at $t = t_2$ is a natural "motion" if, and only if, the variational indicator

$$\text{V.I.} = \int_{t_1}^{t_2} \left[\delta(W_m^* - W_e) + \sum_j e_j \, \delta q_j \right] dt \tag{5-48}$$

vanishes for arbitrary admissible variations in the currents and charges.

In (5–48) the symbol W_m^* represents the *sum* of the magnetic coenergies of the individual inductors in the network, and W_e represents the *sum* of the electrical energies of the individual capacitors. These network statefunctions are functions of currents and charges which, in order to be admissible, must satisfy the first-category requirements of Sec. 5–4; i.e., the currents must satisfy Kirchhoff's current law, and corresponding current and charge variables must satisfy relations of the form

$$\frac{dq_j}{dt} = i_j. \tag{5-49}$$

The summation on j in (5–48) is a *work expression* which accounts for the increments of work done by all nonconservative elements in the network under an admissible variation of the charges. Further discussion of the work expression follows shortly. The agreement in Hamilton's principle that we consider fixed configurations at $t = t_1$ and $t = t_2$ implies that admissible variations δq_j, while arbitrary in the interior of the interval, must vanish at the end points in time. The requirement that the variational indicator (5–48) vanish for arbitrary admissible variations under these circumstances is equivalent to the imposition of the second-category requirements of Sec. 5–4.

When flux-linkage variables are employed, *Hamilton's principle* for an electrical network states that an admissible dynamic "motion" of the flux linkages from a fixed configuration at $t = t_1$ to another fixed configuration at $t = t_2$ is a natural "motion" if, and only if, the variational indicator

$$\text{V.I.} = \int_{t_1}^{t_2} \left[\delta(W_e^* - W_m) + \sum_j i_j \, \delta\lambda_j \right] dt \tag{5-50}$$

vanishes for arbitrary admissible variations in the voltages and flux linkages.

In (5–50) the symbol W_e^* represents the *sum* of the electrical coenergies of the individual capacitors, and W_m represents the *sum* of the magnetic energies of the individual inductors.   These network state-functions are functions of voltages and flux linkages which, in order to be admissible, must satisfy the second-category requirements of Sec. 5–4; i.e., the voltages must satisfy Kirchhoff's voltage law, and corresponding voltage and flux-linkage variables must satisfy relations of the form

$$\frac{d\lambda_j}{dt} = e_j. \tag{5–51}$$

The summation on j in (5–50) is a *work expression* which accounts for the increments of work done by all nonconservative elements in the network under an admissible variation of the flux linkages.   Variations $\delta\lambda_j$ must

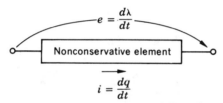

FIG. 5–13.   Work increment delivered to network by element is $e\delta q$ or $i\delta\lambda$.

satisfy the second-category requirements in order to be admissible, and, by agreement, they must vanish at the end points in time: $t = t_1$ and $t = t_2$.   The requirement that the indicator (5–50) vanish for arbitrary admissible variations under these conditions is an indirect alternative to directly imposing the first-category requirements of Sec. 5–4.

WORK EXPRESSIONS

In the variational indicators (5–48) and (5–50) it is necessary to account for the work done by every nonconservative element in the network under an admissible variation.   To clarify these work expressions we consider a single nonconservative network element in Fig. 5–13.   The element might be a resistor or an ideal source.   The current i *through* the element and the potential difference e *across* the element are indicated.   In addition to the voltage and current, let us introduce generalized charge and flux-linkage variables, q and λ, as follows:

$$\frac{dq}{dt} = i \qquad \frac{d\lambda}{dt} = e. \tag{5–52}$$

Now, according to (5–5), the power delivered to the network by the element is *ei*. The work delivered during a time *dt* is

$$ei\,dt = e\,dq = i\,d\lambda. \qquad (5\text{–}53)$$

Thus the increment of work done by the element under a differential change in *q* (or λ) is given by *e dq* (or *i dλ*).

In Hamilton's principle, we consider admissible variations in charges (or flux linkages) which entail infinitesimal variations in the generalized charge *q* (or generalized flux linkage λ) defined by (5–52). The increment of work done by the element is thus *e δq* (or *i δλ*). These work terms should be positive if the element supplies energy to the circuit under the variation, and should be negative if the element absorbs energy under the variation.

For example, suppose the element of Fig. 5–13 were a linear resistor with resistance *R*. Then, for the directions indicated in Fig. 5–13,

$$e = -Ri, \qquad (5\text{–}54)$$

and the contribution to the work expression of (5–48) for such a resistor would be

$$e\,\delta q = -Ri\,\delta q = -R\,\frac{dq}{dt}\,\delta q. \qquad (5\text{–}55)$$

When (5–50) is used, the contribution to the work expression for the same resistor will be

$$i\,\delta\lambda = -\frac{1}{R}\,e\,\delta\lambda = -\frac{1}{R}\frac{d\lambda}{dt}\,\delta\lambda. \qquad (5\text{–}56)$$

As another example, suppose that the element of Fig. 5–13 were an ideal voltage source *E(t)*. Then the contribution to the work expression of (5–48) would be

$$e\,\delta q = E(t)\,\delta q. \qquad (5\text{–}57)$$

It is important to note that when (5–50) is used there is *no contribution* to the work expression from an ideal voltage source. This is because the voltage across such a source at a given instant in time is *prescribed*, and thus cannot be altered in an admissible variation of voltages and flux linkages. This implies that δλ in Fig. 5–13 is identically zero for an ideal voltage source. Voltage sources *do* enter the formulation, using (5–50), as part of the admissibility requirements on voltages and flux linkages, but they do not appear in the work expression.

Similarly, if the element in Fig. 5–13 were an ideal current source *I(t)*, the contribution to the work expression of (5–50) would be

$$i\,\delta\lambda = I(t)\,\delta\lambda. \qquad (5\text{–}58)$$

There is *no contribution* to the work expression of (5–48) from an ideal current source because there can be no variation in the current or the corresponding generalized charge when the current is *prescribed*.

Example 5–3.   To illustrate the application of Hamilton's principle we consider again the circuit of Fig. 5–9.   We shall apply the principle, using flux-linkage variables and the variational indicator (5–50).   Admissible voltages and flux linkages must satisfy Kirchhoff's voltage law, together with relations of the form (5–51).   For the circuit of Fig. 5–9, these requirements are met if all voltages and flux linkages are expressed in terms of a single generalized flux-linkage variable λ, according to (5–32).

We then proceed to construct the variational indicator (5–50).   The electrical coenergy of the capacitor (5–14) is

$$W_e^* = \tfrac{1}{2}Ce_C^2 = \tfrac{1}{2}C\left(\frac{d\lambda}{dt}\right)^2, \tag{5–59}$$

and the magnetic energy of the inductor (5–21) is

$$W_m = \frac{1}{2}\frac{\lambda_L^2}{L} = \frac{1}{2}\frac{\lambda^2}{L}. \tag{5–60}$$

The work expression contains contributions from the resistor and the ideal current source.   Using (5–56) and (5–58), we have

$$\sum_j i_j\,\delta\lambda_j = -\frac{1}{R}\frac{d\lambda}{dt}\,\delta\lambda + I(t)\,\delta\lambda. \tag{5–61}$$

Assembling (5–59) to (5–61), we obtain the variational indicator

$$\text{V.I.} = \int_{t_1}^{t_2}\left\{\delta\left[\tfrac{1}{2}C\left(\frac{d\lambda}{dt}\right)^2 - \frac{1}{2}\frac{\lambda^2}{L}\right] + \left[-\frac{1}{R}\frac{d\lambda}{dt} + I(t)\right]\delta\lambda\right\}\,dt. \tag{5–62}$$

The dynamic equation of "motion" for the network is obtained from the requirement that (5–62) must vanish for arbitrary admissible variations, i.e., arbitrary histories of $\delta\lambda$.   When we carry out the indicated variations in (5–62) and integrate the first term by parts, we find

$$\text{V.I.} = C\frac{d\lambda}{dt}\,\delta\lambda\,\bigg]_{t_1}^{t_2} - \int_{t_1}^{t_2}\left[C\frac{d^2\lambda}{dt^2} + \frac{1}{L}\lambda + \frac{1}{R}\frac{d\lambda}{dt} - I(t)\right]\delta\lambda\,dt. \tag{5–63}$$

Because of the agreement that variations should vanish at the end points in time, the first expression on the right of (5–63) is zero.   The necessary condition for the integral to vanish for arbitrary $\delta\lambda$ is that the contents of the square brackets must be zero.   This is equivalent to requiring that λ must satisfy the following differential equation:

$$C\frac{d^2\lambda}{dt^2} + \frac{1}{R}\frac{d\lambda}{dt} + \frac{1}{L}\lambda = I(t). \tag{5–64}$$

Note that this result of applying Hamilton's principle is identical with (5–37), obtained by direct application of the three categories of requirements listed in Sec. 5–4.

5–7 Generalized coordinates and Lagrange's equations

An important part of any variational method involves the recognition and implementation of the admissibility conditions. For mechanical systems, several techniques for ensuring geometric admissibility are described in Secs. 2–9 and 2–10. Generally, the most convenient technique consists in selecting a complete set of independent generalized coordinates such that the geometric requirements are automatically satisfied by any combination of values of the generalized coordinates. For holonomic systems the variations of these coordinates also provide a complete set of independent variations. In such systems the admissibility requirements are taken care of, once and for all, by the choice of an appropriate set of generalized coordinates.

For electrical networks it is also convenient to employ generalized coordinates which automatically take care of the admissibility requirements. When flux-linkage variables are used, a systematic procedure for selecting generalized coordinates is to take *independent node potentials* to be time derivatives of generalized flux-linkage coordinates λ_k. This technique is illustrated in Example 5–4, to follow. When charge variables are used, a systematic procedure for selecting generalized coordinates is to take *independent loop currents* to be time derivatives of generalized charge coordinates q_k. This technique is also illustrated in Example 5–4.

Once generalized coordinates have been selected, the next step in the variational method consists in evaluating a lagrangian. When flux-linkage coordinates are used, the lagrangian is

$$\mathcal{L}(\dot{\lambda}_k, \lambda_k) = W_e^* - W_m, \tag{5–65}$$

where the electrical coenergy W_e^* of the capacitors is a function of the time derivatives of the generalized coordinates, and the magnetic energy W_m of the inductors is a function of the coordinates themselves. When charge variables are employed, the lagrangian is

$$\mathcal{L}(\dot{q}_k, q_k) = W_m^* - W_e, \tag{5–66}$$

where the magnetic coenergy of the inductors is a function of the time derivatives of the generalized coordinates, and the electrical energy W_e of the capacitors is a function of the coordinates themselves.

When a lagrangian expressed in generalized coordinates is available, the most direct route to the equations of "motion" is by application of Lagrange's equations. In Sec. 2–11 it is shown that these equations are the necessary conditions for the vanishing of the variational indicator of Hamilton's principle. When using n independent generalized flux-linkage coordinates λ_k, *Lagrange's equations* take the form

$$\frac{d}{dt}\left(\frac{\partial \mathcal{L}}{\partial \dot{\lambda}_k}\right) - \frac{\partial \mathcal{L}}{\partial \lambda_k} = I_k \qquad k = 1, \ldots, n, \qquad (5\text{–}67)$$

where the *generalized currents* I_k corresponding to the generalized flux-linkage variables are defined by the work expression

$$\sum_k I_k \, \delta\lambda_k = \sum_j i_j \, \delta\lambda_j \qquad (5\text{–}68)$$

in which the summation over j represents the total work expression for all the nonconservative elements. When using m independent generalized charge coordinates q_k, *Lagrange's equations* take the form

$$\frac{d}{dt}\left(\frac{\partial \mathcal{L}}{\partial \dot{q}_k}\right) - \frac{\partial \mathcal{L}}{\partial q_k} = E_k \qquad k = 1, \ldots, m, \qquad (5\text{–}69)$$

where the *generalized voltages* E_k corresponding to the generalized charge variables are defined by the work expression

$$\sum_k E_k \, \delta q_k = \sum_j e_j \, \delta q_j \qquad (5\text{–}70)$$

in which the summation over j represents the total work expression for all the nonconservative elements.

Example 5–4. To illustrate the selection of generalized coordinates we consider the network of Fig. 5–14a. When using *flux-linkage variables* the admissibility requirements include Kirchhoff's voltage law, together with relations of the form $\lambda_j = e_j$. In Fig. 5–14b, the circuit is redrawn to emphasize that there are three independent nodes. Kirchhoff's voltage law is satisfied if we take one of these nodes to be the datum for potential and assign (unknown) potentials e_1 and e_2 to the other two nodes. Then potential is single-valued and the sum of the voltage drops around any closed loop is zero. The generalized flux-linkage variables λ_1 and λ_2 are then defined by

$$\frac{d\lambda_1}{dt} = e_1 \qquad \frac{d\lambda_2}{dt} = e_2. \qquad (5\text{–}71)$$

In this circuit λ_1 can be identified as the flux linkage of the inductor L_1, and λ_2 can be identified as the flux linkage of the inductor L_2. It is not, however, necessary for generalized flux-linkage coordinates to be always identifiable as flux linkages of actual inductors. The admissibility requirements are automatically satisfied when we express the potential difference across any element in the circuit in terms of λ_1 and λ_2. For example, the voltage across the capacitor C_1 is $\dot{\lambda}_2 - \dot{\lambda}_1$. By using (5–14) and (5–21), we find the lagrangian (5–65) to be

$$\mathcal{L} = \tfrac{1}{2}C_1(\lambda_2 - \lambda_1)^2 + \tfrac{1}{2}C_2\lambda_2^2 + \frac{\lambda_1^2}{2L_1} + \frac{\lambda_2^2}{2L_2}. \tag{5-72}$$

The work expression for the network follows from (5–56) and (5–68).

$$\sum_{k=1}^{2} I_k \, \delta\lambda_k = -\frac{1}{R_1}\dot{\lambda}_1 \, \delta\lambda_1 - \frac{1}{R_2}\dot{\lambda}_2 \, \delta\lambda_2. \tag{5-73}$$

Because of the independence of the generalized coordinates, we can

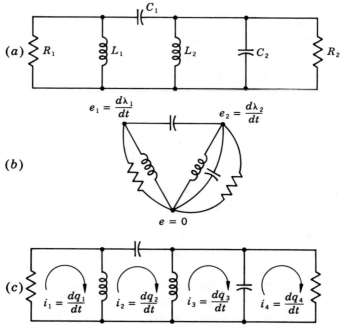

$$e_1 = \frac{d\lambda_1}{dt} \qquad e_2 = \frac{d\lambda_2}{dt}$$

$$e = 0$$

$$i_1 = \frac{dq_1}{dt} \qquad i_2 = \frac{dq_2}{dt} \qquad i_3 = \frac{dq_3}{dt} \qquad i_4 = \frac{dq_4}{dt}$$

Fig. 5–14.   Example 5–4.   (a) Given circuit.   (b) Generalized flux-linkage coordinates.   (c) Generalized charge coordinates.

identify the generalized currents I_1 and I_2 from (5–73).

$$I_1 = -\frac{\lambda_1}{R_1} \qquad I_2 = -\frac{\lambda_2}{R_2}. \tag{5-74}$$

Lagrange's equations (5–67) can now be constructed from (5–72) and (5–74).

$$\frac{d}{dt}\left[-C_1(\lambda_2 - \dot{\lambda}_1)\right] + \frac{\lambda_1}{L_1} = -\frac{\dot{\lambda}_1}{R_1}$$

$$\frac{d}{dt}\left[C_1(\lambda_2 - \dot{\lambda}_1) + C_2\lambda_2\right] + \frac{\lambda_2}{L_2} = -\frac{\dot{\lambda}_2}{R_2}. \tag{5-75}$$

These are a pair of coupled differential equations which describe the dynamics of the circuit. It is interesting to note that the first of (5–75) can be interpreted as a statement of Kirchhoff's current law applied to the node to which we assigned the coordinate λ_1. Similarly, the second of (5–75) represents Kirchhoff's current law applied to the node to which we assigned the coordinate λ_2.

GENERALIZED CHARGE VARIABLES

When using charge coordinates, the admissibility requirements include Kirchhoff's current law, together with relations of the form $\dot{q}_j = i_j$. In Fig. 5–14c, four *loop currents*, i_1, i_2, i_3, and i_4, are indicated. This set of loops is independent and also complete in the sense that any other loop in the network can be built up from combinations of these. The net current flowing in any branch of the network is the sum of the loop currents which pass through that branch. For example, the upward current through R_1 is i_1, and the upward current through L_1 is $i_2 - i_1$. Kirchhoff's current law is automatically satisfied at every node for any set of values of the loop currents. This follows from the fact that, at any node through which a loop current passes, there is automatically an equality between current entering and leaving, on that loop. Since the total currents are simply sums of loop currents, the total current entering a node must be equal to the total current leaving that node.

The generalized charge coordinates q_1, q_2, q_3, and q_4 are then defined by

$$\frac{dq_1}{dt} = i_1 \qquad \frac{dq_2}{dt} = i_2 \qquad \frac{dq_3}{dt} = i_3 \qquad \frac{dq_4}{dt} = i_4. \qquad (5\text{–}76)$$

In this circuit q_2 can be identified as the charge on the capacitor C_1. The other charge coordinates are simply convenient variables defined by (5–76). The charge on the capacitor C_2 can be represented by $q_3 - q_4$. The admissibility requirements are automatically satisfied when we express the current in any branch of the network in terms of the q_k. For example, the current through the inductor L_2 is $\dot{q}_2 - \dot{q}_3$. By using (5–25) and (5–10), we find the lagrangian (5–66) to be

$$\mathcal{L} = \tfrac{1}{2}L_1(\dot{q}_1 - \dot{q}_2)^2 + \tfrac{1}{2}L_2(\dot{q}_2 - \dot{q}_3)^2 - \frac{q_2^2}{2C_1} - \frac{(q_3 - q_4)^2}{2C_2}. \qquad (5\text{–}77)$$

Using (5–55) and (5–70), we find the work expression to be

$$\sum_{k=1}^{4} E_k \, \delta q_k = -R_1\dot{q}_1 \, \delta q_1 - R_2\dot{q}_4 \, \delta q_4. \qquad (5\text{–}78)$$

Because of the independence of the generalized coordinates, we can identify the generalized voltages,

$$E_1 = -R_1\dot{q}_1 \qquad E_2 = 0 \qquad E_3 = 0 \qquad E_4 = -R_2\dot{q}_4. \qquad (5\text{-}79)$$

Lagrange's equations (5–69) can now be constructed from (5–77) and (5–79).

$$\frac{d}{dt}[L_1(\dot{q}_1 - \dot{q}_2)] = -R_1\dot{q}_1$$

$$\frac{d}{dt}[-L_1(\dot{q}_1 - \dot{q}_2) + L_2(\dot{q}_2 - \dot{q}_3)] + \frac{q_2}{C_1} = 0$$

$$\frac{d}{dt}[-L_2(\dot{q}_2 - \dot{q}_3)] + \frac{q_3 - q_4}{C_2} = 0 \qquad\qquad (5\text{-}80)$$

$$-\frac{q_3 - q_4}{C_2} = -R_2\dot{q}_4.$$

This is a set of four coupled differential equations which describe the dynamics of the circuit. These four equations can be interpreted as

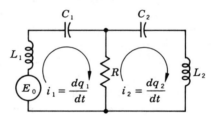

Fig. 5–15. Example 5–5. Network with constant voltage source.

statements of Kirchhoff's voltage law applied to the four loops to which we have assigned the loop currents in Fig. 5–14c. Note that since the network has *four* independent *loops*, there are four generalized charge variables and four dynamic equations (5–80). Because the circuit has *two* independent *nodes*, there are only two generalized flux linkages and two corresponding equations (5–75).

   Example 5–5. As a final example of the application to electrical networks of generalized coordinates and Lagrange's equations, we consider the circuit of Fig. 5–15. A network containing two capacitors, two inductors, and a resistor is connected to an ideal constant-voltage source. The network has three independent nodes and two independent loops. As a consequence, let us analyze the system using the generalized charge variables q_1 and q_2 which are related to the loop currents, as indicated in Fig. 5–15. The lagrangian (5–66) is

$$\mathcal{L} = \tfrac{1}{2}L_1\dot{q}_1^2 + \tfrac{1}{2}L_2\dot{q}_2^2 - \frac{q_1^2}{2C_1} - \frac{q_2^2}{2C_2}, \qquad (5\text{-}81)$$

and the work expression (5–70) is

$$\sum_{k=1}^{2} E_k \, \delta q_k = E_0 \, \delta q_1 - R(\dot{q}_2 - \dot{q}_1)(\delta q_2 - \delta q_1). \qquad (5\text{–}82)$$

Therefore the generalized voltages corresponding to the generalized charges q_1 and q_2 are

$$E_1 = E_0 + R(\dot{q}_2 - \dot{q}_1) \qquad E_2 = -R(\dot{q}_2 - \dot{q}_1). \qquad (5\text{–}83)$$

Lagrange's equations (5–69) follow from (5–81) and (5–83).

$$
\begin{aligned}
L_1 \ddot{q}_1 + \frac{q_1}{C_1} &= E_0 + R(\dot{q}_2 - \dot{q}_1) \\
L_2 \ddot{q}_2 + \frac{q_2}{C_2} &= -R(\dot{q}_2 - \dot{q}_1).
\end{aligned}
\qquad (5\text{–}84)
$$

The dynamic response of the circuit of Fig. 5–15 can be obtained by integrating (5–84) subject to initial conditions. It is interesting to note that these equations are coupled only by the resistance term because the resistor is the only element common to the two loops.

PROBLEMS

5–1. At $t = 0$, with the capacitor uncharged, the ideal constant-voltage source is connected in series with the resistor and capacitor. Find the time history of charge on the capacitor.

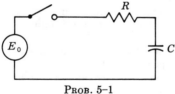

PROB. 5–1

5–2. A resistor and an inductor are connected in parallel with an ideal current source which supplies the current,

$$I(t) = I_0 \sin \omega t.$$

Find the steady-state voltage developed across the resistor.

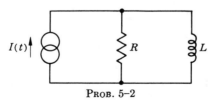

PROB. 5–2

5-3. An ideal voltage source is connected in series with a resistor, a capacitor, and an inductor. Obtain a dynamic equation for the charge on the capacitor.

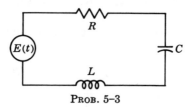

PROB. 5-3

5-4. An ideal current source is connected to a circuit consisting of a pair of capacitors and an inductor.

(a)   Obtain a dynamic equation for the voltage e_1 across the capacitor shown.
(b)   Obtain the steady-state response $e_1(t)$ when $I(t) = I_0 \sin \omega t$.

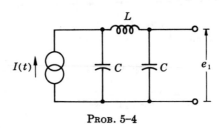

PROB. 5-4

5-5. A linear resistor, a capacitor, and an inductor are connected in series with a source which has the nonlinear characteristic shown.

(a)   Model the source with an ideal constant-voltage source and a nonlinear resistor. Formulate dynamic equations for the system.
(b)   Model the source with an ideal constant-current source and a nonlinear resistor. Formulate dynamic equations for the system.

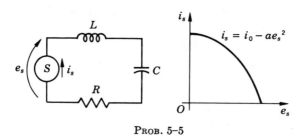

PROB. 5-5

5-6. The circuit shown contains two ideal voltage sources. Select appropriate coordinates and formulate dynamic equations of "motion."

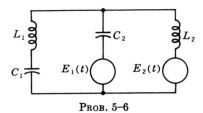

PROB. 5-6

5-7. The lossless circuit shown contains three capacitors and two inductors.

(a) Select a set of generalized flux-linkage coordinates λ_k.
(b) Formulate dynamic equations for the λ_k by direct application of the requirements of Sec. 5-4.
(c) Formulate dynamic equations for the λ_k using Lagrange's equations.
(d) Select a set of generalized charge coordinates q_k.
(e) Formulate dynamic equations for the q_k by direct application of the requirements of Sec. 5-4.
(f) Formulate dynamic equations for the q_k using Lagrange's equations.

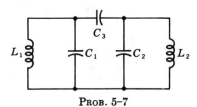

PROB. 5-7

5-8. Eliminate q_R between the two equations of (5-47) and show that q_C must satisfy

$$C\frac{d^2q_C}{dt} + \frac{1}{R}\frac{dq_C}{dt} + \frac{1}{L}q_C = C\frac{dI}{dt}.$$

Compare with (5-37).

5-9. The electric circuit shown contains two inductors and a capacitor.

(a) Formulate dynamic equations for flux-linkage variables.
(b) Formulate dynamic equations for charge variables.
(c) Construct a translational mass-spring system which has analogous equations of motion according to the analogy of Fig. 5-11.
(d) Construct a dual electrical network whose dynamic equations in charge coordinates are analogous to (a) and whose dynamic equations in flux-linkage variables are analogous to (b).

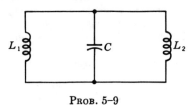

PROB. 5-9

5-10. Answer the questions (*a*) to (*d*) of Prob. 5–9 for the system shown, which contains a resistor and an ideal current source, as well as an inductor and two capacitors.

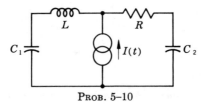

Prob. 5–10

5-11. If mass were strictly analogous to capacitance, as indicated in Fig. 5–11, it would have the following property. When the velocities at its terminals are v_2 and v_1, the tensile force f acting at the terminals would be

$$f = m\left(\frac{dv_2}{dt} - \frac{dv_1}{dt}\right).$$

For a simple mass particle, this holds only when one of the terminals remains fixed in an inertial reference frame. Show that the device indicated in the sketch does have the required property within the limitations imposed by assuming small motions of the rigid massless links.

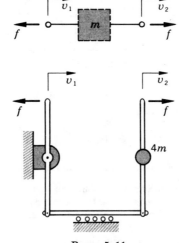

Prob. 5–11

5-12. Construct a symbolic diagram similar to Fig. 5–12 for the mechanical dual analogy which applies to systems with simple translation of masses, springs, and dashpots. Use this diagram as an aid to devise a mechanical system which is the dual of the system of Example 2–9 in Fig. 2–49. Formulate an equation of motion for the dual system and study its relation to (2–136).

5-13. The network shown contains an ideal voltage source.

(*a*) Formulate dynamic equations in terms of flux-linkage variables.
(*b*) Formulate dynamic equations in terms of charge variables.

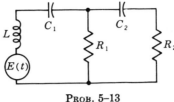

PROB. 5-13

5-14. The network shown contains an ideal voltage source.

(*a*) Formulate dynamic equations in terms of flux-linkage variables.
(*b*) Formulate dynamic equations in terms of charge variables.

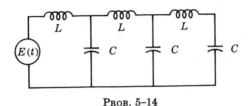

PROB. 5-14

5-15. Formulate dynamic equations for the network shown

(*a*) In terms of flux linkages
(*b*) In terms of charges

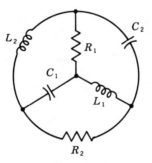

PROB. 5-15

5-16. An ideal *transformer* is a lossless device whose constitutive relations are indicated in the figure on the left. This device is the electrical counterpart of a lever for translational motion or a set of gears for fixed-axis rotations.

(*a*) Formulate dynamic equations for the circuit on the right using direct application of the requirements of Sec. 5-4.
(*b*) Construct a lagrangian for the circuit which yields the same result as (*a*).
(*c*) What is the natural frequency of the oscillations of the circuit?

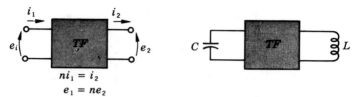

PROB. 5-16

5-17. An *ideal gyrator* is a lossless device whose constitutive relations are indicated in the figure on the left. This device is the electrical counterpart of the mechanical gyrator described in Prob. 4-28.

(a) Formulate dynamic equations for the circuit on the right using direct application of the requirements of Sec. 5-4.

(b) Construct a lagrangian for the circuit which yields the same result as (a). [A mixed lagrangian consisting of (5-65) on one side of the gyrator and (5-66) on the other side is useful here.]

(c) What is the natural frequency of free oscillations of the circuit?

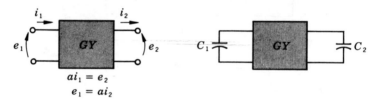

PROB. 5-17

5-18. Derive the equation

$$-d\mathbf{f} \cdot \mathbf{v} = i\,de$$

by using Eqs. (C-2) and (C-3) in the appendix to eliminate the vector $d\mathbf{l} \times \mathbf{B}$ between (5-27) and (5-28). Interpret this result when an external agent forces the conductor to maintain a constant velocity.

CHAPTER SIX

Electromechanical transducers

The preceding chapters have dealt separately with the dynamics of mechanical systems and with the dynamics of electrical networks. In this chapter we consider composite systems in which there is dynamic interaction between mechanical and electrical variables. The elements which convert electrical energy to mechanical energy (or mechanical energy to electrical energy) are called *electromechanical transducers*. Important examples of such transducers include solenoids, microphones, loudspeakers, motors, generators, electrical meters, phonograph pickups, accelerometers, etc.

The chapter begins with a discussion of the constitutive relations of several lossless lumped-parameter transducers. Here energy and coenergy state-functions are particularly helpful. Then we show how a real transducer can be modeled by a system containing auxiliary mechanical and electrical elements in addition to an ideal transducer element. The dynamic behavior of a few representative electromechanical systems is then discussed in some detail. Both direct and variational methods of analysis are illustrated.

6–1 Constitutive relations for transducers

Lossless transducers conserve energy. These may be subdivided into elements which can *store* energy and those which only *transfer* energy. An electromechanical *energy-storage* transducer stores energy when either mechanical work is done on it or when electrical work is done on it. The energy stored can be recovered at a later time. Furthermore, the energy removed can be taken out as either mechanical or electrical work without regard to the form in which the energy was originally stored. Energy conversion is accomplished by putting in energy in one form (mechanical or electrical) and subsequently removing it in the other form.

An *energy-transfer* element is a device for which the instantaneous power out always equals the instantaneous power in. Energy conversion in an ideal electromechanical energy-transfer transducer is accomplished instantaneously.

MOVABLE-PLATE CAPACITOR

As an example of an energy-storage trai......ucer, we consider the ideal movable-plate capacitor shown in Fig. 6–1. The charge on the capacitor is q, and the voltage across the plates is e. The displacement of the movable plate is x, and the external force required to hold the movable plate in equilibrium (against the electrical attraction of the fixed plate)

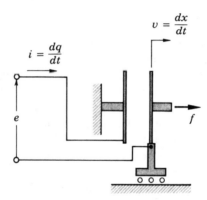

FIG. 6–1. Movable-plate capacitor.

is f. In this ideal transducer we neglect electrical inductance and resistance and mechanical mass and friction.

The constitutive relations for a movable-plate capacitor can be given in the form of equations for the voltage and force in terms of the displacement and charge,

$$e = e(x,q)$$
$$f = f(x,q). \tag{6–1}$$

These relations may be determined experimentally in a real capacitor, or they can be predicted analytically from electrostatic field theory. An important characteristic of the relations (6–1) for a movable-plate capacitor is that when $q = 0$, the force f must be zero for any x. This follows from the fact that when there is no charge, there is no electric field, and hence no attractive force between the plates.

The electrical power delivered to the capacitor is ei, and the mechanical power delivered is fv. The net work done *on* the capacitor in time dt is

$$ei\,dt + fv\,dt = e\,dq + f\,dx. \tag{6–2}$$

For a conservative element this work represents the increase dW_e in the stored electrical energy[1] of the element.   The total *electrical energy* $W_e(x,q)$ with respect to a datum configuration is obtained by integrating (6–2) along *any* path from the datum to (x,q).   Once the energy function $W_e(x,q)$ is known, the constitutive relations (6–1) can be recovered by differentiation.

$$\frac{\partial W_e}{\partial x} = f \qquad \frac{\partial W_e}{\partial q} = e. \tag{6–3}$$

An alternative state-function for the capacitor is the *electrical coenergy* $W_e^*(x,e)$, defined by the Legendre transformation,

$$W_e^*(x,e) = eq - W_e(x,q). \tag{6–4}$$

When the coenergy function is known, the constitutive relations are obtained in the following form:

$$\frac{\partial W_e^*}{\partial x} = -f \qquad \frac{\partial W_e^*}{\partial e} = q. \tag{6–5}$$

ELECTRICALLY LINEAR CASE

For most movable-plate capacitors it is satisfactory to assume that the device is *electrically linear;* i.e., the first of (6–1) can be assumed to have the form

$$e = \frac{q}{C(x)}, \tag{6–6}$$

where $C(x)$ is the capacitance when the movable plate is in the position x.   In general, $C(x)$ will be a nonlinear[2] function of x.   In this case we can obtain explicit representations of the state-functions W_e and W_e^*.   If we integrate (6–2) from a datum configuration with $x = 0$ and $q = 0$ along the path indicated in Fig. 6–2, the electrical energy is, simply,

$$W_e = \int_0^q \frac{q\,dq}{C(x)} = \frac{q^2}{2C(x)}, \tag{6–7}$$

because there is no contribution from the segment along the x axis (where the force f is zero for any x).   The electrical coenergy follows from (6–4) and (6–7) with the aid of (6–6).

$$W_e^* = \tfrac{1}{2}C(x)e^2. \tag{6–8}$$

The constitutive relations for an electrically linear movable-plate capaci-

---

[1] Even though both electrical and mechanical work increments contribute to the stored energy, we use the adjective *electrical* for the energy because the storage mechanism depends on the electric field of the capacitor.

[2] See Prob. 6–13.

tor can be obtained by applying (6–3) to (6–7),

$$f = \frac{\partial W_e}{\partial x} = -\frac{q^2 C'}{2C^2} \qquad e = \frac{\partial W_e}{\partial q} = \frac{q}{C}, \tag{6-9}$$

or by applying (6–5) to (6–8),

$$-f = \frac{\partial W_e^*}{\partial x} = \tfrac{1}{2}C'e^2 \qquad q = \frac{\partial W_e^*}{\partial e} = Ce. \tag{6-10}$$

In these formulas C' represents the derivative,

$$C'(x) = \frac{dC(x)}{dx}. \tag{6-11}$$

It is interesting to note that the assumed form (6–6) of *one* of the constitutive relations (together with the assumption that the device is con-

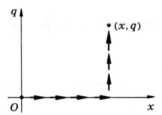

FIG. 6–2. Integration path for electrical energy of movable-plate capacitor.

servative) is sufficient to fix the form of the other constitutive relation. The relations (6–9) and (6–10) for the mechanical force are a direct consequence of the assumption (6–6) regarding the potential difference across the capacitor.

SOLENOID

As a second example of an energy-storing transducer, we consider a solenoid, or movable-core inductor, which is the magnetic counterpart of the movable-plate capacitor. A typical configuration is sketched in Fig. 6–3. The flux linkage of the coil is λ, and the current through it is i. The displacement of the movable iron core is x, and the external force required to hold the core in equilibrium (against the magnetic attraction) is f. In this ideal transducer we neglect electrical capacitance and resistance and mechanical mass and friction.

The constitutive relations for a solenoid can be given in the form of equations for the current and force in terms of the displacement and flux linkage.

$$i = i(x,\lambda) \qquad f = f(x,\lambda). \tag{6-12}$$

In an ideal transducer we take these to be single-valued functions; i.e., we neglect hysteresis. We also assume that the element is conservative. An important characteristic of the relations (6-12) is that, when $\lambda = 0$, there is no magnetic field, and hence no magnetic attraction. The force f is therefore zero for any x when $\lambda = 0$.

The electrical power delivered to the solenoid is ei, and the mechanical power delivered is fv. In time dt the net work done *on* the solenoid is

$$ei\, dt + fv\, dt = i\, d\lambda + f\, dx. \tag{6-13}$$

For a conservative element this work represents the increase dW_m in the stored magnetic energy[1] of the solenoid. The total *magnetic energy* $W_m(x,\lambda)$ with respect to a datum configuration is obtained by integrating (6-13) along *any* path from the datum to (x,λ). Once the energy func-

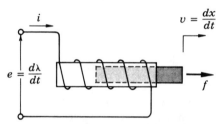

FIG. 6-3.   Movable-core inductor.

tion $W_m(x,\lambda)$ is known, the constitutive relations (6-12) can be recovered by differentiation.

$$\frac{\partial W_m}{\partial x} = f \qquad \frac{\partial W_m}{\partial \lambda} = i. \tag{6-14}$$

An alternative state-function for the solenoid is the *magnetic coenergy* $W_m^*(x,\lambda)$, defined by the Legendre transformation,

$$W_m^*(x,i) = i\lambda - W_m(x,\lambda). \tag{6-15}$$

When the coenergy function is known, the constitutive relations are obtained in the following form:

$$\frac{\partial W_m^*}{\partial x} = -f \qquad \frac{\partial W_m^*}{\partial i} = \lambda. \tag{6-16}$$

ELECTRICALLY LINEAR CASE

When the constitutive relation between current and flux linkage can be assumed to be linear, we can obtain explicit representations for the magnetic energy, the magnetic coenergy, and the constitutive relation for

[1] The stored energy is called magnetic energy here because the storage mechanism depends on the magnetic field of the inductor.

the force. Let the first of (6–12) have the form

$$i = \frac{\lambda}{L(x)},\qquad(6\text{–}17)$$

where $L(x)$ is the inductance of the coil when the core is in the position x. The magnetic energy $W_m(x,\lambda)$ is then obtained by integrating (6–13). Taking $x = 0$, $\lambda = 0$, as the datum configuration and integrating along the path indicated in Fig. 6–4, we find

$$W_m = \int_0^\lambda \frac{\lambda\, d\lambda}{L(x)} = \frac{\lambda^2}{2L(x)},\qquad(6\text{–}18)$$

since there is no contribution from the segment of the path along the x

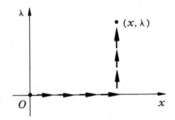

Fig. 6–4.   Integration path for magnetic energy of solenoid.

axis (where the force f is zero). The magnetic coenergy is obtained by substituting (6–17) and (6–18) in (6–15).

$$W_m^* = \tfrac{1}{2}L(x)\, i^2.\qquad(6\text{–}19)$$

The constitutive relations can be recovered from either the energy or the coenergy. Applying (6–14) to (6–18), we have

$$f = \frac{\partial W_m}{\partial x} = -\frac{\lambda^2 L'}{2L^2} \qquad i = \frac{\partial W_m}{\partial \lambda} = \frac{\lambda}{L},\qquad(6\text{–}20)$$

and applying (6–16) to (6–19), we have

$$-f = \frac{\partial W_m^*}{\partial x} = \tfrac{1}{2}L' i^2 \qquad \lambda = \frac{\partial W_m^*}{\partial i} = Li,\qquad(6\text{–}21)$$

where L' represents the derivative,

$$L'(x) = \frac{dL(x)}{dx}.\qquad(6\text{–}22)$$

MOVING-COIL TRANSDUCER

As an example of an energy-transfer transducer, we consider a coil which can move in a uniform magnetic field. The voice coil of a loudspeaker and the meter movement of an ammeter are representative applications. Figure 6–5 indicates a typical geometric configuration for a loudspeaker voice coil.   A permanent magnet maintains a uniform radial magnetic field **B** in the annular gap.   Within the gap is the voice coil, which can move longitudinally with velocity v.   Flexible lead-in wires introduce a current i into the moving coil.   The external force f acting on the coil is required to hold the coil in equilibrium against the magnetic forces. The potential difference across the voice coil is e.   In this ideal transducer we neglect the electrical resistance and self-inductance of the coil, as well as any capacitance.   We also neglect the mechanical mass and friction, as well as any restraint from the lead-in wires.

    The constitutive relations for the ideal moving coil can be obtained by

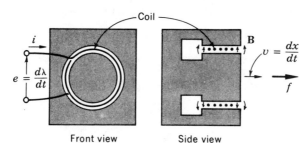

Front view   Side view
FIG. 6–5.   Moving coil in permanent magnetic field.

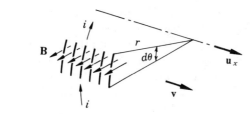

FIG. 6–6.   Element of voice coil.

applying the magnetic-field relations (5–26) and (5–27) to a differential element of the voice coil, and integrating.   In Fig. 6–6 we consider a length $dl = r\,d\theta$ of one of the turns of the coil in which the current i flows. The magnetic field **B** is mutually perpendicular to the direction of current flow and to the velocity **v** of the coil.   Then, according to (5–27), the increment in potential in the direction of current flow is

$$de = \mathbf{v} \times \mathbf{B} \cdot d\mathbf{l}$$

$$= -vBr\,d\theta. \tag{6–23}$$

The total potential difference for n turns is obtained by integrating over θ from zero to $2\pi n$.   This is an increment in the direction of current flow. The potential difference labeled e in Fig. 6–5 represents a drop in potential (in the direction of current flow) within the coil.   Therefore the terminal

298 Electromechanical transducers

voltage e in Fig. 6–5 is

$$e = 2\pi nrBv. \qquad (6\text{--}24)$$

Similarly, the force of the magnetic field on the element of length $r\,d\theta$ in Fig. 6–6 is given by (5–28).

$$d\mathbf{f} = i\,d\mathbf{l} \times \mathbf{B}$$
$$= r\,d\theta\,Bi\mathbf{u}_x. \qquad (6\text{--}25)$$

The total force for n turns is obtained by integrating over θ from zero to $2\pi n$. This is the force of the magnetic field acting on the voice coil. The force labeled f in Fig. 6–5 represents the external force required to balance the magnetic force. Therefore the external force f in Fig. 6–5 is

$$f = -2\pi nrBi. \qquad (6\text{--}26)$$

The constitutive relations can be simplified by introducing the transducer constant T, defined by

$$T = 2\pi nrB. \qquad (6\text{--}27)$$

Then, in place of (6–24) and (6–26), we have

$$e = Tv$$
$$-Ti = f \qquad (6\text{--}28)$$

as the constitutive relations for the moving-coil transducer. The transducer constant T depends on the geometrical configuration of the coil and field, the number of turns in the coil, and the strength of the magnetic field. According to (6–27), T has the dimension of a magnetic-field intensity multiplied by a length. In the first of (6–28), T has the dimension of a quotient of a potential divided by a velocity, and in the second of (6–28), T has the dimension of a force divided by a current. These three dimensions are consistent, and furthermore, if the rationalized mks system is used,[1] the magnitude of T will be the same, whether expressed in units of *webers per meter, volts per meter per second,* or *newtons per ampere.* A simplified symbolic representation of a moving-coil transducer is shown in Fig. 6–7. At the left are the electric terminals through which the current i passes and across which the potential difference e exists. At the right is a massless plunger, attached internally to the moving coil, which moves with the velocity v of the coil and to which

---

[1] When a mixed system of units is employed, it is common practice to express the constitutive relations (6–28) in the form

$$e = T_{em}v$$
$$-T_{me}i = f, \qquad (6\text{--}28a)$$

where the conversion factors, for the units employed, are absorbed in the transducer constants T_{em} and T_{me}. See Prob. 6–18.

the external force f is applied. The constitutive relations (6–28) are displayed underneath the transducer.

STATE-FUNCTIONS

We next consider energy and coenergy functions for the moving-coil transducer. We begin by calculating the work done *on* the transducer. The electrical power delivered is ei, and the mechanical power delivered is fv. In time dt the net work done is, using (6–28),

$$ei\, dt + fv\, dt = Tv\left(-\frac{f}{T}\right) dt + fv\, dt = 0. \qquad (6\text{--}29)$$

There is never *any* work done on the device. There is no accumulation of stored energy. The transducer magnetic[1] energy W_m remains equal to its datum value. The transducer is still conservative, in the sense

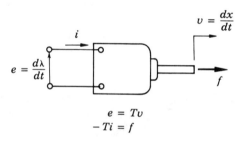

$$e = Tv$$
$$-Ti = f$$

FIG. 6-7. Symbolic representation of moving-coil transducer.

that the instantaneous power in, ei, is always equal to the instantaneous power out, $-fv$. For this reason we call such a device an *energy-transfer* transducer.

We next consider the magnetic coenergy defined by

$$W_m^*(x,i) = \lambda i - W_m(x,\lambda). \qquad (6\text{--}30)$$

If we take the datum value of W_m to be zero, the coenergy W_m^* is λi. A useful representation of W_m^* is obtained by integrating the first of (6–28) to obtain

$$\lambda = T(x - x_0), \qquad (6\text{--}31)$$

where x_0 is a constant of integration. Then, inserting (6–31) in (6–30), we obtain the coenergy as

$$W_m^* = Ti(x - x_0), \qquad (6\text{--}32)$$

when $W_m = 0$. The constitutive relations for the transducer can be

[1] The use of the adjective magnetic here is optional. See Prob. 6–4.

recovered from (6–32) by using (6–16). Thus

$$-f = \frac{\partial W_m^*}{\partial x} = Ti \qquad \lambda = \frac{\partial W^*}{\partial i} = T(x - x_0), \qquad (6\text{–}33)$$

which agree with (6–28) and (6–31).

6–2   Analysis of transducer systems

In the preceding section the constitutive relations for ideal transducers
are discussed. In order to model real transducers, it is necessary to
include additional electrical and mechanical elements to account for
losses, for effects of inductance or capacitance, and for effects of mass or
flexibility. The resulting electromechanical systems are often fairly
complicated. Dynamic analysis of a system of this sort requires a well-
organized plan of attack, with careful attention to many detailed points.
We consider here direct methods of analysis and indirect methods based
on the variational approach, using Hamilton's principle or Lagrange's
equations.

Direct methods

An electromechanical system, being both electrical and mechanical, must
satisfy all the dynamic requirements for electric circuits, plus all the
dynamic requirements for mechanical systems. In addition, the constitutive relations of an ideal transducer in the system serve to couple
electrical and mechanical variables. The direct method of analysis
consists in formulating each and every requirement in mathematical
terms so as to obtain a (large) set of simultaneous relations which must
hold during dynamic changes. Then, by systematic elimination of
variables, it is usually possible to reduce the number of simultaneous
equations to a much smaller number of differential equations of "motion."

The governing requirements for mechanical systems are listed on page
102, and the governing requirements for electric circuits are listed on
page 268. The systematic methods for formulating equations of motion
for mechanical and electrical systems separately can be extended to
composite systems. Mechanical variables are selected which automatically satisfy one of the three categories of mechanical requirements, and
electrical variables are selected which automatically satisfy one of the
three categories of electrical requirements. Suppose there are n_m independent mechanical variables and n_e independent electrical variables.
Then the remaining mechanical requirements are imposed (leading to
n_m differential equations), and the remaining electrical requirements are
imposed (leading to n_e differential equations). Because of transducers in
the system, the n_m mechanical equations will contain some electrical

variables and the n_e electrical equations will contain some mechanical variables. Thus we obtain a set of $n_m + n_e$ coupled differential equations for both the mechanical and electrical variables.

VARIATIONAL METHODS

The alternative indirect procedures for meeting the dynamic requirements of mechanical and electrical systems separately are readily extended to composite systems. For a mechanical system there is a variational indicator (2–134) and a corresponding set of admissibility requirements. Vanishing of the indicator ensures satisfaction of the remaining mechanical requirements. Similarly, for an electrical network there are variational indicators (5–48) and (5–50) and corresponding admissibility requirements. Vanishing of the indicator in each case ensures satisfaction of the remaining electrical requirements.

For a composite system, both the mechanical and the electrical requirements can be met by a composite Hamilton's principle in which the variational indicator is the *sum* of a mechanical indicator and an electrical indicator and the admissibility requirements include *both* the mechanical and the electrical admissibility requirements. An ideal transducer is accounted for in this procedure by including either its energy or coenergy function in the lagrangian of the composite variational indicator.

When displacement and charge variables are employed, the variational indicator for Hamilton's principle is

$$\text{V.I.} = \int_{t_1}^{t_2} \left[\delta(T^* - V + W_m^* - W_e) + \sum_i f_i \, \delta x_i + \sum_j e_j \, \delta q_j \right] dt. \quad (6\text{--}34)$$

The admissibility requirements include geometric constraints, along with kinematic relations between displacements and velocities, *and* Kirchhoff's current law, along with relations between charges and currents.

When displacement and flux-linkage variables are employed, the variational indicator for Hamilton's principle is

$$\text{V.I.} = \int_{t_1}^{t_2} \left[\delta(T^* - V + W_e^* - W_m) + \sum_i f_i \, \delta x_i + \sum_j i_j \, \delta \lambda_j \right] dt. \quad (6\text{--}35)$$

The corresponding admissibility requirements include geometric constraints, along with kinematic relations between displacements and velocities, *and* Kirchhoff's voltage law, along with relations between flux linkages and voltages.

Taking into account these extensions to the variational indicator and to the admissibility requirements, the statement of Hamilton's principle on page 30 still applies. A natural "motion" is an admissible "motion" for which the indicator vanishes (for arbitrary admissible variations).

In general, the most convenient way to meet the admissibility requirements is to select a complete set of independent generalized coordinates.
If the variations of these coordinates are a complete set of independent
variations, the most direct method of implementing the variational
method is to apply Lagrange's equations (2–228).

When employing displacement and charge variables, the generalized
coordinates include generalized geometrical coordinates ξ_i and generalized
charges q_k.   The lagrangian

$$\mathcal{L}(\xi_i, \dot{\xi}_i, q_k, \dot{q}_k) = T^* - V + W_m^* - W_e \qquad (6\text{--}36)$$

accounts for all the conservative elements in the system, and the work
expression

$$\sum_i \Xi_i \, \delta\xi_i + \sum_k E_k \, \delta q_k \qquad (6\text{--}37)$$

defines the generalized forces Ξ_i and the generalized voltages E_k of the
nonconservative elements.   Lagrange's equations then divide into two
sets,

$$\frac{d}{dt}\left(\frac{\partial\mathcal{L}}{\partial\dot{\xi}_i}\right) - \frac{\partial\mathcal{L}}{\partial\xi_i} = \Xi_i \qquad i = 1, 2, \ldots, n_m$$

$$\frac{d}{dt}\left(\frac{\partial\mathcal{L}}{\partial\dot{q}_k}\right) - \frac{\partial\mathcal{L}}{\partial q_k} = E_k \qquad k = 1, 2, \ldots, n_e. \qquad (6\text{--}38)$$

When employing displacement and flux-linkage variables, the generalized coordinates include generalized geometrical coordinates ξ_i and
generalized flux linkages λ_k.   The lagrangian

$$\mathcal{L}(\xi_i, \dot{\xi}_i, \lambda_k, \dot{\lambda}_k) = T^* - V + W_e^* - W_m \qquad (6\text{--}39)$$

accounts for all the conservative elements in the system, and the work
expression

$$\sum_i \Xi_i \, \delta\xi_i + \sum_k I_k \, \delta\lambda_k \qquad (6\text{--}40)$$

defines the generalized forces Ξ_i and the generalized currents I_k of the
nonconservative elements.   Lagrange's equations again[1] split into two
sets,

$$\frac{d}{dt}\left(\frac{\partial\mathcal{L}}{\partial\dot{\xi}_i}\right) - \frac{\partial\mathcal{L}}{\partial\xi_i} = \Xi_i \qquad i = 1, 2, \ldots, n_m$$

$$\frac{d}{dt}\left(\frac{\partial\mathcal{L}}{\partial\dot{\lambda}_k}\right) - \frac{\partial\mathcal{L}}{\partial\lambda_k} = I_k \qquad k = 1, 2, \ldots, n_e'. \qquad (6\text{--}41)$$

The formalism of the variational approach provides an efficient organi-
zational scheme which is helpful in analyzing complex systems.   Every
element in the system must be included once, either in the lagrangian or

[1] The number n_e' of independent flux-linkage variables is generally different from the
number n_e of independent charge variables.

in the work expression.   Note that in the lagrangians (6–36) and (6–39) the coenergies always appear with a plus sign and the energies always appear with a minus sign.

Example 6–1.   To illustrate the analysis of a simple transducer system, we consider the door-chime solenoid shown in Fig. 6–8.   When the door button is pushed, an alternating voltage is connected to the solenoid which should cause the striker to move rapidly to the right.   It should strike the chime once, and then return sufficiently so as not to interfere with the subsequent ringing of the chime.   We shall analyze

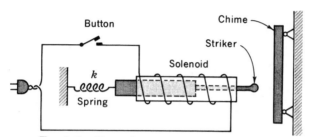

FIG. 6–8.   Example 6–1.   Door-chime solenoid.

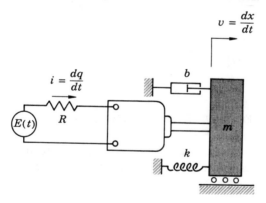

FIG. 6–9.   Model of real transducer by system containing ideal transducer.

the dynamic behavior of the system during the interval in which the button remains depressed.

To model the real solenoid we consider the system of Fig. 6–9, which contains an ideal solenoid.   We take the ideal solenoid to be electrically linear, with constitutive relations of the form given in (6–20) or (6–21). We assume the inductance $L(x)$ to be maximum when the iron core is centered in the solenoid, and to drop off symmetrically as the core is removed from either end.   The functions $L(x)$ and $L'(x)$ are sketched in Fig. 6–10.

On the electrical side of the system of Fig. 6–9 we assume an ideal voltage source and a resistor.   The resistor models the resistance of the solenoid winding plus the source resistance of the actual source (see Fig. 5–8).   On the mechanical side we assume a mass m and a dashpot (with dashpot constant b) in addition to the spring.   The mass models the inertia of the solenoid core and striker, and the dashpot models viscous friction, which retards the motion of the core.

To begin the analysis by the direct method, we dissect the system of Fig. 6–9 into an electric circuit, a mechanical system, and the ideal transducer.   These three subsystems are shown in Fig. 6–11.   In the electric

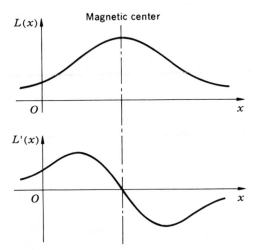

FIG. 6–10.   Inductance function for solenoid.

circuit the electrical terminals of the transducer are included.   In the mechanical system the transducer force f is shown acting on the mass. In Fig. 6–11c the constitutive relations (6–21) for the ideal transducer are displayed.

We next consider the electrical and mechanical subsystems separately. The electric circuit of Fig. 6–11a has a single loop, or mesh.   There is one independent mesh current i and one independent generalized charge q, defined by $\dot{q} = i$.   By using q as our electrical variable, we automatically ensure the satisfaction of Kirchhoff's current law and the charge-current relations.   To obtain an equation for q we impose Kirchhoff's voltage law.   Using Ohm's law for the resistor, we have

$$E(t) = R\frac{dq}{dt} + e, \qquad (6\text{--}42)$$

where e is the voltage across the transducer terminals.

The mechanical system of Fig. 6–11b is of the same form as the system of Example 2–9.   The geometric requirements are automatically satisfied if we take the common displacement x of the mass, spring, and dashpot as a generalized coordinate and set $v = \dot{x}$.   We take the origin of x to correspond with the position in which there is no force in the spring.

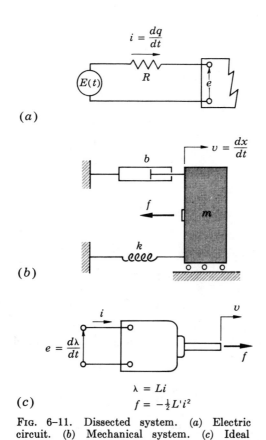

(a)

(b)

(c)

$$\lambda = Li$$
$$f = -\tfrac{1}{2}L'i^2$$

FIG. 6–11.   Dissected system.   (a) Electric circuit.   (b) Mechanical system.   (c) Ideal transducer.

This origin is indicated in Fig. 6–10 as being to the left of the magnetic center of the solenoid.   To obtain an equation for x, we impose the linear-momentum principle.   Using the constitutive relations for the spring and dashpot, we have

$$-f - kx - b\frac{dx}{dt} = m\frac{d^2x}{dt^2}, \tag{6-43}$$

where f is the transducer force.

At this stage we have an electrical equation (6–42), containing the transducer voltage e, and a mechanical equation (6–43), containing the transducer force f. Now turning to the transducer in Fig. 6–11c, we have

$$e = \frac{d\lambda}{dt} = \frac{d}{dt}(Li) = L\frac{d^2q}{dt^2} + L'\frac{dq}{dt}\frac{dx}{dt} \qquad (6\text{–}44)$$

and

$$f = -\tfrac{1}{2}L'\left(\frac{dq}{dt}\right)^2, \qquad (6\text{–}45)$$

on substituting $\dot{q}$ for i. Finally, when (6–44) is inserted in (6–42) and (6–45) is inserted in (6–43), we obtain a pair of coupled dynamic equations for q and x.

$$L\frac{d^2q}{dt^2} + R\frac{dq}{dt} + L'\frac{dq}{dt}\frac{dx}{dt} = E(t)$$

$$m\frac{d^2x}{dt^2} + b\frac{dx}{dt} + kx - \tfrac{1}{2}L'\left(\frac{dq}{dt}\right)^2 = 0. \qquad (6\text{–}46)$$

LAGRANGE'S EQUATIONS

To illustrate the indirect method of obtaining these equations of motion, we reconsider the system of Fig. 6–9. Again we select the generalized coordinates q and x, but now we construct the lagrangian (6–36) and the work expression (6–37). The lagrangian, including the magnetic coenergy (6–19) of the ideal transducer, is

$$\mathcal{L} = T^* - V + W_m^* = \tfrac{1}{2}m\left(\frac{dx}{dt}\right)^2 - \tfrac{1}{2}kx^2 + \tfrac{1}{2}L(x)\left(\frac{dq}{dt}\right)^2, \qquad (6\text{–}47)$$

and the work expression for both the electrical and mechanical nonconservative elements is

$$\left[E(t) - R\frac{dq}{dt}\right]\delta q + \left[-b\frac{dx}{dt}\right]\delta x. \qquad (6\text{–}48)$$

The terms in the square brackets are, respectively, the generalized voltage corresponding to q and the generalized force corresponding to x. Note that every element of the system in Fig. 6–9 is accounted for in either (6–47) or (6–48). Lagrange's equations (6–38) when applied to (6–47) and (6–48) are

$$\frac{d}{dt}\left(L\frac{dq}{dt}\right) = E(t) - R\frac{dq}{dt}$$

$$\frac{d}{dt}\left(m\frac{dx}{dt}\right) + kx - \tfrac{1}{2}L'\left(\frac{dq}{dt}\right)^2 = -b\frac{dx}{dt}. \qquad (6\text{–}49)$$

These are equivalent to (6–46), obtained by the direct method.

INTERPRETATION OF THE EQUATIONS

The equations of motion (6–49) are nonlinear coupled differential equations. We do not attempt an exhaustive study of their properties. We can, however, use them to give a qualitative explanation of how the door chime works when the voltage source is an alternating voltage,

$$E(t) = E_0 \sin \omega_0 t. \tag{6–50}$$

The second of (6–49) describes the motion of the striker. The mass is accelerated by any unbalance between the forces due to the spring, the transducer, and the dashpot. The dashpot always acts against the velocity. The spring and transducer forces, however, depend on position. The spring force acts to move the striker toward $x = 0$, and the transducer force acts to move the striker toward the magnetic center. Note that the transducer force $-\frac{1}{2}L'i^2$ depends on the *square* of the current $i = \dot{q}$. This means that the direction of the transducer force does not depend on the direction of the current. This is why a solenoid can be used on alternating current.

When the frequency ω_0 of the voltage source is high in comparison with the mechanical natural frequency, $\omega_n = \sqrt{k/m}$, we can assume that changes in x occur slowly in comparison with changes in q. We might then consider $L(x)$ as essentially constant during a few cycles of the alternating voltage source. The first of (6–49) could then be written

$$L \frac{di}{dt} + Ri = E_0 \sin \omega_0 t. \tag{6–51}$$

Furthermore, if R/L is large in comparison with ω_0, the electrical transient dies away rapidly and for most times the current i is, approximately,

$$i = \frac{E_0}{R} \sin \omega_0 t. \tag{6–52}$$

The corresponding transducer force would be

$$f = -\frac{1}{2}L'i^2 = -\frac{E_0^2 L'}{4R^2}(1 - \cos 2\omega_0 t), \tag{6–53}$$

which oscillates at twice the source frequency. This high-frequency fluctuation is too rapid for the mechanical system to follow. We therefore make an additional simplification by replacing the oscillating force with a smoothed force, which is essentially a cycle-by-cycle time average of (6–53),

$$f = -\frac{E_0^2 L'(x)}{4R^2}. \tag{6–54}$$

This effective transducer force can be considered to act like a nonlinear spring whose characteristic is determined by the function $L'(x)$ of Fig. 6–10.

With those simplifications, the motion of the striker is described by

$$m\frac{d^2x}{dt^2} + b\frac{dx}{dt} + kx - \frac{E_0^2L'(x)}{4R^2} = 0. \qquad (6\text{--}55)$$

When the door button is first pressed, the striker is at rest, with $x = 0$. Starting from this initial condition, the solution of (6–55) is a damped oscillation about a new equilibrium position x_e, where

$$kx_e = \frac{E_0^2L'(x_e)}{4R^2}. \qquad (6\text{--}56)$$

The door chime will work satisfactorily if the striker hits the chime on its first "overshoot" of the position x_e and then falls back to a damped oscillation about x_e without again touching the chime.

6–3   Frequency response of transducers

In this section we continue to study examples of transducer systems. Models of a loudspeaker and of a microphone are analyzed, and equa-

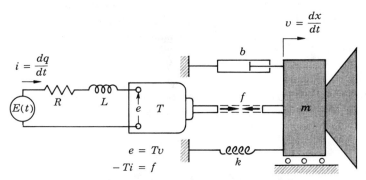

Fig. 6–12.   Example 6–2.   Loudspeaker model.

tions of motion are formulated.   In addition, to illustrate an important application of equations of motion, we obtain frequency-response functions for the two transducer systems.

Example 6–2.   In Fig. 6–12 we consider a model of a permanent-magnet loudspeaker.   In a real loudspeaker the voice coil is attached to a paper cone.   In our model we assume that the voice coil and cone assembly can be modeled by an ideal moving-coil transducer (see Fig. 6–7) plus, on the electrical side, self-inductance L and resistance R, plus, on the mechanical side, mass m.   Furthermore, we assume that the cone suspension and air resistance can be modeled by a simple spring and a dashpot. These hypotheses limit the validity of our model to low frequencies. The dynamic behavior of a loudspeaker at higher frequencies depends importantly on the acoustical properties of the speaker enclosure and on the higher modes of vibration of the cone.

To obtain equations of motion, we select generalized coordinates and apply Lagrange's equations. Since the electrical side of Fig. 6–12 has only a single mesh, we take the generalized mesh charge q as our electrical coordinate. On the mechanical side we take the common displacement x of the mass, spring, dashpot, and transducer as our mechanical coordinate. The lagrangian (6–36) is

$$\mathcal{L} = T^* - V + W_m^*$$

$$= \tfrac{1}{2}m\left(\frac{dx}{dt}\right)^2 - \tfrac{1}{2}kx^2 + \tfrac{1}{2}L\left(\frac{dq}{dt}\right)^2 + T\frac{dq}{dt}\,(x - x_0), \quad (6\text{–}57)$$

on using (6–32) for the transducer coenergy. The work expression for the nonconservative elements is

$$\left[E(t) - R\frac{dq}{dt}\right]\delta q + \left[-b\frac{dx}{dt}\right]\delta x. \quad (6\text{–}58)$$

Lagrange's equations (6–38) applied to (6–57) and (6–58) yield

$$m\frac{d^2x}{dt^2} + kx - T\frac{dq}{dt} = -b\frac{dx}{dt}$$

$$L\frac{d^2q}{dt^2} + T\frac{dx}{dt} = E(t) - R\frac{dq}{dt}, \quad (6\text{–}59)$$

as a pair of coupled equations in q and x. We note that q appears in (6–59) only as a first or second derivative. The equations can be simplified somewhat by reintroducing $i = \dot{q}$. Thus, in terms of i and x, we have

$$m\frac{d^2x}{dt^2} + b\frac{dx}{dt} + kx - Ti = 0$$

$$T\frac{dx}{dt} + L\frac{di}{dt} + Ri = E(t). \quad (6\text{–}60)$$

LOUDSPEAKER IMPEDANCE

Since the equations of (6–60) are linear, with constant coefficients, we know that when the exciting voltage $E(t)$ is a stationary sinusoid at frequency ω, there is a steady-state solution in which $i(t)$ and $x(t)$ are also stationary sinusoids at frequency ω. The amplitude and phase of the response sinusoids will, however, depend on the frequency.

To investigate this further, we take

$$E(t) = \text{Re}\,[Z(\omega)\epsilon^{j\omega t}], \quad (6\text{–}61)$$

where Re [   ] stands for "real part of," and assume that in the stationary solution,

$$x(t) = \text{Re}\,[H(\omega)e^{j\omega t}]$$

$$i(t) = \text{Re}\,[e^{j\omega t}].$$

(6-62)

In postulating (6-61) and (6-62), we have arbitrarily taken the current amplitude to be *unity* and allowed the (complex) amplitudes of the voltage and displacement to be unknown functions of frequency. The function $Z(\omega)$ can be considered to represent the quotient of the complex voltage amplitude divided by the current amplitude. It is called the *impedance* of the loudspeaker at the terminals of the voltage source. This impedance is relatively easy to measure for a real loudspeaker. Steady-state conditions at a frequency ω are established, and then the voltage and current amplitudes are measured, along with their phase difference. It is common practice to describe a loudspeaker's characteristics by means of a plot of $|Z(\omega)|$ versus ω.

To obtain the impedance function for our loudspeaker model we substitute

$$i(t) = e^{j\omega t}$$

$$E(t) = Z(\omega)e^{j\omega t}$$

(6-63)

$$x(t) = H(\omega)e^{j\omega t}$$

in (6-60). Since the equations are linear, the real parts of (6-63) will satisfy (6-60) whenever (6-63) satisfies (6-60). The complex amplitudes thus must satisfy

$$[-\omega^2 m + j\omega b + k]H(\omega) - T = 0$$

$$[j\omega T]H(\omega) + [j\omega L + R] = Z(\omega).$$

(6-64)

Eliminating $H(\omega)$, we obtain

$$Z(\omega) = (j\omega L + R) + \frac{T^2}{b + j(\omega m - k/\omega)} = Z_e(\omega) + Z_m(\omega),$$

(6-65)

where Z_e stands for the electrical contribution to the total impedance, and Z_m for the mechanical contribution. The general behavior of $Z(\omega)$ can be visualized by considering the behavior of Z_e and Z_m separately and superposing. In Fig. 6-13 we sketch these functions of ω in the complex Z plane. The electrical impedance in Fig. 6-13a starts from $Z_e = R$ at $\omega = 0$, and its imaginary part increases linearly with frequency. The mechanical impedance in Fig. 6-13b can be shown[1] to describe a clockwise circle with increasing frequency. The frequency ω_n marked on the sketches is the mechanical natural frequency, $\omega_n = \sqrt{k/m}$.

[1] See Prob. 6-8.

When these two curves are added, point by point, in the complex plane, the total impedance sketched in Fig. 6-13c is obtained.

In Fig. 6-13c the magnitude $|Z(\omega)|$ is represented by the distance from the origin to the point on the curve corresponding to frequency ω. A

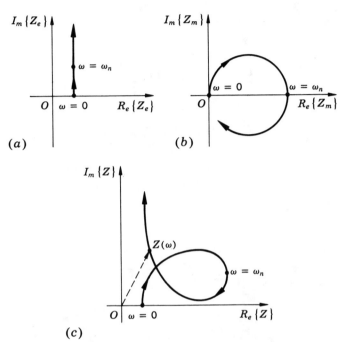

(a)

(b)

(c)

FIG. 6-13.   Complex impedance plots.   (a) Electrical impedance.
(b) Mechanical impedance.   (c) Total impedance.

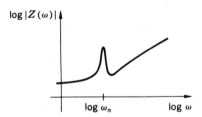

FIG. 6-14.   Impedance magnitude as a
function of frequency.

logarithmic plot of $|Z(\omega)|$ against ω is sketched in Fig. 6-14.   For very low frequencies, $|Z(\omega)|$ is nearly equal to R, and for large frequencies, $|Z(\omega)|$ tends to grow linearly with ω.   In between there is a pronounced peak and valley in the neighborhood of the mechanical resonance at $\omega = \omega_n$.

Example 6-3. We consider in Fig. 6-15 a condenser microphone. A flexible diaphragm is mounted in a case, parallel to a rigid back plate. These two surfaces form a parallel-plate condenser which is charged by a battery with potential E_0. Acoustic pressures cause the membrane to move, thereby altering the capacitance and causing a minute current to flow in response to the sound. An electronic amplifier raises the level

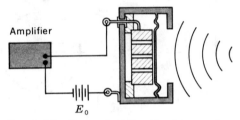

FIG. 6-15.   Example 6-3.   Condenser microphone.

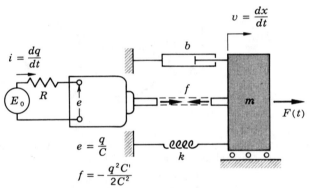

FIG. 6-16.   Model of condenser microphone.

of the signal to a point where it can be used to drive a loudspeaker or recording device.

To study the dynamical properties of the condenser microphone, we consider the system of Fig. 6-16 as a model. The real condenser microphone is replaced by an ideal movable-plate capacitor and several auxiliary electrical and mechanical elements. The constitutive relations (6-9) for an ideal electrically linear transducer are displayed under the transducer in Fig. 6-16. The battery is represented by an ideal constant-voltage source, and the input impedance of the amplifier is modeled by a resistance R. The source resistance of the battery is neglected in comparison with the amplifier impedance (the former is usually a fraction of one ohm, while the latter is of the order of a megohm).

On the mechanical side, the effective mass of the diaphragm is modeled by a mass m, and the effective stiffness of the membrane is modeled by a linear spring with spring constant k. This kind of modeling is satisfactory for low frequencies but is inadequate for frequencies appreciably higher than the first resonant frequency of the diaphragm. In addition, the damping of the membrane is modeled by a dashpot, with dashpot constant b. The major source of this damping is from a set of holes drilled in the back plate. When the diaphragm moves, it forces air (in laminar flow) in and out of these passages. Finally, the resultant force on the diaphragm due to the acoustic pressures is represented by a time history of force $F(t)$. The aim of our analysis is to investigate the relation between the acoustic input $F(t)$ and the voltage developed across the amplifier input terminals, which is the electrical output.

To obtain equations of motion, we select generalized coordinates and apply Lagrange's equations. Except for the transducer characteristics, the system of Fig. 6–16 is quite similar to the loudspeaker system of Fig. 6–12. The same generalized coordinates, q and x, can be used. We shall take the origin of x to correspond with the position in which there is no force in the spring. This corresponds with the natural position of the microphone diaphragm when there is no electrical field. Using (6–7) for the electrical energy of the transducer, the lagrangian (6–36) is

$$\mathcal{L} = T^* - V - W_e = \tfrac{1}{2}m\left(\frac{dx}{dt}\right)^2 - \tfrac{1}{2}kx^2 - \frac{q^2}{2C(x)}. \qquad (6\text{–}66)$$

The work expression for the nonconservative elements is

$$\left(E_0 - R\frac{dq}{dt}\right)\delta q + \left(-b\frac{dx}{dt}\right)\delta x. \qquad (6\text{–}67)$$

When we apply Lagrange's equations (6–38) to (6–66) and (6–67), we obtain

$$m\frac{d^2x}{dt^2} + b\frac{dx}{dt} + kx - \frac{q^2 C'}{2C^2} = F(t)$$

$$R\frac{dq}{dt} + \frac{q}{C} = E_0 \qquad (6\text{–}68)$$

as a pair of coupled differential equations of motion for q and x.

LINEARIZATION OF THE EQUATIONS OF MOTION

Because of the nonlinear capacitance function $C(x)$ and the appearance of q^2 in the first of (6–68), the equations (6–68) for the microphone are *nonlinear*. This is in contrast to the loudspeaker, where the corresponding equations (6–60) are linear. Now, although the equations (6–68) are nonlinear, if the input $F(t)$ is small enough, the deviations of q and

x from their equilibrium values will be sufficiently small so that they can be predicted from *linearized* approximations to the equations (6–68).

In order to perform the linearization, it is first necessary to determine the equilibrium configuration of the microphone due to the voltage source E_0. To permit an explicit solution here, we shall make a particular choice for the capacitance function $C(x)$. The capacitance of an ideal parallel-plate condenser, with plates having area A and separation d, is

$$C = \frac{\epsilon A}{d}, \tag{6-69}$$

where ϵ is the dielectric constant of the medium between the plates. We therefore assume that the capacitance function $C(x)$ has the form

$$C(x) = C_0 \frac{d_0}{d_0 + x}, \tag{6-70}$$

where d_0 is the separation between diaphragm and back plate when $x = 0$, and C_0 is the capacitance of the condenser microphone when $x = 0$. Although $C(x)$ as given by (6–70) is nonlinear in x, the reciprocal of $C(x)$ is linear. As a consequence, the expression

$$-\frac{C'(x)}{C^2(x)} = \frac{1}{C_0 d_0} \tag{6-71}$$

is independent of x. When we insert (6–70) and (6–71) in (6–68), we obtain

$$m \frac{d^2x}{dt^2} + b \frac{dx}{dt} + kx + \frac{q^2}{2C_0 d_0} = F(t)$$

$$R \frac{dq}{dt} + \frac{q(d_0 + x)}{C_0 d_0} = E_0 \tag{6-72}$$

as the equations of motion corresponding to the assumed capacitance function (6–70).

The equilibrium position of the uncharged microphone is $x = 0$. To determine the equilibrium position when the microphone is charged by the voltage source E_0, we set $F(t) = 0$ in (6–72) and seek the time-independent solution. Deleting the time-derivative terms from (6–72), we find

$$kx + \frac{q^2}{2C_0 d_0} = 0$$

$$\frac{q(d_0 + x)}{C_0 d_0} = E_0 \tag{6-73}$$

as nonlinear algebraic equations for the equilibrium values of x and q.

Eliminating q, we obtain a single equation for x,

$$-kx = \frac{C_0 E_0^2}{2d_0(1 + x/d_0)^2},\qquad (6\text{--}74)$$

which can be interpreted as a statement of mechanical equilibrium for the diaphragm. In equilibrium, the spring force on the left of (6–74) must balance the attractive force between plates on the right. The nature of the solutions to (6–74) is portrayed graphically in Fig. 6–17. The two sides of (6–74) are plotted separately as functions of x, and solutions are indicated by intersections of the curves.

Several interesting conclusions can be drawn from Fig. 6–17. For small values of E_0, there are *two* intersections, A and B, and for large[1] values of E_0, there are *no* real intersections. When there are two inter-

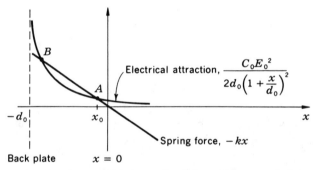

FIG. 6–17. For equilibrium, the spring force must balance the electrical attraction.

sections, as in Fig. 6–17, the intersection at A is stable and the intersection at B is unstable. This can be seen by considering small displacements from equilibrium. The spring force acts to bring the diaphragm toward $x = 0$, while the electrical attraction acts to move the diaphragm toward the back plate, $x = -d_0$. If the diaphragm moves to the right of point A, where these forces balance, the electrical attraction becomes larger than the spring force, which tends to drive the diaphragm back toward A. Also, if the diaphragm moves to the left of point A, the spring force becomes larger than the electrical attraction, and the diaphragm is again driven back toward A. The equilibrium at A is said to be *stable*.

At point B we have the opposite situation. If the diaphragm moves in either direction from the equilibrium position at B, the unbalanced force tends to drive the diaphragm farther away from B. The equilibrium at B is said to be *unstable*.

[1] See Prob. 6–9.

We shall designate the stable-equilibrium solution of (6–74) by the symbol x_0. The corresponding equilibrium value of charge is called q_0; i.e., x_0 and q_0 constitute a (stable) solution to (6–73). We now return to the dynamic equations (6–72) and assume solutions of the form

$$
\begin{aligned}
x(t) &= x_0 + x_1(t) \\
q(t) &= q_0 + q_1(t),
\end{aligned}
\tag{6–75}
$$

where $x_1(t)$ and $q_1(t)$ represent dynamic deviations from the equilibrium values x_0 and q_0. Furthermore, we assume that the excitation $F(t)$ is so small that x_1/x_0 and q_1/q_0 can be treated as first-order infinitesimals, in the sense that products and powers of these ratios can be neglected in comparison with the ratios themselves. Then, when we substitute (6–75) into (6–72), we can arrange the result as follows:

$$
\left(kx_0 + \frac{q_0^2}{2C_0 d_0} \right) + \left(m\frac{d^2 x_1}{dt^2} + b\frac{dx_1}{dt} + kx_1 + \frac{q_0 q_1}{C_0 d_0} \right) + (\cdot \cdot \cdot) = F(t)
$$

$$
\left(\frac{q_0(d_0 + x_0)}{C_0 d_0} - E_0 \right) + \left(R\frac{dq_1}{dt} + \frac{q_1(d_0 + x_0)}{C_0 d_0} + \frac{q_0 x_1}{C_0 d_0} \right) + (\cdot \cdot \cdot) = 0.
$$

$$
\tag{6–76}
$$

The first parentheses on the left of (6–76) contain terms involving only x_0 and q_0, the second parentheses contain terms which are linear in x_1 and q_1, and the third parentheses contain terms which involve products and powers of x_1 and q_1. Now, since the equilibrium values x_0 and q_0 satisfy (6–73), the first parentheses on the left of (6–76) are equal to zero. If we agree to neglect the higher-order terms in the third parentheses, we are left with the *linearized approximation* to the equations of motion,

$$
m\frac{d^2 x_1}{dt^2} + b\frac{dx_1}{dt} + kx_1 + \frac{q_0}{C_0 d_0}q_1 = F(t)
$$

$$
R\frac{dq_1}{dt} + \frac{d_0 + x_0}{C_0 d_0}q_1 + \frac{q_0}{C_0 d_0}x_1 = 0.
$$

$$
\tag{6–77}
$$

A solution to the linearized equations of (6–77) does not provide an exact solution to the nonlinear equations of (6–72), but if x_1/x_0 and q_1/q_0 are small, the deviations[1] will be of the order of magnitude of the squares and products of these quantities. Note that the coefficients in the linear differential equations (6–77) depend on the equilibrium values x_0 and q_0, which can be obtained only by solving the nonlinear algebraic equations (6–73).

---

[1] If the accuracy of the linear approximation is in question, it is possible to go back and estimate distortions in the linear solution due to the omitted higher-order terms. See F. V. Hunt, "Electroacoustics," Harvard University Press, Cambridge, Mass., and John Wiley & Sons, Inc., New York, 1954, p. 192.

FREQUENCY RESPONSE

We shall use the linearized equations (6–77) to estimate the amplitude of the response voltage which appears across the terminals of the amplifier in Fig. 6–15 due to a steady sinusoidal oscillation of the acoustic force at frequency ω. The voltage across the amplifier terminals is $R\, dq_1/dt$. When the exciting force $F(t)$ is taken as the real part of $e^{j\omega t}$, the response voltage $R\, dq_1/dt$ can be taken as the real part of $H(\omega)e^{j\omega t}$, where $H(\omega)$ is the (unknown) complex amplitude of the response. In order to obtain relations between complex amplitudes, it is sufficient to assume that all time-varying quantities have the form of a complex amplitude multiplied by $e^{j\omega t}$. Thus, putting

$$x_1 = X(\omega)e^{j\omega t} \qquad q_1 = Q(\omega)e^{j\omega t} \qquad F(t) = e^{j\omega t} \qquad (6\text{–}78)$$

in (6–77), we obtain

$$\begin{bmatrix} -\omega^2 m + j\omega b + k & \dfrac{q_0}{C_0 d_0} \\[2ex] \dfrac{q_0}{C_0 d_0} & j\omega R + \dfrac{d_0 + x_0}{C_0 d_0} \end{bmatrix} \begin{Bmatrix} X \\ Q \end{Bmatrix} = \begin{Bmatrix} 1 \\ 0 \end{Bmatrix} \qquad (6\text{–}79)$$

as a pair of equations for the amplitudes $X(\omega)$ and $Q(\omega)$. Writing the response voltage in the same way, we have

$$R\frac{dq_1}{dt} = H(\omega)e^{j\omega t} = j\omega R Q(\omega)e^{j\omega t}, \qquad (6\text{–}80)$$

or
$$H(\omega) = j\omega R Q(\omega). \qquad (6\text{–}81)$$

Thus the desired complex amplitude of the response, $H(\omega)$, is obtained by solving (6–79) for $Q(\omega)$ and substituting in (6–81). Carrying out this calculation, we find

$$H(\omega) = \frac{-j\omega R q_0}{(j\omega R C_0 d_0 + d_0 + x_0)(-\omega^2 m + j\omega b + k) - q_0^2/C_0 d_0}. \qquad (6\text{–}82)$$

We shall not enter into a complete analysis of the effects of the individual parameters in (6–82). We can, however, obtain a qualitative estimate of the behavior of (6–82) by considering some limiting cases. Since the amplifier input resistance R is usually very high, let us look at the limiting form taken by (6–82) when $R \to \infty$:

$$\lim_{R \to \infty} H(\omega) = \frac{-q_0/C_0 d_0}{-\omega^2 m + j\omega b + k}. \qquad (6\text{–}83)$$

In Fig. 6–18 the solid curve shows a logarithmic plot of the magnitude of (6–83) as a function of frequency. Within the limitations of this approximation we see that the microphone has substantially uniform response over a wide band of frequencies which lie below the natural frequency,

$\omega_n = \sqrt{k/m}$, of the diaphragm's first mechanical resonance. Thus, for a high-fidelity instrument, ω_n should be made as large as possible. In addition, this approximation also suggests that the amplitude of the resonant peak at $\omega = \omega_n$ is principally controlled by the mechanical damping constant b; i.e., increasing the damping should flatten out the peak and thereby increase the frequency range of the microphone.

A more extensive analysis[1] of (6–82) shows that when R is finite but large, the most important deviations from (6–83) occur at very low fre-

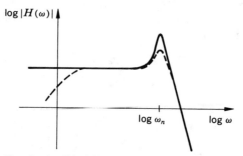

FIG. 6–18.   Limiting response of microphone as $R \rightarrow \infty$. Dotted curve indicates response for large, but finite, R.

quencies and in the neighborhood of the resonant peak.   The resulting response is indicated by the dotted curve in Fig. 6–18.

6–4   Variable mutual inductance

The transducers treated in the preceding sections have all permitted characterization in terms of two independent coordinates, one generalized mechanical coordinate and one generalized electrical coordinate.   An alternative description in terms of power flows is to say that these transducers are *two-port* devices.[2]   Power enters (or leaves) at one pair of electrical terminals (which constitutes one port) and at one moving mechanical junction (which constitutes the other port).   At each port the power flow is computed as a product of power-related variables.   At a pair of electrical terminals, power flow is ei.   At a translating mechanical junction, power flow is fv.

In this section we consider an example of a *three-port* device, a transducer with two pairs of electrical terminals and one mechanical port.

[1] See Probs. 6–11 and 6–12.

[2] The *ideal gyrator* of Prob. 4–28 is an example of an all-mechanical two-port device, and the *ideal transformer* of Prob. 5–16 is an example of an all-electrical two-port device.

The fundamental mechanism involved is variation in mutual inductance due to mechanical motion. We first examine the phenomenon of mutual inductance without mechanical motion and then go on to study transducer properties when motion is possible. The theory is illustrated by a model of a single-phase synchronous generator.

MUTUAL INDUCTANCE

In Fig. 6–19 we consider two fixed inductors placed close to one another so that there is interaction between their magnetic fields. The flux

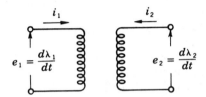

FIG. 6–19. Pair of interacting inductors.

linkage in one coil depends on the currents flowing in both coils. The constitutive relations would have the form

$$\lambda_1 = \lambda_1(i_1, i_2)$$
$$\lambda_2 = \lambda_2(i_1, i_2). \tag{6–84}$$

We assume that λ_1 and λ_2 are single-valued functions of i_1 and i_2 which can be inverted to give

$$i_1 = i_1(\lambda_1, \lambda_2)$$
$$i_2 = i_2(\lambda_1, \lambda_2) \tag{6–85}$$

as alternative constitutive relations. The power flowing into the pair of coils at any instant is $e_1 i_1 + e_2 i_2$. The work done during a time interval dt is

$$e_1 i_1\, dt + e_2 i_2\, dt = i_1\, d\lambda_1 + i_2\, d\lambda_2. \tag{6–86}$$

If the pair of coils is *conservative*, this work equals the increase dW_m of stored magnetic energy, and $W_m(\lambda_1, \lambda_2)$ is obtained by integrating (6–86) along *any* path in the $\lambda_1\lambda_2$ plane from a datum to the point (λ_1, λ_2). In order for the integral to be path-independent, it is necessary that the constitutive relations (6–85) satisfy[1] the *integrability condition*

$$\frac{\partial i_1}{\partial \lambda_2} = \frac{\partial i_2}{\partial \lambda_1}. \tag{6–87}$$

[1] See p. 16. See also (2–110).

If (6–87) is satisfied, $W_m(\lambda_1,\lambda_2)$ exists and the constitutive relations (6–85) can be recovered by differentiation.

$$\frac{\partial W_m}{\partial \lambda_1} = i_1 \qquad \frac{\partial W_m}{\partial \lambda_2} = i_2. \tag{6–88}$$

An alternative state-function is the *magnetic coenergy* $W_m^*(i_1,i_2)$ defined by the Legendre transformation,

$$W_m^*(i_1,i_2) = \lambda_1 i_1 + \lambda_2 i_2 - W_m(\lambda_1,\lambda_2). \tag{6–89}$$

The constitutive relations (6–84) can be recovered by differentiating (6–89).

$$\frac{\partial W_m^*}{\partial i_1} = \lambda_1 \qquad \frac{\partial W_m^*}{\partial i_2} = \lambda_2. \tag{6–90}$$

Furthermore, since the second mixed partial derivatives of any (sufficiently smooth) function are equal, independently of the order of differentiation, the constitutive relations (6–84) must also satisfy an integrability requirement,

$$\frac{\partial \lambda_1}{\partial i_2} = \frac{\partial \lambda_2}{\partial i_1}, \tag{6–91}$$

in the conservative case.

When the constitutive relations of a conservative pair of coils is *linear*, we can give explicit formulas for W_m and W_m^*. In the linear case (6–84) would take the form

$$\begin{aligned} \lambda_1 &= L_{11}i_1 + L_{12}i_2 \\ \lambda_2 &= L_{21}i_1 + L_{22}i_2, \end{aligned} \tag{6–92}$$

where L_{11} and L_{22} are called *self-inductance* coefficients, and L_{12} and L_{21} are called *mutual-inductance* coefficients. In this case the integrability condition (6–91) reduces to the requirement that

$$L_{12} = L_{21}. \tag{6–93}$$

The constitutive relations (6–92) can thus be written in matrix form,

$$\begin{Bmatrix} \lambda_1 \\ \lambda_2 \end{Bmatrix} = \begin{bmatrix} L_{11} & L_{12} \\ L_{12} & L_{22} \end{bmatrix} \begin{Bmatrix} i_1 \\ i_2 \end{Bmatrix}, \tag{6–94}$$

where the *inductance matrix*, or L matrix, is *symmetric*. The inverse of (6–94) has the form

$$\begin{Bmatrix} i_1 \\ i_2 \end{Bmatrix} = \begin{bmatrix} L_{11} & L_{12} \\ L_{12} & L_{22} \end{bmatrix}^{-1} \begin{Bmatrix} \lambda_1 \\ \lambda_2 \end{Bmatrix} = \begin{bmatrix} \Gamma_{11} & \Gamma_{12} \\ \Gamma_{12} & \Gamma_{22} \end{bmatrix} \begin{Bmatrix} \lambda_1 \\ \lambda_2 \end{Bmatrix}, \tag{6–95}$$

where the Γ matrix, which is the inverse[1] of the L matrix, is also symmetric.

[1] See Prob. 6–15.

The magnetic energy W_m is the integral of (6–86) along any path. Taking $\lambda_1 = 0$, $\lambda_2 = 0$, as the datum, following the path indicated in Fig. 6–20, and using the constitutive relations (6–95), we have

$$
\begin{aligned}
W_m(\lambda_1,\lambda_2) &= \int_0^{\lambda_1} \Gamma_{11}\lambda_1 \, d\lambda_1 + \int_0^{\lambda_2} (\Gamma_{12}\lambda_1 + \Gamma_{22}\lambda_2) \, d\lambda_2 \\
&= \tfrac{1}{2}(\Gamma_{11}\lambda_1^2 + 2\Gamma_{12}\lambda_1\lambda_2 + \Gamma_{22}\lambda_2^2) \\
&= \tfrac{1}{2}\{\lambda_1 \ \lambda_2\}
\begin{bmatrix} \Gamma_{11} & \Gamma_{12} \\ \Gamma_{12} & \Gamma_{22} \end{bmatrix}
\begin{Bmatrix} \lambda_1 \\ \lambda_2 \end{Bmatrix}.
\end{aligned}
\tag{6–96}
$$

The magnetic coenergy (6–89) is evaluated by inserting (6–96) and using (6–93) to eliminate λ_1 and λ_2 in favor of i_1 and i_2.

$$
\begin{aligned}
W_m^*(i_1,i_2) &= \tfrac{1}{2}(L_{11}i_1^2 + 2L_{12}i_1i_2 + L_{22}i_2^2) \\
&= \tfrac{1}{2}\{i_1 \ i_2\}
\begin{bmatrix} L_{11} & L_{12} \\ L_{12} & L_{22} \end{bmatrix}
\begin{Bmatrix} i_1 \\ i_2 \end{Bmatrix}.
\end{aligned}
\tag{6–97}
$$

The parallel between the magnetic energy (6–96) for a pair of coils and

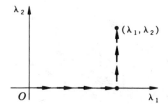

Fig. 6–20. Integration path for magnetic energy.

the potential energy (4–45) of an elastic cantilever beam should be noted. The coils store energy from two electrical ports, and the beam stores energy from two mechanical ports (a translational port and a rotational port).

MUTUAL-INDUCTANCE TRANSDUCER

We next consider a pair of coils in which the relative orientation of the interacting coils can be changed. In Fig. 6–21 coil 1 is stationary, and is called the *stator*. Coil 2 is attached to a shaft which can turn in fixed bearings, and is called the *rotor*. The angular displacement θ of the rotor with respect to the stator can change with time. The external torque τ acting on the rotor is the torque required to balance the magnetic torque. When the rotor turns with angular velocity, $\omega = d\theta/dt$, mechanical power $\tau\omega$ is delivered to the device.

In the ideal transducer we neglect electrical resistance and capacitance and mechanical friction and inertia. These effects must be introduced by auxiliary elements when we model a real device.

We shall take the transducer of Fig. 6–21 to be conservative and electrically linear. We assume that the effect of the varying geometry can be represented by taking the inductance coefficients to be functions of θ; i.e., we take the constitutive relations to have the form

$$\begin{Bmatrix} \lambda_1 \\ \lambda_2 \end{Bmatrix} = \begin{bmatrix} L_{11}(\theta) & L_{12}(\theta) \\ L_{12}(\theta) & L_{22}(\theta) \end{bmatrix} \begin{Bmatrix} i_1 \\ i_2 \end{Bmatrix}$$

$$\begin{Bmatrix} i_1 \\ i_2 \end{Bmatrix} = \begin{bmatrix} \Gamma_{11}(\theta) & \Gamma_{12}(\theta) \\ \Gamma_{12}(\theta) & \Gamma_{22}(\theta) \end{bmatrix} \begin{Bmatrix} \lambda_1 \\ \lambda_2 \end{Bmatrix}. \tag{6–98}$$

There are now three ports through which power can be delivered to the

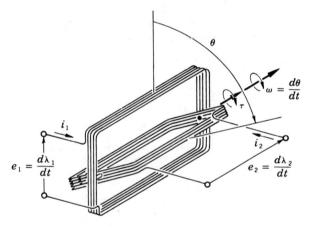

FIG. 6–21. Ideal mutual-inductance transducer.

transducer, two electrical and one mechanical. The work done on the transducer in an interval dt is

$$e_1 i_1\, dt + e_2 i_2\, dt + \tau \omega\, dt = i_1\, d\lambda_1 + i_2\, d\lambda_2 + \tau\, d\theta. \tag{6–99}$$

The total stored energy $W_m(\lambda_1, \lambda_2, \theta)$ is the integral of (6–99), along any path in λ_1, λ_2, θ space, from a datum point to the operating point. Now this transducer has an important property in common with the movable-plate capacitor and the movable-core inductor. When there is no magnetic field, there is no magnetic torque, and hence no external torque τ is required. Thus, when the magnetic field is zero, we can displace the rotor at will without storing any energy.

Taking $\lambda_1 = 0$, $\lambda_2 = 0$, $\theta = 0$, as the datum configuration, the integration of (6–98) is simplest if we take the first leg of our path along the θ axis. We can reach the operating geometric position without introducing any energy.   Then, with θ fixed, the L matrix and Γ matrix become constant matrices, and we can proceed to integrate along the path of Fig. 6–20 to obtain the stored magnetic energy,

$$W_m(\lambda_1,\lambda_2,\theta) = \tfrac{1}{2}\{\lambda_1\ \lambda_2\} \begin{bmatrix} \Gamma_{11}(\theta) & \Gamma_{12}(\theta) \\ \Gamma_{12}(\theta) & \Gamma_{22}(\theta) \end{bmatrix} \begin{Bmatrix} \lambda_1 \\ \lambda_2 \end{Bmatrix}, \qquad (6\text{--}100)$$

as in (6–96).   The magnetic coenergy $W_m^*(i_1,i_2,\theta)$, defined by the Legendre transformation,

$$W_m^*(i_1,i_2,\theta) = i_1\lambda_1 + i_2\lambda_2 - W_m(\lambda_1,\lambda_2,\theta), \qquad (6\text{--}101)$$

then follows from (6–97) as

$$W_m^*(i_1,i_2,\theta) = \tfrac{1}{2}\{i_1\ i_2\} \begin{bmatrix} L_{11}(\theta) & L_{12}(\theta) \\ L_{12}(\theta) & L_{22}(\theta) \end{bmatrix} \begin{Bmatrix} i_1 \\ i_2 \end{Bmatrix}. \qquad (6\text{--}102)$$

The state-functions (6–100) and (6–102) are concise representations of the properties of the transducer.   The constitutive relations for the electrical variables are recovered by differentiation, according to (6–88) and (6–90).   The mechanical torque τ is given by the derivatives

$$\frac{\partial W_m}{\partial \theta} = \tau \qquad \frac{\partial W_m^*}{\partial \theta} = -\tau, \qquad (6\text{--}103)$$

which follow from (6–99) and (6–101).   Because the transducer is conservative, we are able to derive explicit expressions for the mechanical torque corresponding to any magnetic state by using only the fact that no torque is required when there is no magnetic field.   Evaluating (6–103), we obtain

$$\begin{aligned} \tau &= \tfrac{1}{2}\{\lambda_1\ \lambda_2\} \begin{bmatrix} \Gamma_{11}'(\theta) & \Gamma_{12}'(\theta) \\ \Gamma_{12}'(\theta) & \Gamma_{22}'(\theta) \end{bmatrix} \begin{Bmatrix} \lambda_1 \\ \lambda_2 \end{Bmatrix} \\ &= -\tfrac{1}{2}\{i_1\ i_2\} \begin{bmatrix} L_{11}'(\theta) & L_{12}'(\theta) \\ L_{12}'(\theta) & L_{22}'(\theta) \end{bmatrix} \begin{Bmatrix} i_1 \\ i_2 \end{Bmatrix}, \end{aligned} \qquad (6\text{--}104)$$

where the prime on the inductance coefficients represents differentiation with respect to θ.

Example 6–4.   We consider in Fig. 6–22 a model for a single-phase synchronous generator, or alternator.   This device converts mechanical energy of rotation into electrical energy available at the slip rings.   The basic element in Fig. 6–22 is an ideal mutual-inductance transducer of the type illustrated in Fig. 6–21.   To simulate a real alternator, we also include the effects of electrical resistance and mechanical inertia in our model.

In the stator circuit we have an ideal constant-voltage source E_1 and a resistance R_1 which models the resistance of the stator winding and the internal resistance of a real source.   The rotor leads are brought out via

slip rings and connected to a resistance R_2, which models the electrical
load that absorbs the power generated by the alternator. The resistance
of the rotor winding is taken to be negligible in comparison with the load
resistance.

On the mechanical side the rotor is taken to be a rigid body. The axis
of rotation is assumed to be a principal direction, and the moment of
inertia about this axis is J. The shaft connecting the rotor with the
driver is assumed to be elastic. The elasticity is represented by a torsional spring constant k. Finally, the shaft is driven at the far right by
a constant-speed motor, so that the angular speed is ω_0. We shall obtain
equations of motion for this system by applying Lagrange's equations.

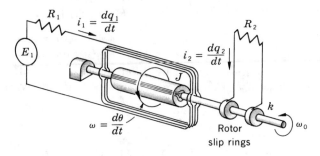

Fig. 6–22. Example 6–4. Single-phase alternator.

First we select generalized coordinates. There are two independent
single-loop electric circuits. The current-charge requirements can be
met by taking the charges q_1 and q_2 as generalized variables and setting
$i_1 = \dot{q}_1$ and $i_2 = \dot{q}_2$, as indicated in Fig. 6–22. The geometric requirements for the mechanical part of the system can be met by taking the
angular displacement of the rotor to be θ, the twist in the shaft to be
$\theta - \omega_0 t$, and the angular velocity of the rotor to be $\omega = \dot{\theta}$. The coordinates q_1, q_2, and θ constitute a complete set of independent generalized
coordinates.

The lagrangian (6–36) is

$$\mathcal{L} = T^* - V + W_m^*$$

$$= \tfrac{1}{2}J\dot{\theta}^2 - \tfrac{1}{2}k(\theta - \omega_0 t)^2 + \tfrac{1}{2}\{\dot{q}_1\ \dot{q}_2\} \begin{bmatrix} L_{11}(\theta) & L_{12}(\theta) \\ L_{12}(\theta) & L_{22}(\theta) \end{bmatrix} \begin{Bmatrix} \dot{q}_1 \\ \dot{q}_2 \end{Bmatrix}, \quad (6\text{–}105)$$

on using (6–102) for the transducer coenergy. The work expression
(6–37) for the nonconservative elements is

$$(E_1 - R_1\dot{q}_1)\,\delta q_1 + (-R_2\dot{q}_2)\,\delta q_2. \quad (6\text{–}106)$$

Applying Lagrange's equations (6–38) to (6–105) and (6–106), we obtain

$$J \frac{d^2\theta}{dt^2} + k\theta - \tfrac{1}{2}\{\dot{q}_1 \ \dot{q}_2\} \begin{bmatrix} L'_{11}(\theta) & L'_{12}(\theta) \\ L'_{12}(\theta) & L'_{22}(\theta) \end{bmatrix} \begin{Bmatrix} \dot{q}_1 \\ \dot{q}_2 \end{Bmatrix} = k\omega_0 t$$

$$\frac{d}{dt}[L_{11}(\theta)\dot{q}_1 + L_{12}(\theta)\dot{q}_2] + R_1\dot{q}_1 = E_1 \qquad (6\text{–}107)$$

$$\frac{d}{dt}[L_{12}(\theta)\dot{q}_1 + L_{22}(\theta)\dot{q}_2] + R_2\dot{q}_2 = 0$$

as equations of motion for the alternator system. We do not enter into an exhaustive study of (6–107). We can, however, use (6–107) to give a simple explanation of the usual mode of operation.

We assume that the major effect of orientation on the inductance coefficients is represented by a variation in *mutual inductance* proportional to the cosine of θ. Thus we put

$$L_{12} = L_0 \cos \theta \qquad (6\text{–}108)$$

and assume that variations in self-inductance can be neglected; i.e., we take

$$L'_{11} = \frac{dL_{11}}{d\theta} = 0 \qquad L'_{22} = \frac{dL_{22}}{d\theta} = \theta. \qquad (6\text{–}109)$$

Under this hypothesis, (6–107) takes the form

$$J \frac{d^2\theta}{dt^2} + k\theta + L_0 i_1 i_2 \sin \theta = k\omega_0 t$$

$$\frac{d}{dt}(L_{11}i_1 + L_0 i_2 \cos \theta) + R_1 i_1 = E_1 \qquad (6\text{–}110)$$

$$\frac{d}{dt}(L_0 i_1 \cos \theta + L_{22}i_2) + R_2 i_2 = 0$$

when we reintroduce the symbols i_1 and i_2 for the charge derivatives $\dot{q}_1$ and $\dot{q}_2$.

Now if the shaft is very stiff, the magnetic torque in the first of (6–110) will have little effect, and we can take $\theta \approx \omega_0 t$ and $d\theta/dt \approx \omega_0$. Suppose, also, that in the second of (6–110), the voltage developed across the stator is small compared with the voltage across the resistance R_1. If this is the case, we can take $i_1 \approx E_1/R_1$. Finally, in the third of (6–110), we suppose that the major rate of change of flux is due to the mutual inductance rather than the self-inductance. Then we may take

$$R_2 i_2 \approx -\frac{d}{dt}(L_0 i_1 \cos \theta). \qquad (6\text{–}111)$$

Inserting the approximations $d\theta/dt \approx \omega_0$ and $i_1 \approx E_1/R_1$, we obtain

$$R_2 i_2 \approx \frac{E_1 L_0}{R_1} \omega_0 \sin \omega_0 t \qquad (6\text{-}112)$$

as the approximate voltage developed across the load resistor R_2.   When the alternator operates in a manner consistent with our approximations, the voltage developed is an alternating voltage in synchronism with the driving speed.   For given electrical characteristics, the amplitude of the voltage varies directly with the speed of operation.

PROBLEMS

6-1. The capacitance of a butterfly condenser varies with rotor position according to the relation

$$C(\theta) = C_0 + C_1 \cos 2\theta,$$

where C_0 and C_1 are constants.   The rotor has axial moment of inertia J and turns without friction in fixed bearings.   Obtain the equation of motion for the rotor when an ideal constant-voltage source E_0 is connected across the condenser.

(a)   Determine all equilibrium positions for the rotor.   Which are stable?
(b)   What is the natural frequency of small oscillations in the neighborhood of a stable equilibrium position?
(c)   What is the maximum electrical torque available if

$$C_0 = 15 \times 10^{-12} \text{ farad}$$

$$C_1 = 10 \times 10^{-12} \text{ farad}$$

$$E_0 = 1{,}000 \text{ volts}?$$

(d)   If the rotor shaft is 0.25 in. in diameter and the rotor weighs 0.1 lbf, estimate the minimum coefficient of friction required to prevent motion.

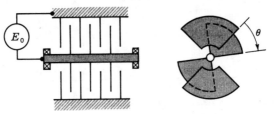

PROB. 6-1

6-2. The electrical energy W_e stored in a configuration consisting of a point charge q separated a distance d from a flat plate with charge $-q$ is

$$W_e = \frac{q^2}{16\pi\epsilon d},$$

where ϵ is the dielectric constant of the intervening medium.   Use this fact to determine the equations of motion of the system shown.   When the spring is unstretched and the pendulum is pointing straight down, the spacing between the mass particle m (with charge q) and the plate of mass M (with charge $-q$) is a.

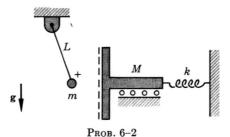

PROB. 6–2

6–3. The capacitor shown has a movable plate of mass m which is bonded to an elastic dielectrical material. When the capacitor is uncharged, the equilibrium spacing of the plates is d_0. When the spacing of the plates is $d_0 + x$, the restoring force acting on the movable plate is kx, where k is the spring constant of the dielectric and the capacitance of the element is

$$C(x) = \frac{C_0 d_0}{d_0 + x}.$$

(a) Obtain an equation of motion when the ideal constant-voltage source E_0 is connected across the capacitor.

(b) What is the stable equilibrium position, and what is the natural frequency of small oscillations in the neighborhood of equilibrium?

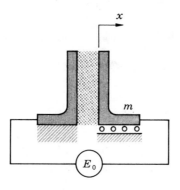

PROB. 6–3

6–4. We may call the stored energy of a transducer potential energy V, if we wish. For the energy-transfer transducer with constitutive relations (6–29), show that if the datum value of V is zero, the *potential coenergy* V^* is given by

$$V^* = \frac{f}{T} (\lambda - \lambda_0).$$

6-5. The capacitance C and inductance L in the system shown depend on the position x of the single moving part of mass m as follows:

$$C(x) = \frac{C_0 d_0}{d_0 + x}$$

$$L(x) = L_0\left(1 - \frac{x}{x_0}\right),$$

where C_0, L_0, d_0, and x_0 are constants. Assume that the spring is unstretched when $x = 0$. Take the charge on the capacitor (due to the ideal current source) to be

$$q = \frac{I_0}{\omega}\sin\omega t.$$

(a) Obtain a differential equation of motion for $x(t)$ valid for arbitrary ω.
(b) Find the frequency ω_0 for which the resultant of the electrical and magnetic forces, acting on the moving member, does not change with time.
(c) Solve for $x(t)$ when $\omega = \omega_0$.

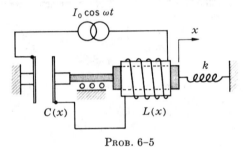

PROB. 6-5

6-6. The movable plate of a movable-plate capacitor is connected mechanically by a linear spring with spring constant k to the movable core of a movable-core inductor. The two transducers are also connected electrically, in series with an ideal constant-voltage source E_0. Let

$$L(x_1) = \frac{L_0}{1 + (x_1/x_0)^2}$$

$$C(x_2) = \frac{C_0 d_0}{d_0 - x_2},$$

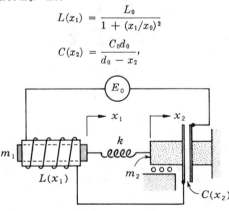

PROB. 6-6

where L_0, C_0, x_0, and d_0 are constants. Assume that when $x_1 = 0$ and $x_2 = 0$, the spring is unstretched.

(a) How many degrees of freedom does the system have?
(b) Obtain the equations of motion.

6-7. A model of an electrostatic loudspeaker is shown. $I(t)$ is an ideal current source with a prescribed signal. The ideal transducer is completely described by its electrical-energy function,

$$W_e(x,q) = \frac{q^2}{2C_0}\left(1 + \frac{x}{d_0}\right).$$

The transducer moves an actuator plate of mass m, and the resistance of the surrounding air is modeled by the dashpot. Obtain equations of motion for $q(t)$ and $x(t)$.

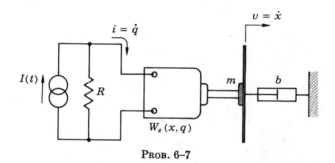

PROB. 6-7

6-8. Consider the admittance function $Y_m(\omega)$ which is the reciprocal of the function $Z_m(\omega)$ defined by (6-65). Show that in the complex Y plane, $Y_m(\omega)$ plots as a straight line. Use this to show that $Z_m(\omega)$ plots as a circle in the complex Z plane.

6-9. For the condenser microphone of Example 6-3, show that the spring force cannot balance the electrical attraction in any position when the charging voltage E_0 is greater than E_{max}, where

$$E_{max}^2 = \frac{4}{27}\frac{kd_0^2}{C_0}.$$

What happens to the diaphragm when $E_0 > E_{max}$? What is the magnitude of E_{max} when the parameters have the following values: $k = 10^3$ lbf/in., $d_0 = 0.010$ in., $C_0 = 10 \times 10^{-12}$ farad?

6-10. A model of a vibration instrumentation scheme is shown. A vibratory system represented by a mass m, supported by a spring and a dashpot, is attached to a moving foundation (shaker table) which is driven with a prescribed displacement history $x_0(t)$. A capacitance pickup utilizes the mass as one plate of a movable-plate capacitor. Take the capacitance to be

$$C(x) = \frac{C_0 d_0}{d_0 - x}$$

and assume that the spring is unstretched when $x_0 = x$. The capacitance pickup is charged by the ideal constant-voltage source E_0, and the resistor R models the input resistance of a signal amplifier. The capacitance C_1 models the stray capacitance of a shielded coaxial cable running from the pickup to the amplifier.

(a)   Obtain equations of motion for the complete electromechanical system.

(b)   Suppose that the force of electrical origin acting on the mass can be neglected. Then the motion of the mass can be obtained independently of the electrical variables. Take $x_0 = \text{Re } [e^{j\omega t}]$ and $x = \text{Re } [G(\omega)e^{j\omega t}]$ under these circumstances. Find $G(\omega)$.

(c)   Under the conditions of (b), obtain a linearized differential equation connecting $x(t)$ and the voltage $e(t)$ across R (assume $x/d_0 \ll 1$). When $x_0 = \text{Re } [e^{j\omega t}]$, assume $e(t) = \text{Re } [H(\omega)e^{j\omega t}]$. Find $H(\omega)$.

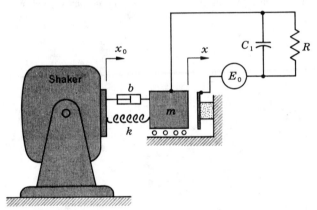

PROB. 6–10

6–11. Examine the behavior of (6–82) for very low frequencies by retaining in the denominator only those terms which are independent of ω. Repeat the analysis, retaining in addition those terms which are linear in ω. Compare these results with the low-frequency behavior of (6–83).

6–12. Study the behavior of (6–82) in the neighborhood of $\omega = \omega_n$. Obtain an exact expression for $H(\omega_n)$ and an approximate result for large R. Compare these results with the corresponding value obtained from (6–83).

6–13. A movable-plate capacitor which is electrically linear *and* mechanically linear would have constitutive relations of the form

$$e(x,q) = \frac{q}{C(x)} \qquad f(x,q) = xk(q),$$

where the capacitance is a function of x and the spring constant is a function of q.

(a)   Find the state-functions $W_e(x,q)$ and $W_e^*(x,e)$ for such a device.

(b)   What must be the functional forms of $C(x)$ and $k(q)$?

(c)   Show that when the *voltage* across such a capacitor is held constant, the force is no longer a strictly linear function of x.

6–14. A movable-core inductor which is electrically linear *and* mechanically linear would have constitutive relations of the form

$$i(x,\lambda) = \frac{\lambda}{L(x)} \qquad f(x,\lambda) = xk(\lambda),$$

where the inductance is a function of x and the spring constant is a function of λ.

(a) Find the state-functions $W_m(x,\lambda)$ and $W_m^*(x,i)$ for such a device.
(b) What must be the functional forms of $L(x)$ and $k(\lambda)$?
(c) Show that when the current through the inductor is held constant, the force is no longer a strictly linear function of x.

6–15. Show that the Γ matrix and L matrix of (6–95) satisfy the following equation:

$$\begin{bmatrix} \Gamma_{11} & \Gamma_{12} \\ \Gamma_{12} & \Gamma_{22} \end{bmatrix} = \frac{1}{L_{11}L_{22} - L_{12}^2} \begin{bmatrix} L_{22} & -L_{12} \\ -L_{12} & L_{11} \end{bmatrix}.$$

6–16. Show that the common value of (6–96) and (6–97) is $\frac{1}{2}(\lambda_1 i_1 + \lambda_2 i_2)$.

6–17. Consider the alternator of Example 6–4, with the following modification. Instead of a constant-speed drive at the right end of the shaft, assume that the motor drive delivers a *constant torque* τ_0. Formulate the equations of motion for the modified system, and show, under circumstances similar to those assumed in deriving (6–111) and (6–112), that the voltage developed across the load resistor alternates with an approximate *average* frequency

$$\omega_1 = 2\tau_0 R_2 \left(\frac{R_1}{L_0 E_1} \right)^2.$$

Estimate this frequency when the parameters have the following values: $E_1 = 12$ volts, $R_1 = 4$ ohms, $R_2 = 100$ ohms, $L_0 = 2$ henrys, $\tau_0 = 10$ lbf-ft.

6–18. An ideal moving-coil transducer is described by the constitutive relations (6–28a). If the transducer constant T_{em} is 10 millivolts per inch per second, what is the magnitude of the constant T_{me} in pounds force per ampere?

CHAPTER SEVEN

Dynamics of continuous systems

In this chapter we consider mechanical and electrical systems for which one-dimensional continuous media are used as dynamic models. We examine the dynamic behavior of bars, strings, beams, and electric transmission lines. Both direct and variational methods of analysis are extended to obtain equations of motion which include *partial* differential equations. To illustrate an important application of the equations of motion, we consider some examples of wave propagation.

7–1   Continuum models

Consider a uniform bar subject to longitudinal forces and capable of longitudinal motion. Figure 7–1 indicates a variety of dynamic models for the bar. When the bar is light and stiff in comparison with the major elements of a dynamic system, we may model the bar as in Fig. 7–1a by a *rigid massless* element. In another situation, where the elasticity of the bar is important but the inertia is not, we may model the bar by an ideal *spring*, as in Fig. 7–1b. Alternatively, in situations where the elasticity can be neglected but the inertia cannot, we may model the bar by a *rigid body*, with mass M, as in Fig. 7–1c.

In systems subject to rapid small motions such as are encountered in vibrations and wave motions, it is necessary to account for both the elasticity and the inertia of the bar. Figure 7–1d shows a possible lumped-parameter model which includes both properties. Such a model would be useful for describing low-frequency vibrations or long-wavelength motions. Figure 7–1e and f shows models with increasingly finer subdivisions of the inertia and elasticity which would be useful for higher frequencies and shorter wavelengths.

A continuum model for the bar can be considered heuristically as a limit of the subdivision process indicated in Fig. 7–1, in which the number of elements goes to infinity while the size of an individual element shrinks

332

to zero. Strictly speaking, a continuum is a medium having no micro-structure and possessing assigned properties, such as inertia and elasticity, which are *continuously* (but not necessarily uniformly) distributed throughout the medium.

For a bar with cross-sectional area A, we consider a continuum model with mass density ρ and modulus of elasticity E. This implies, as indi-

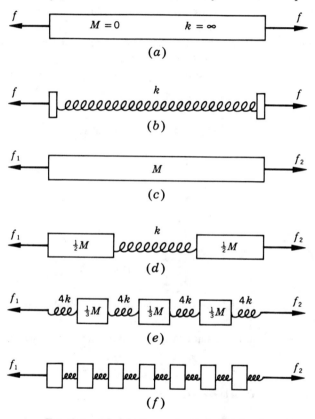

FIG. 7–1. Models for an elastic bar in tension.

cated in Fig. 7–2, that for any segment Δx, there is always some *mass* Δm, and the limit of the mass divided by the volume as the volume shrinks to zero is the mass density ρ. Also, for any segment Δx, there is always some *elasticity;* i.e., the elongation $\Delta \xi$ is proportional to the tensile force f. The limiting form of this relation is Hooke's law,[1]

$$\sigma = E \frac{\partial \xi}{\partial x}, \tag{7-1}$$

[1] See, for example, S. H. Crandall and N. C. Dahl (eds.), "Mechanics of Solids," McGraw-Hill Book Company, New York, 1959, p. 52.

where σ is the axial *stress* (here taken to be uniformly distributed across the area A), and $\partial\xi/\partial x$ is the axial *strain*. Depending on the application, the properties ρ and E may vary with position along the length of the bar.

The continuum model may represent the bar of Fig. 7–1 with great accuracy over a wide frequency range. The continuum model is still a model, however, and it has its limitations. The properties assigned to the model may not in fact be adequate to describe completely the real phenomenon; e.g., it might be necessary to assign a distributed damping to model the internal friction of the bar material. In addition, for very high frequencies and very short wavelengths, continuum models are

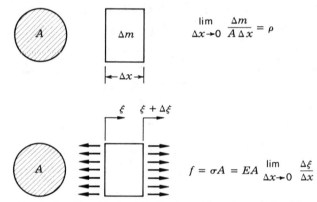

$$\lim_{\Delta x \to 0} \frac{\Delta m}{A\,\Delta x} = \rho$$

$$f = \sigma A = EA \lim_{\Delta x \to 0} \frac{\Delta \xi}{\Delta x}$$

FIG. 7–2.   Continuous distribution of inertia and elasticity.

faulty, because real materials actually do have microstructure (molecules, crystal lattices, etc.).

7–2   Analysis of continuous systems

Dynamic equations of motion for continuous models can be formulated by using extensions of the methods employed for lumped-parameter models. The direct method involves writing explicit relations representing all the governing requirements and using algebraic elimination to obtain a concise formulation. When the direct method is applied to a continuous medium whose assigned properties change smoothly, it is possible to write down all the governing requirements in terms of one representative differential volume element. Algebraic elimination leads to one or more *partial* differential equations which must be satisfied throughout the interior of the medium. In general, a continuous medium is only part of a system, so that in addition to the partial differential equations describing the continuum, it is necessary to obtain equations describing the behavior of the remainder of the system. These equations act as *boundary conditions* for the partial differential equations.

The governing requirements of a lumped-parameter model provide a heuristic guide to the corresponding requirements for a continuum model. In Fig. 7–3 we consider a lumped-parameter model side by side with a continuous model for the bar of Fig. 7–1. In the lumped-parameter model, i is an index which distinguishes one section of the model from another. In the continuous model x is distance from a reference station measured under a reference condition (usually equilibrium). The continuous position variable x identifies a particular section of the continuous model, just as the discrete index variable i identifies a particular section of the lumped-parameter model. The coordinate $\xi_i(t)$ is the dis-

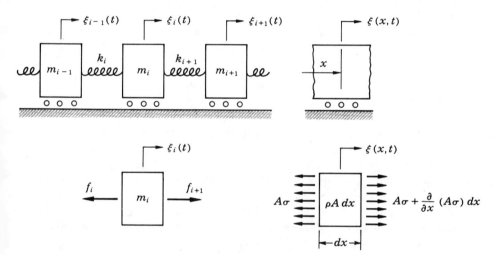

FIG. 7–3.   Analogy between lumped-parameter model and continuous model.

placement of the mass m_i in the lumped-parameter model, and the coordinate $\xi(x,t)$ is the displacement of the section with position variable x.

Geometric compatibility for the lumped-parameter model requires that the ξ_i be independent and that the elongation of the ith spring be taken as $\xi_i - \xi_{i-1}$. For the continuous model the corresponding requirement is that $\xi(x,t)$ should be a single-valued function with an x-derivative (7–1). The linear-momentum principle applied to m_i requires that

$$f_{i+1} - f_i = m_i \frac{d^2\xi_i}{dt^2}, \tag{7–2}$$

while the corresponding result for a differential length dx of the continuous model is

$$\frac{\partial}{\partial x}(A\sigma) = \rho A \frac{\partial^2 \xi}{\partial t^2}. \tag{7-3}$$

The constitutive relations for the springs in the lumped-parameter model of Fig. 7–3 permit us to eliminate the forces f_i and f_{i+1} from (7–2) as follows:

$$k_{i+1}(\xi_{i+1} - \xi_i) - k_i(\xi_i - \xi_{i-1}) = m_i \frac{d^2 \xi_i}{dt^2}. \tag{7-4}$$

Equation (7–4) is an ordinary differential equation of motion for ξ_i with coupling to ξ_{i-1} and ξ_{i+1}. In a long chain of similar elements the corresponding equations are obtained by simply advancing the index variable i. Special attention would, however, have to be given to the first and last sections, where the bar either ends or is coupled to other parts of a larger system.

The constitutive relations for the continuous model of Fig. 7–3 are represented by Hooke's law (7–1). Eliminating the stress σ between (7–1) and (7–3), we obtain

$$\frac{\partial}{\partial x}\left(EA \frac{\partial \xi}{\partial x}\right) = \rho A \frac{\partial^2 \xi}{\partial t^2} \tag{7-5}$$

as a single partial differential equation for $\xi(x,t)$. Note that (7–5) is valid in this form if E, A, and ρ vary with the position x. Equation (7–5) has been derived for a particular differential element dx in Fig. 7–3, but it can be applied to any other element within the continuum by simply advancing the position variable x. The analogy between (7–4) and (7–5) should be noted. Here, also, special attention must be given to the end points of the continuum in order to obtain *boundary conditions*. This is illustrated in the examples which follow.

7–3   Hamilton's principle for continuous systems

The variational method of formulating equations of motion for continuous models is a natural extension of the corresponding method for lumped-parameter models. The state-functions and work expressions which enter the variational indicator (2–222) of Hamilton's principle must now include *integrals* over the continuous portions of the system. Only minor alterations in the calculations are required to handle this extension.

For the bar of Fig. 7–1 we can use the analogy of Fig. 7–3 as a heuristic guide to the proper formulation for the continuous model. Here we consider only the interior of the bar, and defer the formulation for a complete system (which includes boundary conditions) to the examples which follow.

The kinetic coenergy for the lumped-parameter model of Fig. 7–3 would have the form

$$\sum_i \tfrac{1}{2} m_i \dot{\xi}_i^2, \tag{7-6}$$

where the summation includes all the masses in the model of the bar. If in the continuous model the bar is modeled by a segment extending from $x = 0$ to $x = L$, the corresponding kinetic coenergy has the form

$$T^* = \int_0^L \tfrac{1}{2} \rho A \left(\frac{\partial \xi}{\partial t} \right)^2 dx. \tag{7-7}$$

The potential energy for the lumped-parameter model due to the stretching of the springs is

$$V = \sum_i \tfrac{1}{2} k_i (\xi_i - \xi_{i-1})^2, \tag{7-8}$$

where again the summation includes all the springs in the model of the bar. For the continuous model the potential energy can be calculated as follows: For an elastic element with uniaxial strain ϵ subject to uniaxial stress σ, the work done per unit volume, in straining the element, is

$$\int_0^\epsilon \sigma \, d\epsilon = \int_0^\epsilon E\epsilon \, d\epsilon = \tfrac{1}{2} E\epsilon^2, \tag{7-9}$$

when Hooke's law, $\sigma = E\epsilon$, applies. This is the potential energy per unit volume in the strained element. In the continuous model of Fig. 7–3, the strain ϵ is $\partial \xi / \partial x$ and the volume corresponding to a length dx is $A \, dx$. The total potential energy for the bar of length L is

$$V = \int_0^L \tfrac{1}{2} EA \left(\frac{\partial \xi}{\partial x} \right)^2 dx. \tag{7-10}$$

The analogy between (7–8) and (7–10) is evident when we consider that the effective spring constant of length Δx in Fig. 7–2 is

$$k = \frac{f}{\Delta \xi} = \frac{EA}{\Delta x}. \tag{7-11}$$

The integrals (7–7) and (7–10) are typical of the integrals which enter into the variational indicator when Hamilton's principle is applied to continuous systems. In the examples which follow, the actual technique of the variational method is illustrated for several mechanical and electrical systems.

Example 7–1.   The bar in Fig. 7–4 has an equilibrium length L. We consider a one-dimensional continuous model with mass density ρ, elastic modulus E, and cross-sectional area A. We shall formulate equations and boundary conditions to describe the motion when a prescribed time history of force $f(t)$ is applied to the right-hand end.

The longitudinal displacement of the section whose equilibrium position is x is denoted by $\xi(x,t)$. The kinetic coenergy T^* of the bar is expressed by (7–7), and the potential energy V is expressed by (7–10). The work expression due to the end force is $f(t)\ \delta\xi(L,t)$. The variational indicator (2–222) for Hamilton's principle is thus

$$\int_{t_1}^{t_2} dt\ \left\{ \delta \int_0^L \left[\tfrac{1}{2}\rho A \left(\frac{\partial \xi}{\partial t}\right)^2 - \tfrac{1}{2}EA \left(\frac{\partial \xi}{\partial x}\right)^2 \right] dx + f(t)\ \delta\xi(L,t) \right\}, \quad (7\text{–}12)$$

and according to Hamilton's principle the natural motions are admissible motions for which (7–12) vanishes, for arbitrary geometrically admissible variations. In the interior of the bar any single-valued differentiable

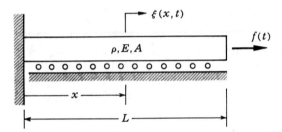

Fig. 7–4.   Example 7–1.   Longitudinal motion of a
continuous elastic bar.

function $\xi(x,t)$ represents an admissible motion, and any single-valued differentiable variation $\delta\xi$ is admissible. At the left end of the bar, $x = 0$, the bar is taken to be rigidly clamped, which requires that $\xi(0,t) = 0$ and

$$\delta\xi(0,t) = 0 \qquad\qquad (7\text{–}13)$$

for geometric admissibility. There are no geometric restrictions on the motion of the right-hand end.

The manipulation of (7–12) required to draw out the dynamic equations for the natural motion is a simple extension of the manipulations used in the previous chapters. Carrying out the indicated variations, we encounter terms of the form

$$\delta\left(\frac{\partial \xi}{\partial t}\right) \qquad \text{and} \qquad \delta\left(\frac{\partial \xi}{\partial x}\right). \qquad (7\text{–}14)$$

It can be shown[1] that the variation operation commutes with both space

[1] See Appendix A.

and time differentiation, so that these can be written

$$\frac{\partial(\delta\xi)}{\partial t} \quad \text{and} \quad \frac{\partial(\delta\xi)}{\partial x}, \tag{7-15}$$

respectively. The derivatives of the variations are then eliminated by integrating by parts timewise, for the first, and spacewise, for the second. In this way we transform (7–12) into

$$\int_{t_1}^{t_2} dt \left\{ \int_0^L \left[-\rho A \frac{\partial^2 \xi}{\partial t^2} + \frac{\partial}{\partial x}\left(EA \frac{\partial \xi}{\partial x}\right) \right] \delta\xi \, dx - EA \frac{\partial \xi}{\partial x} \delta\xi \Big|_0^L + f(t)\, \delta\xi(L,t) \right\}, \tag{7-16}$$

where the integrated portion for the space integration-by-parts is indicated but the corresponding integrated portion for the time integration-by-parts has vanished because of the agreement in Hamilton's principle that the variations should vanish at t_1 and t_2.

Now, at $x = 0$, we have the geometric admissibility requirement (7–13), which means that the integrated portion vanishes at the lower limit. The integral (7–16) can thus be rewritten

$$\int_{t_1}^{t_2} dt \left\{ \int_0^L \left[-\rho A \frac{\partial^2 \xi}{\partial t^2} + \frac{\partial}{\partial x}\left(EA \frac{\partial \xi}{\partial x}\right) \right] \delta\xi(x,t) \, dx \right.$$
$$\left. - \left[\left(EA \frac{\partial \xi(L,t)}{\partial x}\right) - f(t) \right] \delta\xi(L,t) \right\}. \tag{7-17}$$

The necessary conditions that (7–17) vanish for arbitrary values of $\delta\xi(x,t)$ throughout $0 < x < L$ and at the end $x = L$ are the partial differential equation

$$\rho A \frac{\partial^2 \xi}{\partial t^2} - \frac{\partial}{\partial x}\left(EA \frac{\partial \xi}{\partial x}\right) = 0 \qquad 0 < x < L \tag{7-18}$$

and the boundary condition

$$EA \frac{\partial \xi}{\partial x} = f(t) \qquad \text{at } x = L. \tag{7-19}$$

These are the results of applying Hamilton's principle. Note that the partial differential equation (7–18) is equivalent to (7–5) obtained by direct application of the governing requirements to a differential element. In addition to the partial differential equation, we have two boundary conditions: at $x = 0$, we have $\xi(0,t) = 0$, and at $x = L$, we have (7–19), which can be interpreted as the requirement that the force at the end of the beam as computed via Hooke's law and the strain should equal the prescribed force.

It is interesting to observe the difference in the nature of these two boundary conditions from the point of view of the variational method.

The condition at $x = 0$ is part of the geometric admissibility requirements, and thus forms part of the *input* to Hamilton's principle. The
condition (7–19) at $x = L$ is a force-balance condition, which is part of
the *output* from Hamilton's principle. Boundary conditions such as
(7–19) which are derived by applying a variational principle are usually
called *natural boundary conditions.*

The partial differential equation (7–18) and the two boundary conditions represent the extension to the continuous model of a set of ordinary
differential equations of motion for a lumped-parameter model. In
either case, in order to describe a particular motion, it would be necessary
to incorporate *initial conditions;* i.e., the initial displacements and

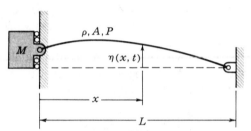

FIG. 7–5.   Example 7–2.   Transverse motion of
a tightly stretched string.

velocities of the model would have to be prescribed. For the continuous model this would require specification of

$$\xi(x,0) \qquad \text{and} \qquad \frac{\partial \xi(x,0)}{\partial t} \qquad\qquad (7\text{–}20)$$

for $0 < x < L$.

Example 7–2.   In this example we use a continuum model for a
tightly stretched flexible string. The system shown in Fig. 7–5 consists
of a rigid mass M which can slide up and down without friction and the
continuous string which has mass density ρ, cross-sectional area A, and
elastic modulus E. To model the effect of a viscous-fluid environment
for the string, we postulate a retarding force proportional to the velocity;
i.e., we assume that when the string has a transverse velocity $\partial \eta / \partial t$, it
experiences a retarding force $b \, (\partial \eta / \partial t)$ per unit length, where b is a viscous-
damping coefficient. We also assume that when the system is in equilib-
rium (gravity is neglected), the (large) tensile force in the undeflected
string is P. We shall formulate the dynamic equations for *small* transverse motions of the system.

We assume that longitudinal motions can be neglected and take the
transverse displacement at the position x to be $\eta(x,t)$. A geometrically
admissible motion is represented by an arbitrary differentiable function

$\eta(x,t)$ provided that, at $x = L$,

$$\eta(L,t) = 0 \tag{7-21}$$

and that, at $x = 0$, the displacement of the mass M is taken to be equal to $\eta(0,t)$.

The kinetic coenergy of the system is the sum of contributions from the mass M and the continuous string,

$$T^* = \tfrac{1}{2}M \left[\frac{\partial \eta (0,t)}{\partial t} \right]^2 + \int_0^L \tfrac{1}{2}\rho A \left(\frac{\partial \eta}{\partial t} \right)^2 dx. \tag{7-22}$$

To evaluate the potential energy of the deflected string, we begin by sketching the constitutive relation for the elastic string in Fig. 7–6. The

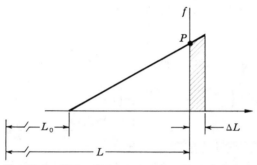

FIG. 7–6. String tension as a function of elongation.

tensile force f in the string is plotted against the elongation ΔL. We assume the unstretched length of the string to be L_0. Considerable work has been done in stretching it to length L. We establish our datum for energy at this point since we are interested only in additional elongations ΔL due to small transverse deflections $\eta(x,t)$. The potential energy V associated with such a deflection is the shaded area in Fig. 7–6. The base ΔL of the shaded trapezoid is the increase in length of the deflected string,

$$\Delta L = \int_0^L ds - L = \int_0^L \left[\sqrt{1 + \left(\frac{\partial \eta}{\partial x} \right)^2} - 1 \right] dx. \tag{7-23}$$

If we expand (7–23) in powers of the slope $\partial \eta / \partial x$, we obtain

$$\Delta L = \int_0^L \frac{1}{2} \left(\frac{\partial \eta}{\partial x} \right)^2 \left[1 - \frac{1}{4} \left(\frac{\partial \eta}{\partial x} \right)^2 + \cdots \right] dx. \tag{7-24}$$

Now, if the slope $\partial \eta / \partial x$ is a small quantity of first order, an approximation to ΔL which is correct to second order is

$$\Delta L = \int_0^L \frac{1}{2} \left(\frac{\partial \eta}{\partial x} \right)^2 dx. \tag{7-25}$$

Furthermore, since ΔL is of second order in the slope, the shaded trapezoidal area in Fig. 7–6 is given correctly to second order by the rectangular approximation

$$V = P \, \Delta L = P \int_0^L \frac{1}{2} \left(\frac{\partial \eta}{\partial x} \right)^2 dx. \tag{7-26}$$

This approximation is adequate[1] for the purpose of deriving a linear equation of motion for $\eta(x,t)$.

The lagrangian of our system is made up from (7–22) and (7–26). The work expression due to the damping force per unit length $b(\partial \eta / \partial t)$ is

$$\int_0^L - b \frac{\partial \eta}{\partial t} \, \delta \eta \, dx. \tag{7-27}$$

The variational indicator (2–222) of Hamilton's principle is thus

$$\int_{t_1}^{t_2} dt \left\{ \delta \int_0^L \left[\tfrac{1}{2}\rho A \left(\frac{\partial \eta}{\partial t} \right)^2 - \tfrac{1}{2}P \left(\frac{\partial \eta}{\partial x} \right)^2 \right] dx \right.$$
$$\left. + \delta \frac{M}{2} \left[\frac{\partial \eta(0,t)}{\partial t} \right]^2 - \int_0^L b \frac{\partial \eta}{\partial t} \, \delta \eta \, dx \right\}. \tag{7-28}$$

The equations of motion for a natural motion follow from the necessary conditions that (7–28) vanish for arbitrary admissible variations $\delta \eta$. Because of (7–21), admissible variations must vanish at $x = L$. Elsewhere there is no restriction on $\delta \eta$ (beyond differentiability).

To draw out the conditions for the vanishing of (7–28), we perform the variation of the lagrangian, interchange the orders of variation and differentiation, and integrate by parts timewise and spacewise, as appropriate, to obtain

$$\int_{t_1}^{t_2} dt \left[\int_0^L \left(-\rho A \frac{\partial^2 \eta}{\partial t^2} - b \frac{\partial \eta}{\partial t} + P \frac{\partial^2 \eta}{\partial x^2} \right) \delta \eta \, dx \right.$$
$$\left. - P \frac{\partial \eta}{\partial x} \, \delta \eta \, \Big|_0^L - M \frac{\partial^2 \eta(0,t)}{\partial t^2} \, \delta \eta(0,t) \right]. \tag{7-29}$$

At the limit $x = L$, the integrated portion of the spacewise integration by parts vanishes because $\delta \eta(L,t) = 0$ for geometric admissibility. The contribution at the limit $x = 0$ has the factor $\delta \eta(0,t)$, and can be associated with the other term at $x = 0$ having the same factor. The necessary

[1] See p. 134.

conditions that (7–29) vanish for arbitrary $\delta\eta(x,t)$ are then the *partial differential equation*

$$\rho A \frac{\partial^2 \eta}{\partial t^2} + b \frac{\partial \eta}{\partial t} - P \frac{\partial^2 \eta}{\partial x^2} = 0 \qquad 0 < x < L \qquad (7\text{–}30)$$

and the *natural boundary condition*

$$M \frac{\partial^2 \eta}{\partial t^2} - P \frac{\partial \eta}{\partial x} = 0 \qquad \text{at } x = 0. \qquad (7\text{–}31)$$

These two statements are the dynamic equations furnished by Hamilton's principle. For a complete formulation of a particular motion of the string it would be necessary to use the geometrical boundary condition $\eta = 0$ at $x = L$, together with prescribed initial velocity and displacement distributions. The formulations (7–30) and (7–31) can be alternatively derived[1] by the direct method. An advantage of the variational method when applied to systems involving continuous members and lumped elements is that, once the lagrangian and the work expression have been constructed in terms of admissible coordinates, then all equations and boundary conditions for a consistent model are automatically furnished by a routine procedure. In the direct method the same consistency is not automatic, particularly in a long analysis, where consistency depends on the compatibility of the analyst's decisions at many stages.

Example 7–3. In this example we use a one-dimensional continuum model for an elastic beam undergoing small transverse motions. The model we use[2] incorporates two forms of inertia and two mechanisms of elastic deformation. A preliminary appreciation of the model can be obtained from an examination of the lumped-parameter version sketched in Fig. 7–7.

The discrete model consists of a number of rigid rectangular bodies with mass M_i connected by rigid massless links and massless elastic springs. The joints are frictionless. Small transverse displacements η_i and small rotations φ_i in the plane of the sketch are considered. Longitudinal motions are neglected. The system is stiffened (so that it tends to remain straight) by bending springs (indicated by k_b) and shearing springs (indicated by k_s). The bending springs tend to keep the rectangular masses parallel, and the shearing springs tend to keep the connecting links perpendicular to the rectangles. The configuration in Fig. 7–7 involves mostly bending deformation àt the left and mostly shearing

---

[1] See Prob. 7–4.

[2] S. Timoshenko, On the Correction for Shear in the Differential Equation for Transverse Vibration of Prismatic Bars, *Phil. Mag.*, 41(6):744–746 (1921).

deformation at the right.   Potential energy is stored in the bending springs *and* in the shearing springs.   The kinetic coenergy of each rigid rectangle is due to translation $\dot{\eta}_i$ *and* rotation $\dot{\varphi}_i$; i.e., in addition to the mass of the rectangle, we need to know its moment of inertia with respect to an axis through the centroid perpendicular to the plane of motion.

Before considering the particular system of this example, let us develop the general form of the lagrangian for the continuous Timoshenko-beam model.   Figure 7–8 shows a segment of continuous beam which corresponds to the lumped-parameter model of Fig. 7–7.   The displacement

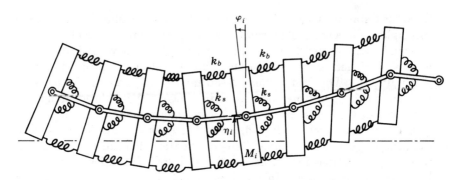

FIG. 7–7.   Lumped-parameter Timoshenko-beam model.

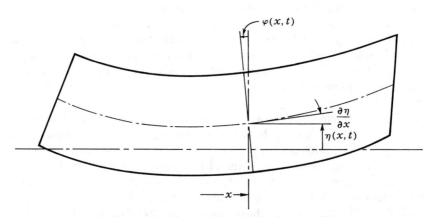

FIG. 7–8.   Continuous Timoshenko-beam model.

coordinates $\eta(x,t)$ and $\varphi(x,t)$ corresponding to $\eta_i(t)$ and $\varphi_i(t)$ are shown in Fig. 7–8.   When the beam is undeformed, the cross section is perpendicular to the neutral axis.   After deformation, the neutral axis has been rotated through the (small) angle $\partial\eta/\partial x$, while the cross section has been rotated through the angle φ.   The shearing angle γ is the net decrease in angle,

$$\gamma = \frac{\partial \eta}{\partial x} - \varphi. \tag{7-32}$$

The continuous constitutive relations corresponding to the bending and shearing springs in the discrete approximation are indicated in Fig. 7-9.   The bending moment–curvature relation (a) is a basic result[1] in mechanics of solids.   Here E is the tension modulus of the material, and I is the second moment of the cross-sectional area about the neutral axis.   The shear-force-shear angle relation (b) is somewhat less satisfactory.   If the element were in a state of uniform shear, the constant of proportionality would be[2] GA, where G is the shear modulus of the material and A is the cross-sectional area.   The element cannot, however,

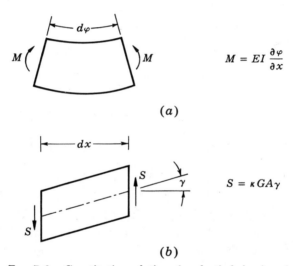

$$M = EI \frac{\partial \varphi}{\partial x}$$

(a)

$$S = \kappa GA\gamma$$

(b)

FIG. 7-9.   Constitutive relations for elastic behavior of Timoshenko beam.

be in a state of uniform shear because the top and bottom faces are stress-free.   This implies that the shear forces S must be distributed nonuniformly (e.g., a parabolic distribution) and that there is not a single angle γ for the section.   In order to retain a one-dimensional model, we must use some kind of average over the section to represent an equivalent uniform shear.   The slope of the shear-force-shear angle relation is thus indicated as κGA, where κ is a numerical factor which depends on the beam cross-sectional shape and on the rationale adopted in the averaging process.   Fortunately, there is not very much spread in the values of κ

[1] See, for example, Crandall and Dahl, *op. cit.*, p. 289.

[2] See Prob. 7-17.

obtained by different averaging processes.[1] The influence of shape is noticeable, and is indicated[2] in Table 7–1.

The linear constitutive relations of Fig. 7–9 for a differential beam element lead[3] to the integral for the elastic potential energy of a finite segment of a Timoshenko beam,

$$V = \int \left[\tfrac{1}{2} EI \left(\frac{\partial \varphi}{\partial x} \right)^2 + \tfrac{1}{2} \kappa G A \left(\frac{\partial \eta}{\partial x} - \varphi \right)^2 \right] dx, \qquad (7\text{--}33)$$

where in any particular application the appropriate limits must be inserted. Note that the geometric-compatibility relation (7–32) has been employed to eliminate γ. If the beam material has mass density ρ, the mass per unit length is ρA and the mass moment of inertia per unit

Table 7–1
Shear coefficient for Timoshenko beams

Shape of cross section	κ
Rectangle	0.85
Circle	0.89
Typical *I*-beam	0.31

length is ρI. The kinetic coenergy for a finite beam segment is then given by the integral

$$T^* = \int \left[\tfrac{1}{2} \rho A \left(\frac{\partial \eta}{\partial t} \right)^2 + \tfrac{1}{2} \rho I \left(\frac{\partial \varphi}{\partial t} \right)^2 \right] dx. \qquad (7\text{--}34)$$

The state-functions (7–33) and (7–34) describe the elastic and inertia properties of the Timoshenko beam in compact form. We shall use them with Hamilton's principle to formulate the dynamic equations for the system of Fig. 7–10.

The Timoshenko model is not the simplest one-dimensional continuum model for beam dynamics. A widely used model is the Bernoulli-Euler model,[4] which assigns only transverse inertia and bending elasticity to the continuum. The state-functions for the Bernoulli-Euler model are obtained by neglecting the shear-deformation term in (7–33) and neglecting the rotational-inertia term in (7–34). The Bernoulli-Euler model is

[1] See Probs. 7–18 and 7–19.

[2] For a more extensive table see G.R. Cowper, The Shear Coefficient in Timoshenko's Beam Theory, *J. Appl. Mechs.*, 33 : 335-340 (1966). The shear coefficient varies slightly with Poisson's ratio and with frequency. The values in Table 7-1 are for $\nu = 0.3$ and for low frequencies.

[3] See Prob. 7–10.

[4] See Prob. 7–6.

satisfactory for low frequencies and long wavelengths. The Timoshenko model provides accurate simulation over a much wider range of frequencies and wavelengths. There are also two intermediate models[1] between the Bernoulli-Euler and the Timoshenko models which have occasionally been used.

TIMOSHENKO CANTILEVER

The beam in Fig. 7–10 is modeled by a Timoshenko continuum built in at the left end and by a rigid body at the right end which has mass M_L and centroidal moment of inertia I_L. The parameters of the continuum

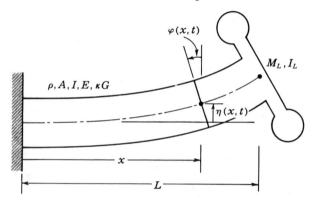

FIG. 7–10. Example 7–3. Cantilever beam with lumped inertia at end.

are the mass density ρ, the cross-sectional area A, the second moment of the cross-sectional area about the neutral axis, I, the tension modulus E, and the effective shear modulus κG. These parameters are taken to be known functions of the position x.

The linear and angular displacements $\eta(x,t)$ and $\varphi(x,t)$ are taken as generalized coordinates for the beam. Any combination of differentiable functions $\eta(x,t)$ and $\varphi(x,t)$ represents a geometrically admissible motion provided that, at $x = 0$, $\eta(0,t)$ and $\varphi(0,t)$ both vanish and that, at $x = L$, we take the linear and angular displacements of the rigid body to be $\eta(L,t)$ and $\varphi(L,t)$. The lagrangian for the system is obtained by adding the kinetic coenergy of the rigid body to the lagrangian for the continuum. Using (7–34) and (7–33), we have

$$\mathcal{L} = \int_0^L \left[\tfrac{1}{2}\rho A \left(\frac{\partial \eta}{\partial t} \right)^2 + \tfrac{1}{2}\rho I \left(\frac{\partial \varphi}{\partial t} \right)^2 - \tfrac{1}{2}EI \left(\frac{\partial \varphi}{\partial x} \right)^2 - \tfrac{1}{2}\kappa GA \left(\frac{\partial \eta}{\partial x} - \varphi \right)^2 \right] dx$$
$$+ \tfrac{1}{2}M_L \left[\frac{\partial \eta(L,t)}{\partial t} \right]^2 + \tfrac{1}{2}I_L \left[\frac{\partial \varphi(L,t)}{\partial t} \right]^2. \quad (7\text{–}35)$$

[1] See Probs. 7–22 and 7–23.

Since there are no nonconservative elements in the system, there is no work expression.

When (7–35) is inserted in the variational indicator (2–222) for Hamilton's principle, and the usual variation calculus is performed, we obtain, after timewise and spacewise integrations-by-parts,

$$
\int_{t_1}^{t_2} dt \left[\int_0^L \left(\left\{ -\rho A \frac{\partial^2 \eta}{\partial t^2} + \frac{\partial}{\partial x} \left[\kappa GA \left(\frac{\partial \eta}{\partial x} - \varphi \right) \right] \right\} \delta \eta \right. \right.
$$
$$
+ \left\{ -\rho I \frac{\partial^2 \varphi}{\partial t^2} + \frac{\partial}{\partial x} \left(EI \frac{\partial \varphi}{\partial x} \right) + \kappa GA \left(\frac{\partial \eta}{\partial x} - \varphi \right) \right\} \delta \varphi \right) dx
$$
$$
- \left\{ M_L \frac{\partial^2 \eta(L,t)}{\partial t^2} + \kappa GA \left[\frac{\partial \eta(L,t)}{\partial x} - \varphi(L,t) \right] \right\} \delta \eta(L,t)
$$
$$
\left. - \left\{ I_L \frac{\partial^2 \varphi(L,t)}{\partial t^2} + EI \frac{\partial \varphi(L,t)}{\partial x} \right\} \delta \varphi(L,t) \right], \quad (7\text{–}36)
$$

where we have already used the fact that $\delta \eta$ and $\delta \varphi$ must both vanish at $x = 0$ for geometrically admissible variations. The necessary conditions for (7–36) to vanish for arbitrary independent variations of η and φ are the pair of coupled *partial differential equations*

$$
\left. \begin{array}{l}
\dfrac{\partial}{\partial x} \left[\kappa GA \left(\dfrac{\partial \eta}{\partial x} - \varphi \right) \right] = \rho A \dfrac{\partial^2 \eta}{\partial t^2} \\[3mm]
\dfrac{\partial}{\partial x} \left(EI \dfrac{\partial \varphi}{\partial x} \right) + \kappa GA \left(\dfrac{\partial \eta}{\partial x} - \varphi \right) = \rho I \dfrac{\partial^2 \varphi}{\partial t^2}
\end{array} \right\} \quad 0 < x < L \quad (7\text{–}37)
$$

and the *natural boundary conditions*

$$
\left. \begin{array}{l}
\kappa GA \left(\dfrac{\partial \eta}{\partial x} - \varphi \right) + M_L \dfrac{\partial^2 \eta}{\partial t^2} = 0 \\[3mm]
EI \dfrac{\partial \varphi}{\partial x} + I_L \dfrac{\partial^2 \varphi}{\partial t^2} = 0
\end{array} \right\} \quad \text{at } x = L. \quad (7\text{–}38)
$$

These same results can be obtained by application of the direct method. The first of (7–37) is essentially a statement of the linear-momentum principle for an element of the continuum expressed in terms of the geometric coordinates. The first of (7–38) is likewise a statement of the linear-momentum principle for the rigid body at the end. Similarly, the second of (7–37) and the second of (7–38) can be interpreted as statements of the angular-momentum principle applied, respectively, to an element of the continuum and to the rigid body at the end.

Example 7–4. In this example we use a continuum model for an electric transmission line. The system under consideration is sketched in Fig. 7–11. An ideal voltage source is connected to one end of a long transmission line, and a load resistance R_l is connected to the other end. The transmission line is modeled by a continuous distribution of series inductance $L(x)$ per unit length, series resistance $R(x)$ per unit length, and shunt capacitance $C(x)$ per unit length.

A differential length of the line is sketched in Fig. 7–12, showing series inductance $L\,dx$, series resistance $R\,dx$, and shunt capacitance $C\,dx$. Current is continually leaking from the top conductor, through the distributed shunt capacitance, to the bottom conductor. In a length dx the total current which flows through the capacitance $C\,dx$ must be $-(\partial i/\partial x)\,dx$, as indicated in Fig. 7–12, if charge is to be conserved.

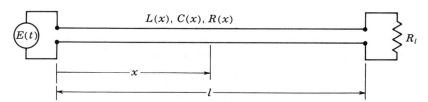

FIG. 7–11. Example 7–4. Electrical transmission line.

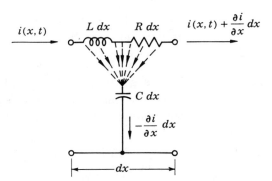

FIG. 7–12. Schematic diagram of current flow through differential element of continuum model.

A convenient generalized coordinate for the continuous line is the charge $q(x,t)$, defined by

$$\frac{\partial q(x,t)}{\partial t} = i(x,t). \qquad (7\text{–}39)$$

An arbitrary (differentiable) function $q(x,t)$ represents an admissible "motion" of the system provided that we take the charge dq_C on the shunt capacitance $C\,dx$ in Fig. 7–12 to be

$$dq_C = -\frac{\partial q}{\partial x}\,dx. \qquad (7\text{–}40)$$

Under this circumstance charge will be conserved, and the current flow will satisfy the relations indicated in Fig. 7–12. The total magnetic coenergy in the series inductances is

$$W_m^* = \int_0^l \tfrac{1}{2}(L\,dx)i^2 = \int_0^l \tfrac{1}{2}L\left(\frac{\partial q}{\partial t}\right)^2 dx, \qquad (7\text{--}41)$$

on using (5–25) and (7–39). The total electrical energy in the shunt capacitance is

$$W_e = \int_0^l \frac{1}{2}\frac{(dqc)^2}{C\,dx} = \int_0^l \frac{1}{2}\frac{\left[-\left(\dfrac{\partial q}{\partial x}\right)dx\right]^2}{C\,dx} = \int_0^l \frac{1}{2C}\left(\frac{\partial q}{\partial x}\right)^2 dx, \quad (7\text{--}42)$$

on using (5–10) and (7–40). The work expression (5–70) for the system of Fig. 7–11 includes contributions from the voltage source $E(t)$, the load resistance R_l, and the distributed series resistance of the line,

$$\sum_k E_k\,\delta q_k = E(t)\,\delta q(0,t) - i(l,t)R_l\,\delta q(l,t) - \int_0^l i(x,t)R\,dx\,\delta q(x,t)$$

$$= E(t)\,\delta q(0,t) - R_l\frac{\partial q(l,t)}{\partial t}\,\delta q(l,t) - \int_0^l R\frac{\partial q}{\partial t}\,\delta q\,dx. \qquad (7\text{--}43)$$

At this stage the analysis is essentially complete. What remains is the routine manipulation of the variational method. The state-functions (7–41) and (7–42) and the work expression (7–43) are all expressed in terms of an admissible charge coordinate $q(x,t)$. When we insert them in the variational indicator (5–48) for Hamilton's principle and perform the usual calculus of variations, we find that the necessary conditions for the indicator to vanish for arbitrary variations δq are a *partial differential equation*

$$L\frac{\partial^2 q}{\partial t^2} + R\frac{\partial q}{\partial t} - \frac{\partial}{\partial x}\left(\frac{1}{C}\frac{\partial q}{\partial x}\right) = 0 \qquad 0 < x < l \qquad (7\text{--}44)$$

and a pair of *natural boundary conditions*,

$$-\frac{1}{C}\frac{\partial q}{\partial x} = E(t) \qquad \text{at } x = 0$$

$$-\frac{1}{C}\frac{\partial q}{\partial x} = R_l\frac{\partial q}{\partial t} \qquad \text{at } x = l. \qquad (7\text{--}45)$$

These same results can also be obtained[1] by the direct method. The equation (7–44) can be interpreted as a statement of Kirchhoff's voltage law applied to a differential loop and expressed in terms of charges. The boundary conditions (7–45) ensure unique potentials at the ends of the line.

It is of interest to observe the similarity between (7–44) and the corresponding partial differential equation (7–30) for the transverse motion

---

[1] See Prob. 7–14.

of a taut string in a viscous medium. When the shunt capacitance $C(x)$ is independent of x, there is an exact analogy[1] between the equations.

7-4   The wave equation

The previous sections in this chapter have been addressed to the problem of formulating dynamic equations of motion for systems containing continua.   To illustrate an important application of these equations we enter into some simple aspects of the theory of wave propagation.   In this section we study the solutions of one particular equation, and in the following section we investigate the nature of sinusoidal waves in a wider class of systems.

We consider the longitudinal motions of an elastic bar with the partial differential equation (7–5) or (7–18).   In the case of a *uniform* bar in which the parameters ρ, E, and A do not vary with position, the equation can be written

$$\frac{\partial^2 \xi}{\partial x^2} = \frac{1}{c^2} \frac{\partial^2 \xi}{\partial t^2}, \tag{7-46}$$

where c is a constant (with the dimensions of a velocity) defined by

$$c^2 = \frac{E}{\rho}. \tag{7-47}$$

We note that in the case of transverse vibrations of a uniform string, the partial differential equation (7–30) reduces to the same form as (7–46), provided that the viscous damping is removed and that the constant c in (7–46) is defined by

$$c^2 = \frac{P}{\rho A}. \tag{7-48}$$

We also note that in the case of a *uniform* electric transmission line, if the series resistance is omitted, then (7–44) takes the same form as (7–46), provided that the constant c is defined by

$$c^2 = \frac{1}{LC}. \tag{7-49}$$

Equation (7–46) is called the classical one-dimensional *wave equation* because it has wave-type solutions..   Consider a function $f(u)$, where the argument u is equal to $x - ct$.   Then

$$\begin{aligned}
\frac{\partial f}{\partial x} &= \frac{df}{du} \frac{\partial u}{\partial x} = \frac{df}{du} \\
\frac{\partial f}{\partial t} &= \frac{df}{du} \frac{\partial u}{\partial t} = -c \frac{df}{du},
\end{aligned} \tag{7-50}$$

[1] See Prob. 7–15.

and on substitution in (7–46), we see that $f(x - ct)$ is a solution regardless of the shape of the function. In the same way we find that $g(x + ct)$, where g is an equally arbitrary function, is also a solution. The general solution to (7–46) can be represented as a superposition of these two types of solutions,

$$\xi(x,t) = f(x - ct) + g(x + ct). \qquad (7\text{–}51)$$

Solutions of the form $\xi = f(x + ct)$ may be considered as *waves* which propagate to the right with velocity c, since, if after an interval Δt we move to the right a distance $\Delta x = c\,\Delta t$, the argument of the function will be unchanged. A possible solution to the wave equation is a wave of arbitrary shape which moves, without change of shape, to the right with velocity c. Similarly, solutions of the form $\xi = g(x + ct)$ represent waves of arbitrary shape which travel to the left with velocity c. The

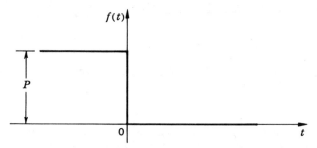

FIG. 7–13. Example 7–4. Tensile force P is suddenly removed at $t = 0$.

general solution (7–51) consists of a wave of arbitrary shape traveling to the right plus a wave of arbitrary shape traveling to the left. The parameter c in the wave equation (7–46) is thus identified as the *velocity of propagation* of waves in the medium.

Example 7–4. To illustrate the application of (7–51) to a particular example, we consider that the bar of Fig. 7–4 is at rest while in a state of tension under a constant force P until $t = 0$, when, suddenly, the force is removed. The time history of the prescribed force $f(t)$ is thus as sketched in Fig. 7–13.

To completely specify the problem mathematically, we need boundary conditions and initial conditions for the wave equation (7–46). At $t = 0$ the bar is at rest ($\partial \xi / \partial t = 0$) in a state of uniform tension. This implies the uniform strain

$$\frac{\partial \xi}{\partial x} = \frac{P}{AE} = \epsilon_0, \qquad (7\text{–}52)$$

say, which follows, for example, from (7–19). Thus the initial conditions are

$$\left.\begin{array}{l} \dfrac{\partial \xi}{\partial t} = 0 \\[2mm] \dfrac{\partial \xi}{\partial x} = \epsilon_0 \end{array}\right\} \quad \text{for } 0 < x < L \text{ at } t = 0. \tag{7-53}$$

The boundary conditions are the geometric condition $\xi = 0$ at the left and the natural boundary condition (7–19) at the right. Taking into account Fig. 7–13, the boundary conditions which apply for $t > 0$ are

$$\left.\begin{array}{l} \xi(0,t) = 0 \\[2mm] \dfrac{\partial \xi(L,t)}{\partial x} = 0 \end{array}\right\} \quad \text{for } t > 0. \tag{7-54}$$

It remains to evaluate the unknown functions f and g in (7–51) in such a manner that (7–53) and (7–54) will be satisfied.

Inserting (7–51) in the first of (7–54), we find

$$\xi(0,t) = f(-ct) + g(ct) = 0, \tag{7-55}$$

or that $g(u) = -f(-u)$, so that we can eliminate the function g by writing

$$\xi(x,t) = f(x - ct) - f(-x - ct). \tag{7-56}$$

Next, inserting (7–56) in the second of (7–54), we find

$$\frac{\partial \xi(L,t)}{\partial x} = f'(L - ct) + f'(-L - ct) = 0. \tag{7-57}$$

Now, on setting $u = -L - ct$ in (7–57), we see that

$$f'(u) = -f'(u + 2L), \tag{7-58}$$

which means that f' is periodic, repeating itself with sign reversed after an interval of $2L$. After an interval of $4L$, the sign will again be reversed, so that f' has period $4L$. This further implies that the function f will also be periodic, with period $4L$. The two boundary conditions (7–54) have thus reduced the general solution (7–51), with two unknown functions to the form (7–56), with only one unknown function f, which must be periodic, with period $4L$.

Next we insert (7–56) into the initial conditions (7–53) to find

$$f'(x) = f'(-x) = \frac{\epsilon_0}{2} \quad \text{for } 0 < x < L, \tag{7-59}$$

which means that $f'(x)$ remains constant over the interval $-L < x < L$. The condition (7–58) can then be used to define $f'(x)$ outside of this region, as indicated in Fig. 7–14a. The function $f(x)$ must then have the shape sketched in Fig. 7–14b.   Note that an arbitrary constant of integration could be added to this $f(x)$, but it would not affect the solution (7–56), since $\xi(x,t)$ is the difference between two f values.

The complete solution for $\xi(x,t)$ is provided by (7–56) and the diagram of Fig. 7–14b.   Further insight into this solution is provided by returning to the f and g notation, using $g(x + ct) = -f(-x - ct)$.   As indicated in Fig. 7–15, the solution $\xi(x,0)$ in the range $(0 < x < L)$ is obtained by adding the f and g waves.   Now as time goes on, the f wave travels to the right with velocity c, and the g wave travels to the left with velocity

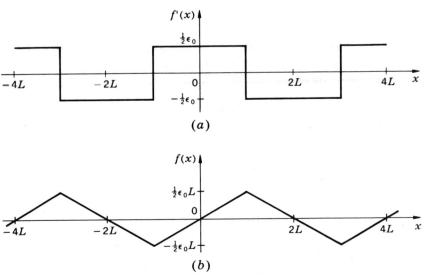

(a)

(b)

FIG. 7–14.   Wave shapes for Example 7–4.

c.   The solution at any moment is the sum of f and g.   For example, at time $t = L/2c$, we have the situation sketched in Fig. 7–16.   The waves have moved half the length of the bar in their respective directions. Their sum provides the displacements in the bar at this time.   Note that there has been no change in the state of the left half of the bar.   The g wave coming in from the suddenly released boundary brings with it a change in displacement, but this wave travels at the finite velocity c.

As time goes on, the solution can be visualized by imagining the f and g waves to move steadily to the right and left, respectively.   The successive displacement patterns obtained by summing f and g are sketched in Fig. 7–17.   This sequence of displacements repeats in time with a period of $T = 4L/c$.

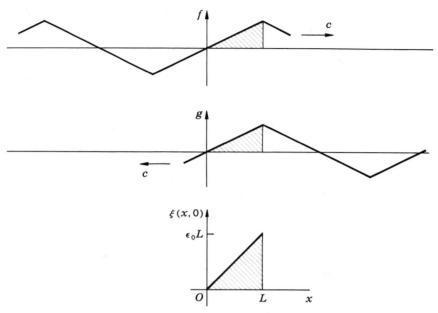

FIG. 7–15.   Initial displacement $\xi(x,0)$ decomposed into waves f and g.

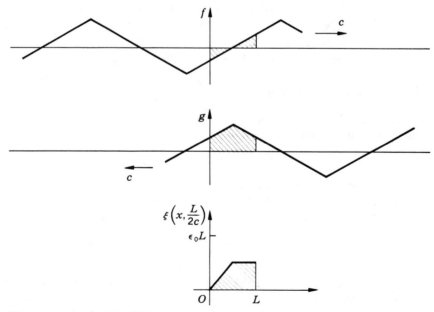

FIG. 7–16.   At $t = L/2c$ displacement $\xi(x, L/2c)$ is obtained by summing f and g waves.

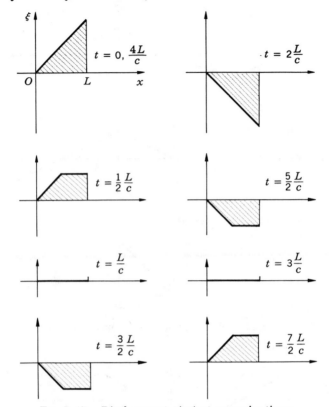

FIG. 7–17.   Displacement $\xi(x,t)$ at successive times.

7–5   Sinusoidal waves

A particular shape for the progressive wave $f(x - ct)$ which is of special importance is the sinusoidal shape

$$f(x - ct) = a \sin (kx - \omega t), \tag{7-60}$$

in which a is called the *amplitude*, k is called the *wave number*, and ω is called the *frequency*.   In order for the argument of the sinusoid actually to be a function of $x - ct$, it is necessary that ω and k satisfy the relation

$$\frac{\omega}{k} = c. \tag{7-61}$$

For the sinusoid (7–60), the velocity c given by (7–61) is called the *phase velocity*, because at this velocity the argument of the sinusoid, which can be thought of as a phase angle, remains constant.   In other words, a fixed phase of the sinusoid travels with the velocity c.

At a fixed location x, the function (7–60) represents a simple harmonic motion at frequency ω and *period T*, where

$$T = \frac{2\pi}{\omega}. \qquad (7\text{–}62)$$

At a fixed time t the function (7–60) represents a sinusoidal space variation with wave number k and *wavelength λ*, where

$$\lambda = \frac{2\pi}{k}. \qquad (7\text{–}63)$$

For media described by the classical wave equation, the propagation velocity c is a fundamental property of the medium. See, for example, (7–47) to (7–49). In these cases the phase velocity of a sinusoidal wave is independent of the frequency or wavelength. There are more general

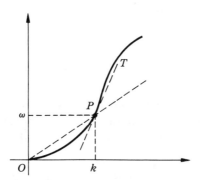

Fig. 7–18. Dispersion relation. Phase velocity is given by slope of chord OP. Group velocity is given by slope of tangent PT.

linear media, which, although they admit sinusoidal-wave solutions of the form (7–60), do not have an intrinsic propagation speed. In these media the phase velocity depends on the frequency or wavelength of the sinusoidal solution. We shall see that the Timoshenko-beam continuum is an example of such a medium.

In studying such a medium we assume a solution of the form (7–60) and find that it will satisfy the dynamic equations of the medium only if the parameters ω and k satisfy a certain equation,

$$\omega = \omega(k), \qquad (7\text{–}64)$$

called the *dispersion relation*. The reason for this name will appear shortly. A possible dispersion relation is sketched in Fig. 7–18. If we pick a particular wave number k, there is a sinusoidal wave solution only if the corresponding frequency ω satisfies (7–64). This means that in Fig. 7–18 the frequency ω is obtained from k by running up from k to a point P on the dispersion relation to determine the ordinate ω. Now, in such a sinusoidal wave there is a phase velocity given by (7–61). This phase velocity is represented graphically by the slope of the chord OP.

We now can see that as the wave number k is varied, the corresponding phase velocity c will generally change. This means that waves of different lengths have different phase velocities. A consequence of this is that if two or more waves of different wave numbers superpose to form a certain complex wave shape at one instant, they will not in general superpose to form the same complex shape at a later time. A fixed phase in each component sinusoid travels at its own speed, and phases which coincide at one instant will separate in space as the motion continues. This separation is called *dispersion*. A medium which supports sinusoidal waves, but for which the phase velocity depends on the wave number, is called *dispersive*. In a dispersive medium the only wave forms that are transmitted without change in wave shape are sinusoids. A disturbance or pulse of any other shape will suffer distortion as it propagates.

GROUP VELOCITY

We consider a linear dispersive medium and a solution to the dynamic equation which consists of the superposition of two sinusoidal wave solutions that differ only slightly in frequency and wave number,

$$\xi(x,t) = \sin\left[(k + \Delta k)x - (\omega + \Delta\omega)t\right]$$
$$+ \sin\left[(k - \Delta k)x - (\omega - \Delta\omega)t\right]. \quad (7\text{–}65)$$

This combination suffers distortion as it progresses. Nevertheless, when Δk and $\Delta\omega$ are small, there is an easily observed characteristic of the disturbance pattern which is preserved. To see this we employ the trigonometric identities for sums and differences of angles to rewrite (7–65) as

$$\xi(x,t) = 2 \cos\left(\Delta k\, x - \Delta\omega\, t\right) \sin\left(kx - \omega t\right). \quad (7\text{–}66)$$

At fixed time t, (7–66) appears as a sinusoid of wave number k, with an amplitude which varies like a cosine with small wave number Δk. In Fig. 7–19a, the two components of (7–65) are plotted separately, and their sum is shown in Fig. 7–19b. As time goes on, the individual cycles with wave number k move with a phase velocity

$$c = \frac{\omega}{k}. \quad (7\text{–}67)$$

The dotted envelope of the groups of sinusoids also moves as time goes on. Since the envelope is represented by the amplitude function $2 \cos\left(\Delta k\, x - \Delta\omega\, t\right)$, a point of fixed phase *on the envelope* travels with velocity $\Delta\omega/\Delta k$. The limiting value of this quotient is called the *group velocity* V_ϱ, associated with the wave number k,

$$V_\varrho = \lim_{\Delta k \to 0} \frac{\Delta\omega}{\Delta k} = \frac{d\omega}{dk}. \tag{7–68}$$

The group velocity is represented graphically in Fig. 7–18 by the slope of the tangent to the dispersion relation.

In Fig. 7–19, as time goes on, the eye can follow the envelope, which appears to move with a velocity close to the limiting value (7–68). The

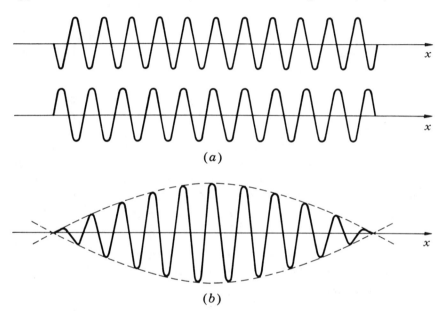

(a)

(b)

Fig. 7–19. Superposition of wave trains with slightly different wave numbers produces groups.

eye can also fasten on an individual cycle which appears to move at the phase velocity (7–67). In general, these two velocities are quite different. This means that the makeup of any one group is constantly changing. If the group velocity is faster than the phase velocity, individual cycles appear to enter at the front of a group and move steadily toward the rear.

A common example which illustrates phase and group velocities is provided by the wake of a motorboat moving steadily on a calm surface. Some distance behind and to one side of the boat the wake appears to be made up of a few wavelengths of relatively smoothly undulating water. This section of wake can be considered to constitute a group. The group moves, in a direction oblique to the course of the boat, with the group velocity appropriate to the wavelength of the undulations. The undulations do not, however, stand still with respect to the group. Individual crests appear to be created at the rear of the group. They then advance to the front of the group and disappear. This indicates that, for surface waves in deep water, the phase velocity is greater[1] than the group velocity.

When a steady sinusoidal-disturbance wave passes through a dispersive medium, the medium acquires energy, and there is a power flow through the medium. In a mechanical case the energy per unit volume is the sum of the kinetic and potential energies per unit volume. The magnitude of the energy fluctuates with time and position, but in the steady state the energy in a unit volume will have a fixed time average $\bar{E}$. The power flow across a section of unit area normal to the direction of wave propagation can be established by examining the forces and velocities of the section. The magnitude of the power flow fluctuates with time, but in the steady state the power flow per unit area will have a fixed time average $\bar{P}$. The quotient $\bar{P}/\bar{E}$ can be interpreted as the average *velocity of energy transport*. In a lossless dispersive medium this velocity is identical[2] with the *group velocity*.

Example 7-5. To illustrate the foregoing discussion we shall evaluate phase and group velocities for the Timoshenko-beam continuum. When the parameters of the medium do not depend on position, the dynamic equations for the interior of the medium (7-37) can be written

$$a_1^2 \left(\frac{\partial^2 \eta}{\partial x^2} - \frac{\partial \varphi}{\partial x} \right) = \frac{\partial^2 \eta}{\partial t^2}$$

$$a_2^2 \frac{\partial^2 \varphi}{\partial x^2} + a_1^2 k_0^2 \left(\frac{\partial \eta}{\partial x} - \varphi \right) = \frac{\partial^2 \varphi}{\partial t^2}, \tag{7-69}$$

where we have introduced the abbreviations

$$a_1^2 = \frac{\kappa G}{\rho} \qquad a_2^2 = \frac{E}{\rho} \qquad k_0^2 = \frac{A}{I}. \tag{7-70}$$

The parameter k_0 is the reciprocal of the section radius of gyration, and a_1 and a_2 are, as we shall see, limiting velocities of propagation. We assume sinusoidal-wave-train solutions of the form

$$\eta(x,t) = Y \sin (kx - \omega t)$$

$$\varphi(x,t) = \Phi \cos (kx - \omega t), \tag{7-71}$$

where we expect that the frequency ω and the amplitudes Y and Φ will depend on the wave number k. Inserting (7-71) into (7-69), we find

[1] See also Prob. 8-16.

[2] M. A. Biot, General Theorems of Equivalence of Group Velocity and Energy Transport, *Phys. Rev.*, **105**:1129–1137 (1957). See also Prob. 7-29.

that (7–71) is a solution, provided the parameters satisfy

$$\begin{bmatrix} \omega^2 - a_1^2 k^2 & a_1^2 k \\ a_1^2 k_0^2 k & \omega^2 - a_2^2 k^2 - a_1^2 k_0^2 \end{bmatrix} \begin{Bmatrix} Y \\ \Phi \end{Bmatrix} = \begin{Bmatrix} 0 \\ 0 \end{Bmatrix}. \tag{7–72}$$

For a nontrivial solution for Y and Φ, it is necessary that the determinant

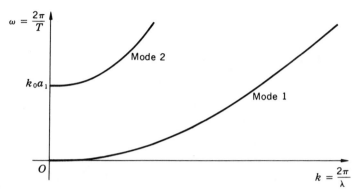

FIG. 7–20. Dispersion relation for Timoshenko beam has two branches.

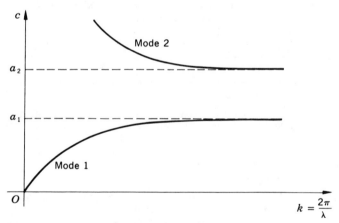

FIG. 7–21. Phase velocities of sinusoidal waves on a Timoshenko beam.

of the coefficient matrix be zero. This requirement yields

$$\omega^4 - \omega^2[a_1^2(k^2 + k_0^2) + a_2^2 k^2] + a_1^2 a_2^2 k^4 = 0, \tag{7–73}$$

which is an implicit form of the dispersion relation (7–64). Because (7–73) is a quadratic in $\omega^2(k^2)$, the dispersion relation has two distinct branches, as shown in Fig. 7–20. The phase velocities (7–61) corresponding to these branches are sketched in Fig. 7–21. The corresponding group velocities (7–68) are sketched in Fig. 7–22.

To complete the determination of the solutions (7–71), it is necessary to return to (7–72) and solve for the amplitudes Y and Φ for those combinations of values of ω and k which satisfy (7–73). In general, most waves involve a combination of transverse amplitude Y and rotational amplitude Φ. We do not make an exhaustive study of the resulting modal

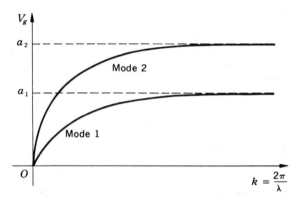

FIG. 7–22. Group velocities of sinusoidal waves on a Timoshenko beam.

FIG. 7–23. Limiting wave configurations for $k \to 0$. (a) Mode 1, rigid translation. (b) Mode 2, thickness shear.

configurations. We can, however, obtain some insight into their general nature by considering the limiting cases $k \to 0$ and $k \to \infty$.

If we put $k = 0$ in (7–72), we obtain the uncoupled equations

$$\begin{bmatrix} \omega^2 & 0 \\ 0 & \omega^2 - a_1^2 k_0^2 \end{bmatrix} \begin{Bmatrix} Y \\ \Phi \end{Bmatrix} = \begin{Bmatrix} 0 \\ 0 \end{Bmatrix}. \qquad (7\text{–}74)$$

One solution (mode 1) is obtained by setting $\omega^2 = 0$ and $\Phi = 0$ and letting Y be an arbitrary constant. This represents a rigid translation of the beam at zero frequency, as indicated in Fig. 7–23a. The other solution (mode 2) is obtained by setting $\omega^2 = a_1^2 k_0^2$ and $Y = 0$ and letting Φ be an arbitrary constant. This represents a thickness shear motion in which all cross sections rotate back and forth in unison at frequency $\omega = a_1 k_0$, as indicated in Fig. 7–23b.

At the other extreme, we divide each term in (7–72) by k^2 and then let

$k \to \infty$ to obtain

$$\begin{bmatrix} c^2 - a_1^2 & 0 \\ 0 & c^2 - a_2^2 \end{bmatrix} \begin{Bmatrix} Y \\ \Phi \end{Bmatrix} = \begin{Bmatrix} 0 \\ 0 \end{Bmatrix}, \qquad (7\text{-}75)$$

where we have put $c = \omega/k$. One solution (mode 1) is obtained by setting $c^2 = a_1^2$ and $\Phi = 0$ and by letting Y be an arbitrary constant. This represents a motion involving only shear, in which the cross sections remain vertical. The limit $k \to \infty$ corresponds to infinite frequency and zero wavelength. The sketch in Fig. 7–24a is suggestive of the mode shape for very large (but finite) k. The frequency would be very high, but both the phase and group velocity would be close to a_1.

The other solution (mode 2) for (7–75) is obtained by setting $c^2 = a_2^2$

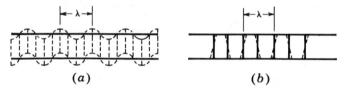

Fig. 7–24. Wave configurations for large $k(\lambda \to 0)$. (a) Mode 1. (b) Mode 2.

and $Y = 0$ and letting Φ be arbitrary. This represents a wave in which the cross sections rotate back and forth, alternately stretching and compressing the longitudinal fibers of the section. Again the limit $k \to \infty$ involves infinite frequency and zero wavelength. The sketch in Fig. 7–24b is suggestive of the mode shape for very large k. Again the frequency would be very high, but both the phase and group velocities would be close to a_2.

The Timoshenko continuum is a one-dimensional model for a truly three-dimensional elastic body. It nevertheless provides surprisingly good results over the entire range of frequencies and wavelengths. When the Timoshenko model is extended to plates (which, in a sense, are two-dimensional beams), it provides a satisfactory tool for predicting the extremely high harmonics of crystal oscillators.

PROBLEMS

7–1. Consider a model for the torsional system shown in which the flywheels at the ends are treated as rigid and the shaft of length L is treated as a uniform continuum with mass density ρ, shear modulus G, and polar second moment of area $I_p = \pi r^4/2$, where r is the radius. The torsional spring at the left can be taken as massless, with a torsional spring constant k.

(a) Select generalized coordinates to describe torsional motion and evaluate the system lagrangian.

(b) Obtain the dynamic equations of motion, including natural boundary conditions.

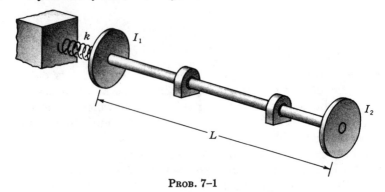

PROB. 7–1

7–2. The system shown consists of a massless spring; a rigid mass M; a continuous elastic bar of length L, density ρ, elastic modulus E, and cross-sectional area A; and an ideal dashpot. Formulate equations and boundary conditions for longitudinal motion.

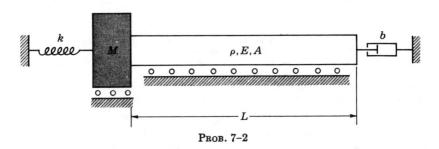

PROB. 7–2

7–3. Consider the taut string to be a uniform continuum with mass density ρ, cross-sectional area A, and large tension P.

(a) Formulate dynamic equations and boundary conditions for small transverse motion.

(b) If the string is plucked at its center point and released from rest at $t = 0$, obtain the subsequent motion as a superposition of waves traveling to the right and to the left.

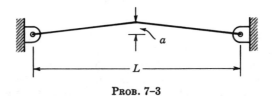

PROB. 7–3

7–4. Reconsider Example 7–2, using the direct method of analysis. Show how (7–30) can be obtained by combining the governing requirements which apply to a differential element of string. Show how (7–31) can be obtained from similar considerations applied to the mass M.

7–5. Consider a taut string similar to that in Prob. 7–3, except that the entire structure rotates steadily about an axis through the end points of the string at an angular velocity ω_0.

(a) Select variables to describe small transverse (not necessarily coplanar) motions of the string. Construct the lagrangian and obtain equations of motion.

(b) Find the value of ω_0 for which the string can rotate in a deflected shape like a jumping rope.

7–6. Consider a model for the system shown in which the inertia at the left is taken as a rigid body of mass M, the spring at the right is taken to be massless, and the flexible member joining them is taken to be a Bernoulli-Euler continuum with mass per unit length ρA and bending modulus EI.

(a) Show that the dynamic equation for the interior of the Bernoulli-Euler continuum is

$$\frac{\partial^2}{\partial x^2}\left(EI\,\frac{\partial^2\eta}{\partial x^2}\right) + \rho A\,\frac{\partial^2\eta}{\partial t^2} = 0.$$

(b) Obtain the complete set of dynamic equations and boundary conditions for the system shown.

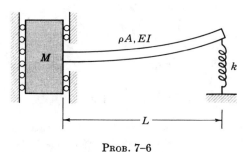

PROB. 7–6

7–7. Consider the beam shown to be a Bernoulli-Euler continuum with frictionless supports at the ends.

(a) Formulate the dynamic equations and boundary conditions for small transverse vibrations.

(b) Obtain the natural frequencies of free vibration.

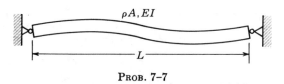

PROB. 7–7

7–8. Consider a one-dimensional Bernoulli-Euler continuum which carries a large longitudinal tensile force P.

(a) Construct the lagrangian for small transverse motions and obtain the dynamic equation for the interior of the continuum.

(b) Derive the dispersion relation for this medium. Determine the phase and group velocities. Examine the limiting values for long and short wavelengths.

7–9. Consider the beam shown to be a Bernoulli-Euler continuum supported by a continuous elastic foundation which exerts a restoring force $K\eta$ per unit length wherever the transverse displacement is η.

(a) Formulate the dynamic equation for the interior of the continuum.

(b) Find the dispersion relation, the phase velocity, and the group velocity. Examine the limiting behavior for long and short wavelengths.

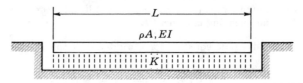

PROB. 7–9

7–10. Consider a uniform elastic beam of length l and bending modulus EI subject to pure bending by a pair of end couples M. Show that the work done on the beam in changing the curvature from zero to $d\varphi/dx = \varphi/l$ is

$$\tfrac{1}{2}EI\left(\frac{d\varphi}{dx}\right)^2 l,$$

and hence that the bending potential energy of a beam of length L with varying curvature is given by

$$\int_0^L \tfrac{1}{2}EI\left(\frac{d\varphi}{dx}\right)^2 dx.$$

7–11. Consider the electric transmission line shown to be a continuum with series inductance $L(x)$ per unit length, shunt capacitance $C(x)$ per unit length, and shunt conductance (reciprocal of resistance) $G(x)$ per unit length.

(a) Select generalized flux-linkage variables to represent dynamic states of the system. Construct the corresponding lagrangian and work expression.

(b) Formulate dynamic equations and boundary conditions for the system.

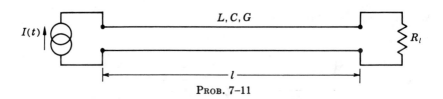

PROB. 7–11

7–12. Consider a long electric transmission line to be a one-dimensional continuum with series inductance $L(x)$ per unit length, series resistance $R(x)$ per unit length, shunt capacitance $C(x)$ per unit length, and shunt conductance (reciprocal of resistance) $G(x)$ per unit length. Formulate dynamic equations for the interior of the continuum. (*Hint:* If current and charge variables are used as in Fig. 7–12, note that the currents through the shunt capacitance $C\,dx$ and the shunt conductance $G\,dx$ are *independent* in an admissible "motion.")

7–13. Consider the electric line to be a uniform lossless line with series inductance L per unit length and shunt capacitance C per unit length. A short circuiting slider has mass m and can slide along the line without friction.

(*a*) Select variables and formulate equations for the motion of the slider and the electrical dynamics when the line is driven by an ideal current source at the left and the line is uncharged to the right of the slider.

(*b*) Consider the same line to be totally without charge and with the slider at rest at $x = x_0$ when at $t = 0$ a capacitor with capacitance C_0 and charge q_0 is suddenly connected to the line at the left. What happens to the slider? Which way does it start to go? Can you predict its ultimate velocity?

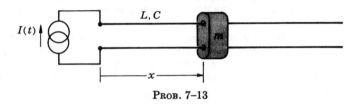

PROB. 7–13

7–14. Apply the direct method of analysis to the differential element, sketched below, for the transmission line of Fig. 7–11. Write Kirchhoff's voltage law and try to reduce the result to the form of (7–44).

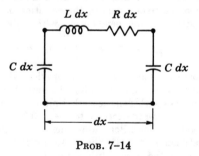

PROB. 7–14

7–15. Study the analogy between (7–30) and (7–44) for the case when all the parameters are independent of x. Show that the velocity $\dot{\eta} = \partial\eta(x,t)/\partial t$ satisfies (7–30) and that the current (7–39) satisfies (7–44). Make a table of analogous quantities.

7–16. A membrane can be taken as a two-dimensional continuum with mass density $\rho(x,y)$ and thickness $h(x,y)$. Take the membrane to be stretched flat in the xy plane by a uniform surface tension S per unit length. Let $\zeta(x,y,t)$ represent small transverse displacements of the membrane. Formulate the equation of motion for ζ within the continuum when the membrane is subjected to a transverse force $f(x,y,t)$ per unit area.

7–17. Consider a rectangular beam cross section of width b and height h subjected to a shear force S. Suppose the shear stress τ_{zy} acting on the cross section were uniformly distributed so that

$$\tau_{xy} = \frac{S}{bh}.$$

Show that the shear coefficient in the relation $S = \kappa G A \gamma$ for this element in a Timoshenko beam would be $\kappa = 1$.

7-18. Consider a rectangular beam cross section of width b and height h subjected to a shear force S. Suppose the shear stress τ_{xy} acting on the cross section is distributed according to the relation

$$\tau_{xy} = \frac{3}{2}\frac{S}{bh}\left(1 - \frac{4y^2}{h^2}\right),$$

where y is measured from the neutral axis. Find the distribution of shear angle γ, and show that if a weighted average of γ is used, where the weighting function has the same distribution as γ itself, the shear coefficient for the Timoshenko beam is $\kappa = \frac{5}{6} = 0.833$.

7-19. An alternative method for fixing the shear coefficient in the Timoshenko theory is to leave κ undetermined and then compare the frequency of thickness shear for mode 2 as $k \to 0$ with the corresponding quantity obtained from a more exact theory. For a rectangular cross section of width b and height h the three-dimensional theory predicts

$$\omega = \frac{\pi}{h}\sqrt{\frac{G}{\rho}}$$

for this thickness shear frequency. Use this to show that the shear coefficient for a rectangular Timoshenko beam according to this method is $\kappa = \pi^2/12 = 0.822$.

7-20. Show that when the parameters in the equation of motion (7-30) for the viscously damped taut string are independent of position, the expression

$$\eta(x,t) = (C_1 e^{\gamma x} + C_2 e^{-\gamma x})e^{-i\omega t}$$

is a solution for arbitrary values of C_1, C_2, and ω, provided that γ is properly chosen. Use this result to show that for light damping the amplitude of a sinusoidal wave traveling to the right decays with distance. Estimate the distance the wave must travel in order for the amplitude to decrease to $1/e = 0.368$ times its original value.

7-21. The dynamic equation for a Bernoulli-Euler continuum is given in Prob. 7-6. Derive the dispersion relation, and obtain the phase and group velocities. Show that for long wavelengths, $k \to 0$, the Bernoulli-Euler model approaches the first mode of the Timoshenko model.

7-22. An intermediate beam model, due to Rayleigh, accounts for both transverse and rotary inertia but considers only bending deformations (i.e., $\varphi \equiv \partial\eta/\partial x$). Formulate a single dynamic equation for $\eta(x,t)$. Derive the dispersion relation and obtain the phase and group velocities. Show that when $k \to 0$ the Rayleigh model approaches the first mode of the Timoshenko model, but that when $k \to \infty$ the Rayleigh model approaches the second mode of the Timoshenko model.

7-23. Another intermediate beam model accounts for both bending and shear deformation but considers only transverse inertia (i.e., the rotary moment of inertia, ρI per unit length, is neglected). Derive the dispersion relation and obtain the phase and group velocities for this model. Show that at both extremes, $k \to 0$ and $k \to \infty$, this model approaches the first mode of the Timoshenko model.

7-24. Show that when the parameters of a Bernoulli-Euler continuum are independent of position, the expression

$$\eta(x,t) = (C_1 \sin kx + C_2 \cos kx + C_3 \sinh kx + C_4 \cosh kx)e^{-i\omega t}$$

is a solution for arbitrary values of C_1, C_2, C_3, C_4, and ω, provided k and ω are related according to the dispersion relation of Prob. 7–21. Use this to obtain the solution for the fundamental mode of free vibration of a cantilever beam clamped at $x = 0$ and free at $x = L$.

7-25. Consider the beam to be a uniform Bernoulli-Euler continuum subject to a uniform longitudinal compressive force P.

(a) Formulate dynamic equations and boundary conditions for small transverse motions.

(b) Formulate equations and boundary conditions for determining the equilibrium deflection mode when P is the static buckling load.

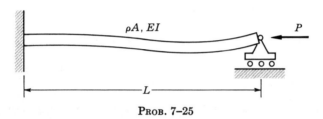

PROB. 7–25

7-26. The uniform lossless electric transmission line has series inductance L per unit length and shunt capacitance C per unit length. Find the natural frequencies of free oscillation of the system.

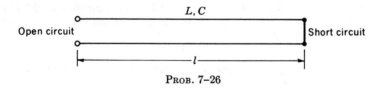

PROB. 7–26

7-27. The uniform lossless electric transmission line has series inductance L per unit length and shunt capacitance C per unit length. Formulate the dynamic equations and boundary conditions for the generalized charge $q(x,t)$ which satisfies (7–39).

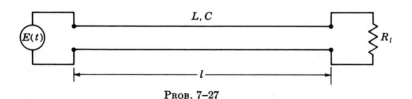

PROB. 7–27

Show that when $E(t) = \sin \omega t$, a solution is given by

$$q(x,t) = C_1 \cos (kx - \omega t) + C_2 \cos (kx + \omega t),$$

provided that the constants C_1 and C_2 and k are properly chosen. Study the behavior of this solution when R_l has the particular value $\sqrt{L/C}$, known as the *characteristic impedance* of the line.

7-28. Consider, as an extension of (7-65), the integral

$$\xi(x,t) = \int_{k_1-\Delta k}^{k_1+\Delta k} F(k) \sin (kx - \omega t)\, dk,$$

where $\omega = \omega(k)$ is given by a dispersion relation (7-64), and $F(k)\, dk$ represents the amplitude of an infinitesimal contribution with wave number k. We suppose that $F(k)$ is so chosen that $\xi(x,0)$ represents a disturbance whose significant effects lie in the neighborhood of $x = 0$. For most values of x and t, the integral tends to be small because of cancellations due to the rapid oscillations of the sine. If, however, x and t have values such that the *phase angle* $\theta = kx - \omega t$ remains nearly constant as k is varied through the range of integration, the integral can be substantial. Show that the x and t values for which the *phase is stationary* at $k = k_1$ satisfy the relation

$$\frac{x}{t} = \left(\frac{d\omega}{dk}\right)_{k=k_1} = V_g(k_1).$$

This means that a disturbance $\xi(x,t)$ which was centered around $x = 0$ at $t = 0$ propagates in such a way that its major consequences appear to travel at the group velocity $V_g(k_1)$.

7-29. Consider a uniform Bernoulli-Euler continuum (see Probs. 7-6 and 7-21) and the solution $\eta = Y \sin (kx - \omega t)$. Show that the instantaneous shear force and bending moment are $EIk^3Y \cos (kx - \omega t)$ and $-EIk^2Y \sin (kx - \omega t)$, that the instantaneous transverse velocity and angular velocity are $-\omega Y \cos (kx - \omega t)$ and $\omega k Y \sin (kx - \omega t)$, and hence that the average power flow across a section is

$$\bar{P} = EIk^3\omega Y^2.$$

Show that the average potential energy in a unit length of beam is $\frac{1}{4}EIk^4Y^2$, that the average kinetic energy in a unit length of beam is $\frac{1}{4}\rho A\omega^2Y^2$, and hence that the average total energy per unit length is

$$\bar{E} = \frac{1}{2}EIk^4Y^2.$$

Show that the average velocity of energy transport is the same as the group velocity.

CHAPTER EIGHT

Dynamics of fluid systems

In this chapter we extend our study of continuous media to include fluid continua. In the early part of the chapter we consider one-dimensional isentropic flow, with particular emphasis on the acoustic approximation. The lagrangian method of description is employed in this discussion. Then the eulerian method of description is introduced and extended to three-dimensional formulations. Finally, we consider mixed systems in which flowing fluids interact dynamically with elastic solids.

8–1 One-dimensional flow

The primary difference between the dynamics of fluids and the dynamics of elastic solids lies in the fact that fluids can undergo unlimited deformation whereas the deformations of solids are usually small. A secondary difference lies in their constitutive relations. The *restoring force* of a fluid due to its compressibility depends only on the fluid *volume* change, whereas the restoring force of an elastic solid depends also on the solid's change in *shape*.

Consider a flow of fluid parallel to the axis of x in which the fluid density, pressure, and velocity do not vary in any plane perpendicular to x, although, in general, all quantities vary with x and with time t. For simplicity of visualization, consider a tube of *unit cross-sectional area* parallel to the axis of x as indicated in Fig. 8–1. We assume that a fluid of (variable) mass density ρ flows within the tube. We shall neglect viscosity.

Let us focus our attention on the fluid system which occupies the space between $x = 0$ and $x = L$ at the time t_0. Figure 8–1 shows the location of this same mass of fluid at a later time t. The motion of the fluid can be described by introducing the position function $\chi(X,t)$, where X is the position of a particular fluid element at $t = t_0$, and $\chi(X,t)$ is the position of the *same* element at the later time t. The coordinate X which identifies the *element* is called a *material* coordinate, or *lagrangian* coordinate. When $t = t_0$, the position χ reduces simply to the lagrangian coordinate X; i.e.,

$$\chi(X,t_0) = X. \tag{8-1}$$

The velocity of a fluid element is the time rate of change of the position *of that element*, so that, for the element identified by X, the fluid velocity is

$$\dot{\chi}(X,t) = \frac{\partial \chi(X,t)}{\partial t}. \tag{8-2}$$

Note that the dot notation is used in this chapter to indicate partial time derivatives taken with fixed values of *lagrangian* coordinates.

When the lagrangian method of description is employed, the fluid density ρ and pressure p are considered to be functions of the lagrangian

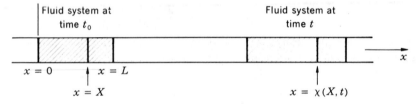

FIG. 8-1.   Same fluid system shown at two instants of time.

coordinates and time. For the one-dimensional case we shall write $\rho(X,t)$ and $p(X,t)$.

As the fluid flows, the mass of any system of fixed identity must be conserved. This imposes an admissibility condition on the density variations. When a fluid system becomes elongated, its density must decrease, and when it becomes shortened, its density must increase. For the fluid system between X and $X + dX$ the mass at time t must be the same as at time t_0. Call the length of this system, at time t, $d\chi$; i.e.,

$$d\chi = \chi(X + dX, t) - \chi(X,t). \tag{8-3}$$

Then the conservation of mass in this system of unit area requires that

$$\rho(X,t)\, d\chi = \rho(X,t_0)\, dX \tag{8-4}$$

or

$$\rho(X,t)\, \frac{\partial \chi}{\partial X} = \rho(X,t_0). \tag{8-5}$$

When geometrically admissible mass-conserving variations of the flow are considered, the same reasoning leads to a constraint between the variations of $\rho(X,t)$ and of $\chi(X,t)$. Applying the variational operator to

(8-5), under the assumption that there is no variation at $t = t_0$, we find

$$\delta\rho \, \frac{\partial\chi}{\partial X} + \rho \, \frac{\partial(\delta\chi)}{\partial X} = 0. \tag{8-6}$$

CONSTITUTIVE RELATION FOR ISENTROPIC COMPRESSIBLE FLUID

In general, the internal energy U per unit mass of a fluid depends on *two* independent state variables such as the density ρ and the entropy S per unit mass. Changes in these quantities are related by the first law of thermodynamics,

$$T \, dS = dU + p \, d\left(\frac{1}{\rho}\right), \tag{8-7}$$

where T is the temperature, and p is the thermodynamic pressure. We shall restrict our considerations to the special case of *isentropic flow* in which all parts of the fluid have the same entropy and there is no change in this entropy with time. This is an idealization which is useful when dissipative effects (e.g., viscosity and heat conduction) are unimportant. In this case there is only *one* independent thermodynamic variable, which we can take to be either the density ρ or the specific volume $1/\rho$. The constitutive relation for a fluid which gives the pressure as a function of the specific volume is commonly called the *equation of state* for the fluid.

In Fig. 8-2 a typical equation of state is sketched. According to (8-7), when dS is zero, the internal energy U can be obtained as the integral,

$$U(\rho) = -\int_{1/\rho_0}^{1/\rho} p \, d\left(\frac{1}{\rho}\right) = \int_{\rho_0}^{\rho} \frac{p}{\rho^2} \, d\rho, \tag{8-8}$$

where ρ_0 is a convenient datum. This integral is represented by the shaded area in Fig. 8-2. If the internal-energy function $U(\rho)$ is known, the constitutive relation, in the form $p = p(\rho)$, can be recovered by differentiating (8-8) to obtain

$$p = \rho^2 \, \frac{dU}{d\rho}. \tag{8-9}$$

Since $U(\rho)$ is stored internal energy *per unit mass*, the potential energy V of a system of fluid, due to its compression, is obtained by integrating ρU over the volume of the system. Thus, at time t in Fig. 8-1, the potential energy in the system which had previously occupied the space between $x = 0$ and $x = L$ at time t_0 is represented by the integral

$$V = \int_{\chi(0,t)}^{\chi(L,t)} U[\rho(X,t)]\rho(X,t) \, d\chi. \tag{8-10}$$

An alternative representation, involving integration with respect to the lagrangian coordinate X, is obtained by introducing the mass-conservation relation (8–4) into (8–10) to get

$$V = \int_0^L U(\rho)\rho_0 \, dX, \tag{8–11}$$

where the abbreviations

$$\rho = \rho(X,t) \qquad \rho_0 = \rho(X,t_0) \tag{8–12}$$

have been employed. Note that ρ_0 describes the density in a reference configuration. It can be taken as a prescribed function of X (in many

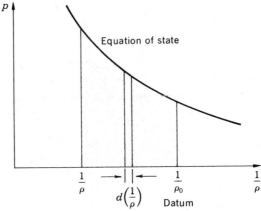

FIG. 8–2. For isentropic flow, internal energy is area under equation of state.

applications the density is uniform in the reference configuration and ρ_0 is a constant).

8–2 Equations of motion

The dynamic equations for one-dimensional flow can be obtained by the direct method or by the variational method. For the direct method we consider the differential element of Fig. 8–3. The mass of this element is $\rho_0 \, dX$ or $\rho \, d\chi$, according to (8–4), and the relation (8–5) between the position $\chi(X,t)$ and the density $\rho(X,t)$ demanded by mass conservation is

$$\rho \frac{\partial \chi}{\partial X} = \rho_0. \tag{8–13}$$

The constitutive relation between pressure $p(X,t)$ and density $\rho(X,t)$ is

$$p = p(\rho). \tag{8–14}$$

Finally, $p(X,t)$ and $\chi(X,t)$ are related by the linear-momentum principle.

The unbalanced force acting on the element of Fig. 8–3 is

$$-dp = -\left(\frac{\partial p}{\partial X}\right) dX.$$

The velocity of the element, according to (8–2), is $\dot{\chi}$; so the momentum is $\rho_0 \dot{\chi} \, dX$. The linear-momentum relation then requires

$$-\frac{\partial p}{\partial X} = \rho_0 \, \ddot{\chi}. \tag{8–15}$$

The three relations (8–13) to (8–15) apply at each point in the interior of a fluid system. These equations are three simultaneous equations for

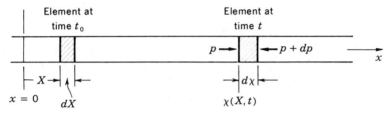

Fig. 8–3.   Differential fluid element.

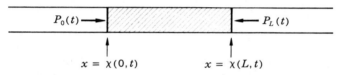

Fig. 8–4.   Prescribed pressure histories $P_0(t)$ and $P_L(t)$ at back and front of fluid system.

$\chi(X,t)$, $p(X,t)$, and $\rho(X,t)$. In principle, it is possible to eliminate two of these variables and obtain a single partial differential equation for the third. In practice, this is generally difficult because of the nonlinearity of (8–13) and (8–14).

In addition to the above relations, which apply in the interior of the fluid continuum, it is usually necessary to formulate boundary conditions at the extremities of the fluid system under consideration. For example, if the fluid system is the system shown in Fig. 8–1, and if the external pressure forces acting at the front and back of the fluid system are prescribed as indicated in Fig. 8–4, the boundary conditions will be

$$p(0,t) = P_0(t)$$
$$p(L,t) = P_L(t). \tag{8–16}$$

APPLICATION OF HAMILTON'S PRINCIPLE

It is instructive to see how these same results are obtained from the variational approach. Here it is necessary to consider the total system at the outset. For the fluid system of Fig. 8–4 the kinetic coenergy per unit mass is $\frac{1}{2}\dot{\chi}^2$, the potential energy per unit mass is simply the internal energy $U(\rho)$ of (8–8), and the work expression is $P_0\,\delta\chi(0,t) - P_L\,\delta\chi(L,t)$. Thus the variational indicator (2–222) for Hamilton's principle is

$$\int_{t_1}^{t_2} dt \left\{ \delta \int_0^L [\tfrac{1}{2}\dot{\chi}^2 - U(\rho)]\rho_0\,dX + P_0\,\delta\chi(0,t) - P_L\,\delta\chi(L,t) \right\}. \quad (8\text{–}17)$$

The dynamic equations to be satisfied by the natural motions are found as the necessary conditions for (8–17) to vanish, for arbitrary admissible variations. In this case, admissible motions must satisfy the conservation-of-mass condition (8–13), and admissible variations must satisfy (8–6). In performing the variation operation we consider that $t = t_0$ lies outside the interval $t_1 < t < t_2$, and hence that $\rho_0 = \rho(X,t_0)$ is not varied. The calculus of variations here is essentially the same as in the preceding chapter. The only term which may require explanation is the potential-energy term. For this term alone we obtain

$$- \int_0^L \delta U\,\rho_0\,dX = - \int_0^L \frac{dU}{d\rho}\,\delta\rho\,\rho_0\,dX$$

$$= \int_0^L \rho^2 \frac{dU}{d\rho} \frac{\partial(\delta\chi)}{\partial X}\,dX, \quad (8\text{–}18)$$

after successive use of (8–6) and (8–5). A spacewise integration-by-parts then yields

$$\int_0^L \rho^2 \frac{dU}{d\rho} \frac{\partial(\delta\chi)}{\partial X}\,dX = \rho^2 \frac{dU}{d\rho}\,\delta\chi \Big|_0^L - \int_0^L \frac{\partial}{\partial X}\left(\rho^2 \frac{dU}{d\rho}\right)\delta\chi\,dX. \quad (8\text{–}19)$$

The kinetic-coenergy term is treated in the standard way with a timewise integration-by-parts. Making use of the agreement that variations should vanish at the end points in time, we find that the complete variational indicator (8–17) is equal to

$$\int_{t_1}^{t_2} dt \left\{ - \int_0^L \left[\rho_0\ddot{\chi} + \frac{\partial}{\partial X}\left(\rho^2 \frac{dU}{d\rho}\right)\right]\delta\chi\,dX + \left(\rho^2 \frac{dU}{d\rho} - P\right)\delta\chi \Big|_0^L \right\}. \quad (8\text{–}20)$$

The necessary conditions that this vanish for arbitrary $\delta\chi$ are that $\chi(X,t)$ and $\rho(X,t)$ satisfy the *partial differential equation*

$$\rho_0\ddot{\chi} = - \frac{\partial}{\partial X}\left(\rho^2 \frac{dU}{d\rho}\right) \qquad 0 < X < L, \quad (8\text{–}21)$$

within the fluid system, and that $\rho(X,t)$ satisfy the *natural boundary conditions*

$$\left(\rho^2 \frac{dU}{d\rho}\right)_{X=0} = P_0(t) \qquad \left(\rho^2 \frac{dU}{d\rho}\right)_{X=L} = P_L(t). \qquad (8\text{–}22)$$

For a complete formulation these must be supplemented with the mass-conservation equation (8–13) relating χ and ρ.

It is interesting to compare these results with the direct formulation. Note that (8–15) and (8–16) reduce to (8–21) and (8–22), respectively, if the constitutive relation (8–9) is used to eliminate the pressure p.

8–3 Acoustic approximation

The constitutive equation $p = p(\rho)$ for fluids is generally nonlinear. If we restrict our considerations to flows in which all density changes are *small*, we can use a *linearized approximation* to the constitutive equation and obtain linearized dynamic equations. This linearized model is useful for describing the propagation of sound in air, and is called the acoustic approximation.

In order to describe changes in density in the neighborhood of a *constant* density ρ_0, it is convenient to introduce the dimensionless *condensation s* defined by

$$\rho - \rho_0 = \rho_0 s. \qquad (8\text{–}23)$$

In the acoustic approximation the condensation is treated as a first-order infinitesimal in the sense that terms proportional to higher powers of s are neglected in comparison with terms which are linear in s. Thus, when the constitutive relation $p = p(\rho)$ is expanded in a Taylor's series about the state where $p = p_0$ and $\rho = \rho_0$,

$$p = p_0 + \rho_0 s \left(\frac{dp}{d\rho}\right)_0 + \frac{(\rho_0 s)^2}{2} \left(\frac{d^2 p}{d\rho^2}\right)_0 + \cdots, \qquad (8\text{–}24)$$

the acoustic approximation is obtained by neglecting all terms beyond the first two on the right. The linearized constitutive relation can be written

$$p - p_0 = \rho_0 c^2 s \qquad (8\text{–}25)$$

if we introduce the symbol c defined by

$$c^2 = \left(\frac{dp}{d\rho}\right)_0. \qquad (8\text{–}26)$$

The quantity c has the dimension of velocity and is a fixed property of an acoustic continuum with reference density ρ_0.

The mass-conservation relation (8–5) can also be linearized by introducing the acoustic approximation. Transforming (8–5), we have

$$\rho \frac{\partial \chi}{\partial X} = \rho_0$$

$$(1 + s) \frac{\partial \chi}{\partial X} = 1 \tag{8–27}$$

$$\frac{\partial \chi}{\partial X} = 1 - s + s^2 - s^3 + \cdots ,$$

after introducing (8–23) and expanding $(1 + s)^{-1}$. The acoustic approximation is obtained by neglecting all but the first two terms on the right of (8–27),

$$\frac{\partial \chi}{\partial X} = 1 - s. \tag{8–28}$$

If we now substitute the linearized constitutive relation (8–25) into the momentum equation (8–15) to get

$$-c^2 \frac{\partial s}{\partial X} = \frac{\partial^2 \chi}{\partial t^2} \tag{8–29}$$

and then eliminate s between (8–28) and (8–29), we obtain a *linear* dynamic equation for $\chi(X,t)$,

$$\frac{\partial^2 \chi}{\partial X^2} = \frac{1}{c^2} \frac{\partial^2 \chi}{\partial t^2}, \tag{8–30}$$

within the acoustic continuum. Alternatively, we can eliminate χ between (8–28) and (8–29) to obtain a linear dynamic equation for $s(X,t)$,

$$\frac{\partial^2 s}{\partial X^2} = \frac{1}{c^2} \frac{\partial^2 s}{\partial t^2}. \tag{8–31}$$

We note that both (8–30) and (8–31) have the form of the classical wave equation (7–46) and that the parameter c can be identified as the velocity of wave propagation in the continuum. In acoustics, c is called the *velocity of sound.*

Since one-dimensional sound waves are described by the same equation as longitudinal strain waves in a uniform elastic bar, there is a strong analogy between these phenomena. The internal character of the disturbance waves are not, however, identical. In an elastic bar the stress is uniaxial, and a disturbance wave usually involves both tensile and compressive phases. In a fluid the stress state (pressure) is triaxial, and a disturbance wave involves only small changes about the reference pressure p_0.

HAMILTON'S PRINCIPLE IN THE ACOUSTIC APPROXIMATION

In order to obtain *linear* equations of motion from a variational indicator, it is necessary for the lagrangian to be *quadratic*.[1] This means that, to obtain the acoustic approximation from Hamilton's principle, it is necessary to retain terms of order s^2 in the lagrangian. This involves using second-order approximations for the internal energy (8–8) and for the mass-conservation relation (8–27).

The internal energy can be evaluated explicitly when the linear acoustic constitutive relation (8–25) is inserted in (8–8). Retaining terms of order s^2, we find

$$U = \frac{p_0}{\rho_0}(s - s^2) + \tfrac{1}{2}c^2 s^2 \tag{8–32}$$

for the internal energy per unit mass in the acoustic approximation. Note that the reference pressure multiplies a term that is linear in s.

The mass-conservation relation (8–27) becomes

$$\frac{\partial \chi}{\partial X} = 1 - s + s^2 \tag{8–33}$$

when second-order terms in s are retained. Applying the variational operator to (8–33), we obtain

$$\delta\left(\frac{\partial \chi}{\partial X}\right) = \frac{\partial(\delta\chi)}{\partial X} = -(1 - 2s)\,\delta s, \tag{8–34}$$

or to the same order in s,

$$\delta s = -(1 + 2s)\frac{\partial(\delta\chi)}{\partial X}, \tag{8–35}$$

as the corresponding restriction on admissible variations. Hamilton's principle will produce the acoustic approximation in a consistent manner if the internal energy is taken in the form (8–32) and the admissibility relation between variations in χ and in s is taken in the form (8–35). The first-order terms involving the reference pressure cancel during the calculus of variations, and the remaining first-order terms give the dynamic equations of the acoustic approximation.

For example, consider the contribution to (8–17) due to the lagrangian alone, in the acoustic case. Using (8–32), the variational indicator is

$$\int_{t_1}^{t_2} dt\,\delta \int_0^L \left[\tfrac{1}{2}\dot{\chi}^2 - \frac{p_0}{\rho_0}(s - s^2) - \tfrac{1}{2}c^2 s^2\right] \rho_0\,dX. \tag{8–36}$$

Performing the variation, we get

$$\int_{t_1}^{t_2} dt \int_0^L \left[\dot{\chi}\frac{\partial}{\partial t}(\delta\chi) - \frac{p_0}{\rho_0}(1 - 2s)\,\delta s - c^2 s\,\delta s\right]\rho_0\,dX. \tag{8–37}$$

[1] See p. 134.

For the variations to be admissible, δs and $\delta \chi$ must satisfy (8–35). Inserting (8–35) and integrating by parts, timewise and spacewise, we find

$$\int_{t_1}^{t_2} dt \int_0^L \left\{ -\ddot{\chi} - \frac{\partial}{\partial X} \left[\frac{p_0}{\rho_0} (1 - 2s)(1 + 2s) + c^2 s(1 + s) \right] \right\} \delta \chi \, \rho_0 \, dX,$$

$$(8\text{–}38)$$

plus boundary terms which we do not show. The interior dynamic equation follows from the requirement that the expression in braces vanish throughout the interior. Retaining only the first-order terms in s, this requirement is

$$\left\{ -\ddot{\chi} - \frac{\partial}{\partial X} \left[c^2 s \right] \right\} = 0 \qquad 0 < X < L. \qquad (8\text{–}39)$$

Note that it is the interaction of (8–32) and (8–35) which results in the cancellation of the first-order contribution due to p_0. The result (8–39) is equivalent to the acoustical equation of motion (8–29).

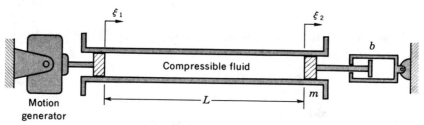

Fig. 8–5.   Example 8–1.   Acoustic-transmission system.

Example 8–1.   To illustrate the application of these ideas to a complete system, we consider the acoustic-transmission system of Fig. 8–5. The left-hand piston is driven by a motion generator. The right-hand piston has mass m, and is connected to a linear dashpot with dashpot constant b. In between is a tube of uniform cross-sectional area A filled with a compressible fluid. We shall investigate the response of the right-hand piston to a prescribed motion of the left-hand piston.

We model the fluid by an acoustic continuum. Furthermore, we suppose that there is a reference configuration with the system at rest, in equilibrium, in which the internal fluid pressure p_0 is equal to the surrounding atmospheric pressure. Under these conditions the fluid occupies a length L of the tube and has a uniform density ρ_0 and velocity of sound c. We use this reference configuration to establish the lagrangian coordinate X, where $0 < X < L$.

Under dynamic conditions the position $\chi(X,t)$ and the condensation $s(X,t)$ of a fluid element will vary. It is convenient in this case to introduce the *relative displacement* function $\xi(X,t)$ defined by

$$\xi = \chi - X. \tag{8-40}$$

For brevity, we designate the displacement $\xi(0,t)$ of the driven piston ξ_1, and designate the displacement $\xi(L,t)$ of the responding piston ξ_2, as indicated in Fig. 8-5.

The equations of motion for the system can be obtained by applying Hamilton's principle. The lagrangian contains the kinetic coenergy and potential energy of the fluid and the kinetic coenergy of the right-hand piston. The work expression contains contributions from the dashpot and the external atmospheric pressure acting on the right-hand piston. We take ξ_1 to be a prescribed time history $\xi_1(t)$; so there is no contribution to either the lagrangian or the work expression from the left-hand piston. Using the acoustic approximation (8-32) for the internal energy, we find that the variational indicator (2-222) of Hamilton's principle is

$$\int_{t_1}^{t_2} dt \left(\delta \left\{ \int_0^L \left[\tfrac{1}{2}\dot{\xi}^2 - \frac{p_0}{\rho_0}(s - s^2) - \tfrac{1}{2}c^2 s^2 \right] \rho_0 A \, dX \right. \right.$$
$$\left. \left. + \tfrac{1}{2}m\dot{\xi}_2^2 \right\} - (b\dot{\xi}_2 + p_0 A)\, \delta\xi_2 \right). \tag{8-41}$$

For a motion to be admissible it is necessary that mass be conserved. In terms of the position χ, this requirement is stated by (8-33). In terms of the displacement $\xi(X,t)$ of (8-40), the second-order mass-conservation relation is

$$\frac{\partial \xi}{\partial X} = -s + s^2. \tag{8-42}$$

The corresponding relation between variations is

$$\delta s = -(1 + 2s)\frac{\partial(\delta \xi)}{\partial X}. \tag{8-43}$$

It is also necessary for geometric compatibility that the pistons in Fig. 8-5 have the same displacements as the adjoining fluid elements. This is automatically ensured by taking the piston displacements ξ_1 and ξ_2 equal, respectively, to $\xi(0,t)$ and $\xi(L,t)$. Since $\xi_1(t)$ is a prescribed motion, we must have $\delta\xi_1 = 0$ in an admissible variation.

What remains is a routine application of the calculus of variations. The variation in (8-41) is performed, the s variation is eliminated by substituting (8-43), and integrations-by-parts, in time and space, are carried out to give

$$\int_{t_1}^{t_2} dt \left(\int_0^L \left\{ -\ddot{\xi} - \frac{\partial}{\partial X}\left[\frac{p_0}{\rho_0}(1 - 4s^2) + c^2 s(1 + 2s) \right] \right\} \delta\xi \, \rho_0 A \, dX \right.$$
$$\left. + [p_0 A(1 - 4s^2) + \rho_0 A c^2 s(1 + 2s) - m\ddot{\xi}_2 - b\dot{\xi}_2 - p_0 A]\, \delta\xi_2 \right). \tag{8-44}$$

In arriving at (8–44), we have utilized the facts that variations vanish at the end points in time and that the variation $\delta\xi_1$ at the driven piston must vanish. The necessary conditions that (8–44) vanish for arbitrary $\delta\xi$ in $0 < X < L$ and arbitrary $\delta\xi_2$ are the *partial differential equation*

$$\ddot{\xi} = -c^2 \frac{\partial s}{\partial X} \qquad 0 < X < L \tag{8–45}$$

and *natural boundary condition*

$$\rho_0 A c^2 s - m\ddot{\xi}_2 - b\dot{\xi}_2 = 0 \qquad \text{at } X = L, \tag{8–46}$$

if at this stage we retain only the first-order terms in s. These dynamic conditions involve both the displacement ξ and the condensation s. Thus, for a complete formulation, we must also restate the mass-conservation relation between ξ and s. At this stage we can use the first-order approximation to (8–42); i.e.,

$$\frac{\partial \xi}{\partial X} = -s \qquad 0 < X < L. \tag{8–47}$$

We also need to state the prescribed motion of the driven piston as a boundary condition,

$$\xi(0,t) = \xi_1(t) \qquad \text{at } X = 0. \tag{8–48}$$

The two equations (8–45) and (8–47) and the two boundary conditions (8–46) and (8–48) constitute a complete set of equations of motion for the acoustic-transmission system.

A somewhat neater formulation is obtained by eliminating the condensation s between (8–45) and (8–47) and between (8–46) and (8–47). In this way we obtain the single partial differential equation

$$\frac{\partial^2 \xi}{\partial X^2} = \frac{1}{c^2} \frac{\partial^2 \xi}{\partial t^2} \qquad 0 < X < L \tag{8–49}$$

and the boundary conditions

$$\xi(0,t) = \xi_1(t) \qquad \text{at } X = 0$$

$$m \frac{\partial^2 \xi}{\partial t^2} + b \frac{\partial \xi}{\partial t} + \rho_0 A c^2 \frac{\partial \xi}{\partial X} = 0 \qquad \text{at } X = L. \tag{8–50}$$

FREQUENCY RESPONSE

To illustrate an application of the equations of motion just obtained, we shall obtain the response of the right-hand piston in Fig. 8–5, when the driven piston executes steady simple harmonic motion at frequency ω. Because our acoustical model is linear, if we know how the system responds to single sinusoids, we can obtain the response to complex signals consisting of superposed sinusoids by superposing the individual responses.

If $\xi_1(t)$ is a unit sinusoid, i.e., if

$$\xi_1(t) = \text{Re } (e^{j\omega t}), \tag{8-51}$$

then in the steady state,

$$\xi(X,t) = \text{Re } [H(X,\omega)e^{j\omega t}] \qquad 0 \leq X \leq L, \tag{8-52}$$

where $H(X,\omega)$ is the (unknown) complex amplitude of the displacement throughout the continuum.

To determine $H(X,\omega)$, we substitute (8–52) into the wave equation (8–49) and the boundary conditions (8–50) to obtain

$$\frac{\partial^2 H}{\partial X^2} + \frac{\omega^2}{c^2} H = 0 \qquad 0 < X < L, \tag{8-53}$$

with the boundary conditions

$$H = 1 \qquad \text{at } X = 0$$

$$(-m\omega^2 + j\omega b)H + \rho_0 A c^2 \frac{\partial H}{\partial X} = 0 \qquad \text{at } X = L. \tag{8-54}$$

The general solution to (8–53) is

$$H(X,\omega) = f_1(\omega) \cos \frac{\omega X}{c} + f_2(\omega) \sin \frac{\omega X}{c}, \tag{8-55}$$

where f_1 and f_2 are functions of integration which can be evaluated by inserting (8–55) into the boundary conditions (8–54). We find

$$f_1(\omega) = 1$$

$$f_2(\omega) = \frac{\sin (\omega L/c) + [(m\omega - jb)/\rho_0 A c] \cos (\omega L/c)}{\cos (\omega L/c) - [(m\omega - jb)/\rho_0 A c] \sin (\omega L/c)}, \tag{8-56}$$

which, when substituted back into (8–55), lead to

$$H(L,\omega) = \frac{1}{[\cos (\omega L/c) - (m\omega/\rho_0 A c) \sin (\omega L/c)] + j(b/\rho_0 A c) \sin (\omega L/c)}, \tag{8-57}$$

for the complex amplitude of the *right-end* displacement response.

The magnitude of (8–57) is sketched as a function of frequency in Fig. 8–6 for a case of light damping ($b \ll \rho_0 A c$). There is a sequence of resonant frequencies[1] at which the response amplitude has peaks. The frequency response is very uneven over any appreciable band of frequencies. This implies that a signal which consists of a superposition of sinusoids of different frequencies will be badly distorted by such a system.

An ideal limiting case occurs when the mass m of the responding piston

[1] See Prob. 8–9.

approaches zero and the dashpot constant takes the particular value

$$b = \rho_0 A c, \tag{8-58}$$

known as the *characteristic impedance.*[1]   Under these circumstances the complex amplitude of the response approaches the value

$$H(L,\omega) = \left(\cos \frac{\omega L}{c} + j \sin \frac{\omega L}{c}\right)^{-1} = e^{-j(\omega L/c)}. \tag{8-59}$$

The magnitude of the response amplitude in this limiting case is sketched

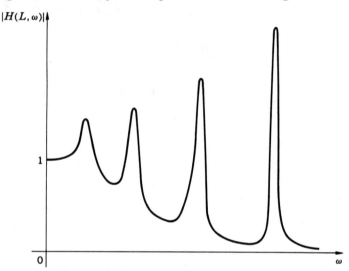

FIG. 8-6.   Frequency response of acoustic-transmission system for light damping.

in Fig. 8-7.   The frequency response is *flat.*   Further interpretation is provided by evaluating (8-55) for this limiting case.   We find

$$H(X,\omega) = e^{-j(\omega X/c)} \qquad 0 \le X \le L \tag{8-60}$$

and, on substituting in (8-52),

$$\xi(X,t) = \text{Re}\left(e^{j\omega[t-(X/c)]}\right) \qquad 0 \le X \le L. \tag{8-61}$$

Thus, when the acoustic-transmission system is terminated by its characteristic impedance, the response to a unit sinusoid at frequency ω is everywhere a unit sinusoid at the same frequency with a phase lag proportional to distance X.   An alternative interpretation is that the response is a progressive wave train moving to the right, with phase velocity c and

[1] See Prob. 8-11.

having frequency ω and wave number $k = \omega/c$. Since a general solution to the wave equation involves waves moving to the right *and* to the left, we can say that the characteristic impedance acts to absorb incoming waves without producing reflections. By using superposition, we can generalize from the result (8–61) for a single frequency to conclude that when the system is terminated by its characteristic impedance, any arbi-

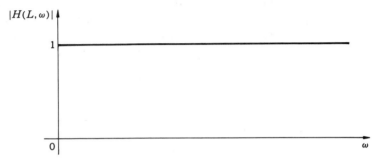

FIG. 8–7. Frequency response when transmission system is terminated by its characteristic impedance.

trary time signal introduced at the left will travel to the right at the velocity of sound without distortion.

8-4   Eulerian description of fluid flow

In the lagrangian method of description the function $\chi(X,t)$ for fixed X describes the motion of an individual fluid element in one-dimensional flow. This method of description can be extended[1] to two- and three-dimensional flows. Its use has, however, been largely restricted to "small motion" situations. For more general three-dimensional fluid flows the alternative eulerian method of description is almost universal. In the eulerian method the observer makes no attempt to follow an individual fluid element in the large, but concentrates his attention on describing how the flow changes at fixed spatial locations. We develop dynamic equations of motion for the eulerian method of description, first for the case of one-dimensional flow, and then for the three-dimensional case.

In Fig. 8–8 we consider one-dimensional isentropic flow in a tube of unit cross section. At a fixed *spatial* coordinate x, the fluid density ρ, pressure p, and instantaneous velocity u will generally change with time. In the eulerian method of description these fluid properties are considered as functions of the spatial coordinate x and the time t; i.e., we write $\rho(x,t)$, $p(x,t)$, and $u(x,t)$. The spatial coordinate x is sometimes called an *eulerian coordinate* to distinguish it from the lagrangian coordinate X.

[1] See Prob. 8–13.

A useful tool for the analysis of fluid flow when using the eulerian method of description is the *control volume*. This is a fixed region in space through which fluid can enter and leave, as indicated in Fig. 8–8. A control volume is a conceptual device established by the analyst for his own convenience. It can be an infinitesimal volume, or it can enclose a complete system.

Now, although in the eulerian description the flow of a particular fluid element *in the large* is not followed, it is possible to follow the flow *infinitesimally*. A fluid particle which has velocity u at the location x at time t will reach the location $x + u\,dt$ at time $t + dt$. The change in a fluid

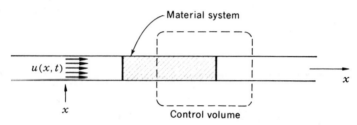

FIG. 8–8.   Spatial coordinate x and control volume fixed in space.

property, for example, the density, experienced by that particle would then be given by

$$d\rho = \frac{\partial \rho}{\partial x}(u\,dt) + \frac{\partial \rho}{\partial t}\,dt, \qquad (8\text{--}62)$$

or the time rate of change, *following the flow*, is

$$\frac{d\rho}{dt} = u\frac{\partial \rho}{\partial x} + \frac{\partial \rho}{\partial t}. \qquad (8\text{--}63)$$

This kind of derivative is called the *material* derivative (also the *substantive* derivative, and *stokesian* derivative). A common notation, which we shall adopt, for the material derivative (8–63) is $D\rho/Dt$.

The dynamic equations for isentropic one-dimensional flow are based on the requirements of mass conservation, the constitutive behavior of the fluid, and the linear-momentum principle. In the interior of the fluid continuum these requirements are represented in the lagrangian description by (8–13) to (8–15). We shall recast these requirements using the eulerian description. First, we note that the constitutive relation (8–14) is independent of the space-time description.

Next, turning to the conservation of mass, we consider the infinitesimal control volume shown in Fig. 8–9. If mass is conserved, the net influx of mass into the control volume during any interval must equal the increase of mass within the control volume. The rate at which mass is carried into the control volume through the tube of unit area at x is ρu. The corresponding rate at which mass is carried out of the control volume at $x + dx$ is $\rho u + [\partial(\rho u)/\partial x]\, dx$. Within the control volume the mass at any instant is $\rho\, dx$. The rate at which mass is increasing within the control surface is $(\partial \rho/\partial t)\, dx$. Thus, for conservation of mass,

$$\frac{\partial \rho}{\partial t}\, dx = \rho u - \left[\rho u + \frac{\partial(\rho u)}{\partial x}\, dx\right], \tag{8–64}$$

or
$$\frac{\partial \rho}{\partial t} + \frac{\partial(\rho u)}{\partial x} = 0. \tag{8–65}$$

Equation (8–65) is called the *continuity equation*. It plays the same role

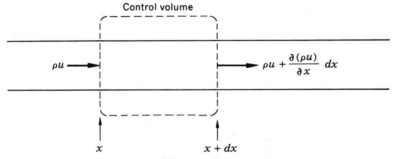

Fig. 8–9. Infinitesimal control volume.

in the eulerian description as the mass-conservation relation (8–13) plays in the lagrangian description.

The continuity equation (8–65) can be rewritten by performing the partial x derivative,

$$\frac{\partial \rho}{\partial t} + u\frac{\partial \rho}{\partial x} + \rho\frac{\partial u}{\partial x} = 0, \tag{8–66}$$

and substituting the material derivative (8–63) to obtain

$$\frac{D\rho}{Dt} + \rho\frac{\partial u}{\partial x} = 0. \tag{8–67}$$

In this form the continuity equation can be interpreted as a conservation-of-mass statement applied to a differential material volume. Equation (8–67) states that a time rate of increase of density in this moving element must be accompanied by a rate of decrease in the element's volume (note that $\partial u/\partial x$ represents a rate of increase of volume per unit volume).

The linear-momentum principle can be expressed in the eulerian description with the help of Fig. 8–10. A differential fluid system is shown acted on by fluid-pressure forces. The mass of this element is conserved throughout the flow, but at the instant depicted in Fig. 8–10, it can be evaluated as $\rho(x,t)\,dx$.   At this same instant the velocity of the fluid system is $u(x,t)$.   The system's acceleration is time rate of change of velocity, following the fluid; i.e., Du/Dt.   Equating the unbalanced pressure force in Fig. 8–10 to the product of the mass and the acceleration of the fluid element, we have

$$\rho\,\frac{Du}{Dt} = -\frac{\partial p}{\partial x} \tag{8–68}$$

as the dynamic equation in the eulerian description which plays the same role as the momentum equation (8–15) plays in the lagrangian description.

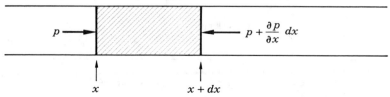

Fig. 8–10.   Material fluid system which occupies the control volume of Fig. 8–9 at time t.

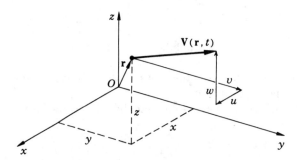

Fig. 8–11.   Fluid velocity is **V** at spatial location **r**.

The constitutive relation (8–14), the continuity equation (8–65) or (8–67), and the momentum equation (8–68) for the three variables p, ρ, and u apply at each point within the fluid continuum in the eulerian description.   Boundary conditions are most conveniently established at fixed locations which act either as walls or as entrances or exits of control volumes.

EXTENSION TO THREE-DIMENSIONAL FLOW

The eulerian formulation is readily generalized to three dimensions by introducing vector notation.   In Fig. 8–11 the vector **r**, with components x, y, and z, represents a fixed *spatial* location in a region through which a fluid flows. The fluid pressure p, density ρ, and velocity $\mathbf{V}$ (with components u, v, and w) are considered as functions of location and time; i.e., we write $p(\mathbf{r},t)$, $\rho(\mathbf{r},t)$, and $\mathbf{V}(\mathbf{r},t)$. A time rate of change following the flow can be indicated by the *material-derivative* operator,

$$\frac{D}{Dt} = \frac{\partial}{\partial t} + u\frac{\partial}{\partial x} + v\frac{\partial}{\partial y} + w\frac{\partial}{\partial z}$$

$$= \frac{\partial}{\partial t} + \mathbf{V} \cdot \boldsymbol{\nabla}, \tag{8-69}$$

which is the extension to three dimensions of the one-dimensional material-derivative operator introduced in (8-63).

The continuity equation in the form (8-65) generalizes[1] to

$$\frac{\partial \rho}{\partial t} + \frac{\partial(\rho u)}{\partial x} + \frac{\partial(\rho v)}{\partial y} + \frac{\partial(\rho w)}{\partial z} = 0$$

$$\frac{\partial \rho}{\partial t} + \boldsymbol{\nabla} \cdot (\rho \mathbf{V}) = 0, \tag{8-70}$$

while the form (8-67) becomes

$$\frac{D\rho}{Dt} + \rho\left(\frac{\partial u}{\partial x} + \frac{\partial v}{\partial y} + \frac{\partial w}{\partial z}\right) = 0$$

$$\frac{D\rho}{Dt} + \rho\boldsymbol{\nabla} \cdot \mathbf{V} = 0. \tag{8-71}$$

The momentum equation (8-68) generalizes into three component equations of the same form, which can be written as the single vector equation

$$\rho\frac{D\mathbf{V}}{Dt} = -\boldsymbol{\nabla} p. \tag{8-72}$$

At each point in the interior of a three-dimensional isentropic fluid continuum the constitutive relation (8-14), the continuity equation (8-70) or (8-71), and the momentum equation (8-72) apply. These are five nonlinear scalar equations for the five quantities p, ρ, u, v, and w. In a completely posed problem it would be necessary to supplement these internal equations with boundary conditions.

HAMILTON'S PRINCIPLE IN THREE-DIMENSIONAL FLOW

The variational method for formulating dynamic equations for one-dimensional fluid systems described in the previous section is easily extended[2] to two- and three-dimensional flows, provided the lagrangian description is maintained. When the eulerian description is used, the application of Hamilton's principle requires considerable delicacy. The

---

[1] See Prob. 8-12.

[2] See Prob. 8-15.

basic difficulty resides in the fact that, in concept, Hamilton's principle applies to a material system of fixed identity during a finite time interval, whereas in the eulerian description it is not possible to follow a material system, except infinitesimally. It is possible, however, to arrange the argument so that Hamilton's principle will produce the dynamic equations for the eulerian description provided that the infinitesimal admissible variations conserve not only mass but the *identity* of the individual fluid elements. A technique[1] for accomplishing this last requirement is to introduce (temporarily) the material coordinates X, Y, and Z and impose as constraints, via Lagrange multipliers, the requirement that the material derivatives of the material coordinates vanish. This technique has proved to be of value in the extension of Hamilton's principle which applies to the flow of electrically charged fluids.[2]

8-5   Interaction of fluids and elastic solids

An important branch of dynamics treats mixed fluid and solid systems in which there are significant dynamic interactions between the fluid and solid subsystems. This branch of dynamics includes *aeroelasticity*, which deals with the dynamic behavior of flexible vehicles as influenced by the air flowing around them. A dramatic example is provided by the phenomenon of "flutter" of aircraft wings. Similar dynamic interactions are responsible for the "galloping" of long transmission wires in the wind and the wind-induced vibrations of tall chimneys. The failure of the Tacoma Narrows bridge was an unfortunate illustration of the potential seriousness of these interactions.

All the examples cited above have been cases of *exterior flow*. There are also some interesting and important dynamic interaction problems in *interior flow*. In these cases the fluid is inside an elastic pipe or container. The problem of blood flow through the arterial system is a good example in which the elasticity of the artery interacts with the flowing fluid inside.[3] Another example is furnished by the sloshing of the comparatively heavy liquid fuel (or oxidizer) in a light elastic tank of a space vehicle. As a brief introduction to problems of this sort we consider a simple case of a fluid flowing in an elastic tube. The model that we use

---

[1] This technique, due to C. C. Lin, is described by J. Serrin, Mathematical Principles of Classical Fluid Mechanics, pp. 125–263 in S. Flügge and C. Truesdell (eds.), "Handbuch der Physik," vol. VIII/1, Springer Verlag OHG, 1959, and also by P. Penfield, Jr., Hamilton's Principle for Fluids, *Phys. Fluids*, **9**:1184–1194 (1966). See Probs. 8–22 and 8–23.

[2] P. Penfield, Jr. and H. A. Haus, "Electrodynamics of Moving Media," The M.I.T. Press, Cambridge, Mass., 1967.

[3] For a simple example of this type, see Prob. 8–24.

has been employed to investigate vibrations in oil pipelines and to study the stability of pipe-nozzle systems.

Example 8–2.   The system is sketched in Fig. 8–12.   A section of pipe acts as an elastic cantilever beam.   Inside the pipe there is a steady flow of fluid.   In the model we consider, the pipe is taken to be a uniform Bernoulli-Euler continuum of length L with bending modulus EI and

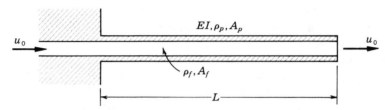

FIG. 8–12.   Example 8–2.   Steady flow of fluid through cantilevered elastic pipe.

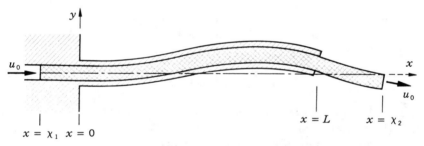

FIG. 8–13.   Total system consists of pipe and moving fluid system which fills pipe and extends beyond it at both ends.

mass per unit length $\rho_p A_p$.   It is clamped at the left and free at the right.   The fluid is taken to be an incompressible fluid with (constant) density ρ_f flowing steadily with velocity u_0 through a tube of constant cross-sectional area A_f.   We neglect friction and gravity and assume the pipe is straight in the reference configuration shown in Fig. 8–12.

We now consider the dynamic behavior of this system for *small* transverse motions in a single plane.   A deformed configuration is shown in Fig. 8–13.   We shall apply Hamilton's principle to obtain the dynamic equations for the system consisting of the pipe and the crosshatched fluid system indicated in Fig. 8–13.   Position along the pipe is denoted by x, where $0 < x < L$.   Transverse displacement of the pipe center line is denoted by $\eta_p(x,t)$.   Longitudinal displacement of the pipe is neglected. We can consider that x acts as a material, or lagrangian, coordinate for the pipe.

For the fluid we introduce a lagrangian coordinate X, which may be thought of as the x coordinate of a fluid element at some earlier reference time t_0 when the entire fluid system was upstream in the uniform straight supply pipe. The present longitudinal position of the fluid element is denoted by $\chi(X,t)$, and the transverse displacement of its center line is denoted by $\eta_f(X,t)$. At present we take the pipe displacement η_p and the fluid displacement η_f to be independent, but eventually we shall have to introduce the constraint that the fluid flows *inside* the pipe.

The lagrangian for the total system of Fig. 8–13 contains the kinetic coenergy of both the pipe and the fluid and the potential energy of bending in the pipe. There is no contribution from the fluid internal energy because of our assumption that the flow is incompressible. We also neglect any nonconservative effects in the system, so that there is no work expression to enter the variational indicator.

The potential energy of bending is (see page 346)

$$V = \int_0^L \tfrac{1}{2}EI \left(\frac{\partial^2 \eta_p}{\partial x^2} \right)^2 dx. \qquad (8\text{–}73)$$

For the kinetic coenergy of the beam we assume that, because of the smallness of the motion, any longitudinal contribution can be neglected in comparison with the transverse contribution.

$$T_p^* = \int_0^L \tfrac{1}{2}\rho_p A_p \left(\frac{\partial \eta_p}{\partial t} \right)^2 dx. \qquad (8\text{–}74)$$

For the fluid, we assume that although the longitudinal velocity is large, it varies so little[1] during the small transverse motions that we can again neglect the longitudinal contribution. Thus

$$T_f^* = \int_{X_1}^{X_2} \tfrac{1}{2}\rho_f A_f \left(\frac{\partial \eta_f}{\partial t} \right)^2 dX. \qquad (8\text{–}75)$$

The total lagrangian for the system is

$$\mathcal{L} = T_p^* + T_f^* - V_p. \qquad (8\text{–}76)$$

If (8–76) is inserted in the variational indicator (2–222) for Hamilton's principle and the usual variations and integrations-by-parts are carried out, the result is

$$\int_{t_1}^{t_2} dt \left[\int_0^L \left(-\rho_p A_p \frac{\partial^2 \eta_p}{\partial t^2} - EI \frac{\partial^4 \eta_p}{\partial x^4} \right) \delta \eta_p \, dx - \int_{X_1}^{X_2} \rho_f A_f \frac{\partial^2 \eta_f}{\partial t^2} \delta \eta_f \, dX \right.$$
$$\left. - EI \frac{\partial^2 \eta_p}{\partial x^2} \delta \frac{\partial \eta_p}{\partial x} \Big|_0^L + EI \frac{\partial^3 \eta_p}{\partial x^3} \delta \eta_p \Big|_0^L \right]. \qquad (8\text{–}77)$$

This is as far as we can go without bringing in the requirements of geometric compatibility. Because the beam is clamped at $x = 0$, it is neces-

[1] See Prob. 8–19.

sary that

$$\delta\eta_p = 0 \qquad \delta\left(\frac{\partial\eta_p}{\partial x}\right) = 0 \qquad \text{at } x = 0. \tag{8-78}$$

Also, the fluid must remain *inside* the pipe. This means that when a fluid element has the same longitudinal position as a pipe element, it must also have the same transverse displacement. Now, for small transverse motions, the longitudinal position at time t of a fluid element with lagrangian coordinate X is

$$\chi = X + u_0(t - t_0), \tag{8-79}$$

and the longitudinal position of a pipe element is x. Thus, when $\chi = x$, it is necessary for $\eta_f(X,t)$ to equal $\eta_p(x,t)$; i.e.,

$$\eta_f(X,t) = \eta_p(X + u_0(t - t_0), t) \qquad \text{for } 0 < \quad X + u_0(t - t_0) \quad < L. \tag{8-80}$$

The relation (8–80) must hold for the actual motions, and also for the admissible variations, which means that

$$\delta\eta_f = \delta\eta_p \qquad \text{for } 0 < x = [X + u_0(t - t_0)] < L. \tag{8-81}$$

To the left of this region (within the supply pipe), $\delta\eta_f = 0$, and to the right of this region (in the free jet), $\delta\eta_f$ is arbitrary. The variational indicator (8–77) can thus be written

$$\int_{t_1}^{t_2} dt \left\{ \int_0^L \left(-\rho_p A_p \frac{\partial^2\eta_p}{\partial t^2} - \rho_f A_f \frac{\partial^2\eta_f}{\partial t^2} - EI \frac{\partial^4\eta_p}{\partial x^4} \right) \delta\eta_p \, dx \right.$$
$$+ \int_{L-u_0(t-t_0)}^{X_2} \rho_f A_f \frac{\partial^2\eta_f}{\partial t^2} \delta\eta_f \, dX - \left[EI \frac{\partial^2\eta_p}{\partial x^2} \delta\left(\frac{\partial\eta_p}{\partial x}\right) \right.$$
$$\left. \left. - EI \frac{\partial^3\eta_p}{\partial x^3} \delta\eta_p \right]_{x=L} \right\}, \tag{8-82}$$

and the dynamic equations obtained as the necessary condition for (8–82) to vanish for arbitrary $\delta\eta_p$ in $0 < x < L$, for arbitrary $\delta\eta_f$ in the free jet, and for arbitrary $\delta\eta_p$ and $\delta(\partial\eta_p/\partial x)$ at $x = L$. We get the *partial differential equation* along the length of the pipe,

$$EI \frac{\partial^4\eta_p}{\partial x^4} + \rho_p A_p \frac{\partial^2\eta_p}{\partial t^2} + \rho_f A_f \frac{\partial^2\eta_f}{\partial t^2} = 0 \qquad 0 < x < L, \tag{8-83}$$

the partial differential equation

$$\rho_f A_f \frac{\partial^2\eta_f}{\partial t^2} = 0 \tag{8-84}$$

in the free jet, and the *natural boundary conditions*

$$EI \frac{\partial^2\eta_p}{\partial x^2} = 0 \qquad EI \frac{\partial^3\eta_p}{\partial x^3} = 0 \qquad \text{at } x = L. \tag{8-85}$$

Equation (8–83) is essentially a statement of the momentum principle. The elastic restoring force acting on an element of pipe changes the transverse momentum of that element of pipe plus the transverse momentum of the element of fluid within the pipe element at that instant. Equation (8–84) states that when the fluid emerges from the pipe, it undergoes no further change in transverse momentum. The natural boundary conditions (8–85) state that there is no external bending moment or shear force applied to the free end of the pipe.

The fluid acceleration in (8–83) can be evaluated in terms of the pipe motion by taking the second time derivative of (8–80). The first derivative of (8–80) yields

$$\frac{\partial \eta_f}{\partial t} = u_0 \frac{\partial \eta_p}{\partial x} + \frac{\partial \eta_p}{\partial t}. \tag{8–86}$$

It is interesting to note here the connection between (8–86) and the material derivative (8–69) used in the eulerian description of fluid flow; i.e., we can write

$$\frac{\partial \eta_f}{\partial t} = \frac{D\eta_p}{Dt}. \tag{8–87}$$

The fluid's transverse velocity is the time rate of change of the pipe's displacement as experienced by an observer moving with the fluid speed u_0. The second derivative of (8–80) is then

$$\frac{\partial^2 \eta_f}{\partial t^2} = \frac{D^2 \eta_p}{Dt^2} = u_0{}^2 \frac{\partial^2 \eta_p}{\partial x^2} + 2u_0 \frac{\partial^2 \eta_p}{\partial x\, \partial t} + \frac{\partial^2 \eta_p}{\partial t^2}. \tag{8–88}$$

The equations of motion for the pipe can then be expressed in terms of the single variable $\eta_p(x,t)$. We have the partial differential equation

$$EI \frac{\partial^4 \eta_p}{\partial x^4} + (\rho_p A_p + \rho_f A_f) \frac{\partial^2 \eta_p}{\partial t^2}$$
$$+ \rho_f A_f u_0 \left(2 \frac{\partial^2 \eta_p}{\partial x\, \partial t} + u_0 \frac{\partial^2 \eta_p}{\partial x^2} \right) = 0 \qquad 0 < x < L, \tag{8–89}$$

the natural boundary conditions (8–85), and the geometric boundary conditions

$$\eta_p = 0 \qquad \text{and} \qquad \frac{\partial \eta_p}{\partial x} = 0 \qquad \text{at } x = 0. \tag{8–90}$$

The formulation just obtained by application of Hamilton's principle can also be obtained directly. Equation (8–83) is essentially a statement of the linear-momentum principle applied to an infinitesimal element of pipe *and* fluid. The further reduction to (8–89) depends only on the kinematic relation (8–88). This may seem relatively obvious to one who has followed the preceding derivation.   Just how obvious this would be to an analyst starting out with the direct method may perhaps be judged from the historical fact that the initial published analysis[1] of this problem was unsuccessful in deriving (8–89).   The first correct derivation of this equation appeared in a subsequent analysis[2] which made use of Hamilton's principle.

The most interesting applications of the above formulation have related to the effect of the fluid speed u_0 on the stability of the straight configuration of Fig. 8–12.   The theory[3] predicts, and experiments[4] have confirmed, that the straight configuration is stable as long as the flow speed u_0 remains less than a certain critical value.   If u_0 is greater than the critical value, the pipe undergoes violent self-excited bending oscillations as it discharges fluid alternately to one side and then to the other.   This kind of instability is similar to "flutter" of airfoils.   A related phenomenon occurs in a pipe which is supported at both ends.[5]   In this case, when the fluid speed becomes greater than a critical value, the pipe buckles and simply diverges from the straight configuration without any tendency to vibrate.

PROBLEMS

8–1. The constitutive relation for a perfect gas undergoing adiabatic changes of state is

$$\frac{p}{p_0} = \left(\frac{\rho}{\rho_0}\right)^{\gamma},$$

where p_0, ρ_0 is a reference state.   Obtain the corresponding internal energy $U(\rho)$.

8–2. Consider an idealized isentropic atmosphere which occupies the half space $z > 0$ under the influence of a uniform gravitational acceleration g in the negative z direction.   Establish a lagrangian coordinate system, and formulate the differential equations and boundary conditions for one-dimensional isentropic dynamic disturbances.   What are the equations determining the distribution of density in the equilibrium state when all dynamic disturbances vanish?   Assume that if the entire atmosphere were brought to the density ρ_0, the atmosphere would have the finite height h.

[1] H. Ashley and G. Haviland, Bending Vibrations of a Pipeline Containing Flowing Fluid, *J. Appl. Mech.*, **17**:229–232 (1950).

[2] G. W. Housner, Bending Vibrations of a Pipeline Containing Flowing Fluid, *J. Appl. Mech.*, **19**:205–208 (1952).

[3] G. Hermann and S. Nemat-Nasser, Instability Modes of Cantilevered Bars Induced by Fluid Flow through Attached Pipes, *Intern. J. Solids and Structures*, **3**:39–52 (1967).

[4] T. B. Benjamin, Dynamics of a System of Articulated Pipes Conveying Fluid: I, Theory, and II, Experiments, *Proc. Roy. Soc. London*, **A261**:457–499 (1961).

[5] See Prob. 8–20.

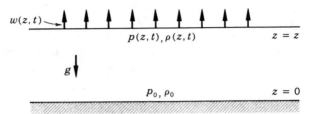

$$w(z,t)$$

$$p(z,t), \rho(z,t) \qquad z = z$$

$$g$$

$$p_0, \rho_0 \qquad z = 0$$

PROB. 8–2

8–3. Consider inviscid isentropic flow of a compressible fluid through a tube whose cross-sectional area $A(x)$ changes slowly with x. It is desired to obtain an approximate one-dimensional description of the flow. Using the lagrangian coordinates of Sec. 8–1, show that the mass-conservation relation is

$$\rho A \frac{\partial x}{\partial X} = \rho_0 A_0$$

and that the governing dynamic equation is

$$\rho_0 A_0 \ddot{x} = -A \frac{\partial p}{\partial X}.$$

$$A(x)$$

$$x$$

PROB. 8–3

8–4. Reformulate the one-dimensional approximation of Prob. 8–3 for isentropic flow in a tube with gradual area change in eulerian coordinates. Show that the continuity equation may be written

$$\frac{\partial}{\partial x}(\rho u A) + A \frac{\partial \rho}{\partial t} = 0.$$

8–5. Formulate the governing equations of spherically symmetric isentropic flow in lagrangian coordinates. Assume that velocity is always directed radially and that spherical symmetry is maintained.

8–6. Reformulate Prob. 8–5 in eulerian coordinates.

8–7. Obtain a formulation in lagrangian coordinates which provides governing equations and boundary conditions for the one-dimensional expansion process in the tube of cross-sectional area A which takes place when, with the gas initially at rest with length L, density ρ_0 and pressure p_0, the right-hand piston is suddenly given a uniform velocity u_2. Assume isentropic expansion and neglect dissipation effects.

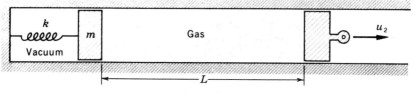

8-8. Reconsider Prob. 8-7 for the case where the right-hand-piston motion is small and the acoustic approximation can be used. If the right-hand-piston velocity is

$$u_2 = \mathrm{Re}\,(u_0 e^{i\omega t})$$

what is the steady-state force reaction of the gas on the right-hand piston?

8-9. Study the response amplitude (8-57) in the resonant case when $b \ll \rho_0 c$. Show that there is an infinite sequence of resonant frequencies ω_n which satisfy the transcendental equation

$$\frac{\rho_0 A L}{m} = \frac{\omega_n L}{c}\tan\frac{\omega_n L}{c}\qquad n = 1, 2, \ldots$$

Obtain a simple asymptotic formula for ω_n valid for large n. Show that the response amplitude has the approximate magnitude

$$\omega_n \frac{m}{b}\sqrt{1 + \left(\frac{\rho_0 A c}{m\omega_n}\right)^2}\qquad \text{when } \omega = \omega_n.$$

8-10. Consider a small dynamic perturbation of a high-speed isentropic compressible one-dimensional flow in eulerian coordinates. Let the velocity be of the form $U + u$, where U is the uniform velocity of the undisturbed flow, and u is a first-order small perturbation. Similarly, let the density be of the form $\rho_0(1 + s)$, where ρ_0 is the uniform density of the undisturbed flow, and the condensation s is a first-order small perturbation. Derive the linearized equation

$$\left(1 - \frac{U^2}{c^2}\right)\frac{\partial^2 s}{\partial x^2} - 2\frac{U}{c^2}\frac{\partial^2 s}{\partial x\,\partial t} - \frac{1}{c^2}\frac{\partial^2 s}{\partial t^2} = 0,$$

and show that this is equivalent to Eq. (8-31).

8-11. Show that $\xi = f(X - ct)$ is a solution of (8-49) which represents a disturbance propagating with phase velocity c. What is the corresponding pressure difference $p - p_0$ and the corresponding particle velocity $\dot{\xi}(X,t)$? Show that the quotient of the dynamic force produced on a unit area by the passage of the disturbance, divided by the particle velocity induced at the same point, is the *characteristic impedance* $\rho_0 c$.

8-12. Consider a fixed infinitesimal rectangular parallelepiped as a control surface in three-dimensional compressible flow. Write approximate expressions for the mass-flow rate across the boundary surfaces and for the rate of increase of mass within the control surface. Derive the three-dimensional continuity equation (8-70).

8-13. Consider small two-dimensional motion of an incompressible inviscid liquid of density ρ_0 in a canal of length L and depth h. Base the analysis on unit thickness normal to the sketch. Let the coordinates X,Y of a particle in the equilibrium position act as lagrangian coordinates for that particle, and take the current coordinates of the particle to be

$$x = X + \xi(X,Y,t)$$

$$y = Y + \eta(X,Y,t).$$

Show that to first order in the spatial derivatives of ξ and η the conservation-of-mass requirement reduces to

$$\frac{\partial \xi}{\partial X} + \frac{\partial \eta}{\partial Y} = 0. \qquad (a)$$

Show that for arbitrary twice-differentiable functions $\psi(X,Y,t)$, Eq. (a) is satisfied provided ξ and η are taken as

$$\xi = \frac{\partial \psi}{\partial Y} \qquad \eta = -\frac{\partial \psi}{\partial X}. \qquad (b)$$

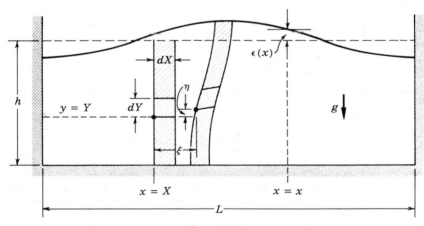

PROB. 8–13

8–14. Consider the potential energy of the fluid in Prob. 8–13 due to gravity. Show that the first-order expression

$$V = \int_0^L dX \int_0^h \rho_0 g \eta \, dY$$

vanishes by using the function ψ introduced in Prob. 8–13. By starting with

$$V = \int_0^L dx \int_0^{h+\epsilon} \rho_0 g y \, dy - \int_0^L dX \int_0^h \rho_0 g Y \, dY,$$

derive the following second-order approximation to the potential energy:

$$V = \int_0^L \tfrac{1}{2} \rho_0 g \eta^2(X,h,t) \, dX.$$

8–15. Apply Hamilton's principle to the small motions of the fluid in Prob. 8–13, neglecting viscosity and interactions at the free surface. Maintain the constraint of Eq. (a) in Prob. 8–13 by introducing the function ψ according to Eq. (b). Verify that Hamilton's principle leads to the governing equation

$$\frac{\partial^2}{\partial t^2}\left(\frac{\partial^2 \psi}{\partial X^2} + \frac{\partial^2 \psi}{\partial Y^2}\right) = 0$$

within the fluid and the natural boundary condition

$$\frac{\partial^3 \psi}{\partial Y \partial t^2} = g \frac{\partial^2 \psi}{\partial X^2}$$

on the free surface $Y = h$. Give an interpretation of the physical significance of these results.

8-16. The classical theory of *deep-water waves* is obtained by assuming a solution to the formulation of Prob. 8-15 in the form

$$\psi = \text{Re}\,[f(Y)e^{i(kX-\omega t)}],$$

where $f(Y)$ and ω are to be determined in terms of the assumed wave number k.

(a) Show that the dispersion relation is

$$\omega^2 = gk \tanh hk.$$

(b) Show that in the limit of extremely deep water $(h \to \infty)$, the phase velocity (7-67) is *twice* the group velocity (7-68).

(c) Show that in the limit of extremely shallow water $(h \to 0)$, the phase velocity and the group velocity both approach $\sqrt{gh}$.

8-17. When $hk \to 0$, the theory of the preceding problems reduces to that of *shallow-water* waves. The formulation for the latter may also be obtained directly by assuming that ξ is independent of Y and that $\dot{\eta}$ is negligible in comparison with $\dot{\xi}$. Show that the conservation-of-mass requirement can then be taken as

$$\eta(X,h,t) + h\frac{\partial \xi}{\partial X} = 0.$$

8-18. Using the assumptions of Prob. 8-17, apply Hamilton's principle to derive the classical wave equation for $\xi(X,t)$ with the phase velocity $c = \sqrt{gh}$.

8-19. Repeat the development in the text for the pipe-and-fluid system of Example 8-2, but include the longitudinal-kinetic-coenergy contribution of the fluid in (8-76). Relate $\delta\chi$ to $\delta\eta_p$, and thus show that the longitudinal term is of higher order than the transverse term.

8-20. Consider a flexible pipe through which liquid flows, as in the system of Example 8-2, except that here both ends of the pipe are hinged, with negligible restraint against rotation. Show that Eq. (8-89) still applies. What are the new natural boundary conditions? By assuming a static deflection of the form $\eta_p = \eta_0 \sin(\pi x/L)$, show that the critical flow velocity for buckling is

$$u_0 = \frac{\pi}{L}\sqrt{\frac{EI}{\rho_f A_f}}.$$

8-21. Use the result of Prob. 8-20 to estimate the critical flow velocity for a steel pipeline 30 in. diameter simply supported at intervals of 70 ft. Take the section moment of inertia to be 0.1 ft⁴, and the fluid to have the density of water. (Normal rates of flow range up to 10 to 15 ft/sec.)

8–22. Hamilton's principle can be applied using the eulerian description in three dimensions by considering the variational integral,

$$\int_{t_1}^{t_2} dt \left\{ \delta \int_{\text{vol}} \left[\tfrac{1}{2}\rho V^2 - \rho U + \lambda_1 \left(\frac{D\rho}{Dt} + \rho \mathbf{\nabla} \cdot \mathbf{V} \right) \right. \right.$$
$$\left. \left. + \rho \boldsymbol{\lambda}_2 \cdot \frac{D\mathbf{R}}{Dt} \right] d\,\text{vol} - \int_S P\mathbf{n} \cdot \delta \boldsymbol{\chi} \, dS \right\},$$

where $\mathbf{R}$ is a lagrangian coordinate with components X, Y, and Z; P is the prescribed pressure acting on the surface S (with unit normal $\mathbf{n}$) of the control volume; and λ_1 and $\rho\boldsymbol{\lambda}_2$ are Lagrange multipliers. By considering independent variations of $\mathbf{V}$, ρ, and $\mathbf{R}$, show that Hamilton's principle requires that

$$\mathbf{V} = \mathbf{\nabla}\lambda_1 + \boldsymbol{\lambda}_2 \cdot (\mathbf{u}_x \mathbf{\nabla}X + \mathbf{u}_y \mathbf{\nabla}Y + \mathbf{u}_z \mathbf{\nabla}Z)$$

$$\frac{D\lambda_1}{Dt} = \tfrac{1}{2}V^2 - U - \frac{p}{\rho}$$

$$\frac{D\boldsymbol{\lambda}_2}{Dt} = 0$$

in the interior of the control volume.

8–23. Verify that if

$$\mathbf{V} = \sum_j a_j \, \mathbf{\nabla}b_j,$$

where a_j and b_j are scalar fields, then

$$\frac{D\mathbf{V}}{Dt} + \mathbf{\nabla}\frac{V^2}{2} = \sum_j \left(a_j \, \mathbf{\nabla}\frac{Db_j}{Dt} + \frac{Da_j}{Dt} \, \mathbf{\nabla}b_j \right).$$

Use this relation to eliminate the Lagrange multipliers from the result of Prob. 8–22 and to verify that these results reduce to (8–72).

8–24. A long light elastic tube contains a heavy compressible fluid. The tube material has tension modulus E, and the tube has a radius r and a wall thickness t, when it is in equilibrium with the fluid inside. The fluid has equilibrium pressure p_0 and density ρ_0. The fluid has sound speed c when in a tube with rigid walls. Estimate the sound speed when the fluid is in the elastic tube.

Consider that internal pressure changes cause the tube radius to change. Consider the effect of radius change on the conservation-of-mass relation, but neglect the transverse contributions to the kinetic coenergy (from both the fluid and the tube). Show that when the effect of the tube compliance is small, the sound speed is reduced,

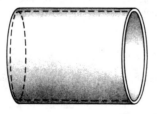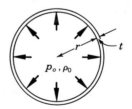

PROB. 8–24

approximately, by the factor

$$1 - \frac{\rho_0 c^2 r}{Et}.$$

8-25. Two light elastic pipes carry heavy incompressible fluid with flow velocities U_1 and U_2. The pipes are long and parallel and coupled together by a continuous elastic medium which exerts a restoring force K per unit length for a unit change in the separation distance between the pipes. Consider small transverse motions of the pipes in the plane of the sketch. Take the fluid to have density ρ, and take both pipes to have uniform bending modulus EI and constant flow area A. Neglect the mass of the pipes in comparison with the fluid mass.

(a) Formulate equations of motion for the transverse displacements of the two pipes, considered as functions of x and t. Take into account the elastic action of the coupling medium and the pipes in bending.

(b) Consider the limiting case where the pipes are so flexible that only the elastic action of the coupling medium has to be taken into account. Show that the dispersion relation, between frequency ω and wave number k, in this case, takes the form

$$\frac{\omega_0^2}{(\omega - kU_1)^2} + \frac{\omega_0^2}{(\omega - kU_2)^2} = 1,$$

where $\omega_0^2 = K/\rho A$.

(c) Show that in case (b) the system is unstable in the range

$$0 < (U_1 - U_2)^2 < \frac{4\omega_0^2}{k^2}.$$

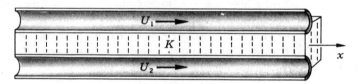

PROB. 8-25

CHAPTER NINE

Prelude to statistical mechanics

The central problem of statistical mechanics is the development of relations between microscopic mechanical models, on the one hand, and macroscopically observed properties of gross material, on the other hand. The key idea connecting the microscopic and macroscopic is the concept that any macroscopic measurement represents an *average* of microscopic behavior. As a result, the dynamics of a microscopic model need only be considered in an average, or statistical, sense.

In this chapter we lay a foundation for subsequent study of statistical mechanics by describing how the dynamic requirements for conservative mechanical systems can be reduced to a convenient form for statistical considerations. We derive Hamilton's canonical equations and Liouville's theorem and give a preliminary outline of how these are incorporated into the framework of statistical mechanics.

9–1 Hamilton's canonical equations

In the latter half of Chap. 2 the dynamic requirements of a general holonomic system are distilled into a compact representation. For a *conservative* system the dynamic equations follow from the lagrangian $\mathcal{L}(\xi, \dot{\xi})$, which is a function of n generalized coordinates ξ_j and n generalized velocities $\dot{\xi}_j$. The dynamic equations are Lagrange's equations (2–228),

$$\frac{d}{dt}\left(\frac{\partial \mathcal{L}}{\partial \dot{\xi}_j}\right) - \frac{\partial \mathcal{L}}{\partial \xi_j} = 0 \qquad j = 1, \ldots, n. \tag{9–1}$$

These equations can be cast into a slightly different form by making use of the generalized momenta π_j of (2–197).

$$\frac{\partial \mathcal{L}}{\partial \dot{\xi}_j} = \frac{\partial T^*}{\partial \dot{\xi}_j} = \pi_j. \tag{9–2}$$

Substituting (9–2) into (9–1), we obtain

$$\frac{\partial \mathcal{L}}{\partial \xi_j} = \frac{d\pi_j}{dt} = \dot{\pi}_j. \tag{9-3}$$

Thus, for a conservative holonomic system, the lagrangian is a function of the n coordinates ξ_j and the n velocities $\dot{\xi}_j$ which has the property that its partial derivatives (9–2) and (9–3) give the n momenta π_j and the n time rates of change of momenta in a natural motion of the system.

We now restate this relationship by means of a device used many times in this book, the Legendre transformation. Here we apply the Legendre transformation to the lagrangian itself. We consider a transformation in which the ξ_j are passive variables, while the $\dot{\xi}_j$ are replaced by the π_j. Denoting the complementary function by $\mathcal{H}(\xi,\pi)$, we have

$$\mathcal{H}(\xi,\pi) = \sum_j \dot{\xi}_j \pi_j - \mathcal{L}(\xi,\dot{\xi}). \tag{9-4}$$

By using (9–2) and (9–3), it is easy to verify that $\mathcal{H}$ is indeed independent of the velocities $\dot{\xi}_j$ and that

$$\left. \begin{array}{l} \dfrac{\partial \mathcal{H}}{\partial \xi_j} = -\dot{\pi}_j \\[2mm] \dfrac{\partial \mathcal{H}}{\partial \pi_j} = \dot{\xi}_j \end{array} \right\} \quad j = 1, \ldots, n. \tag{9-5}$$

The $2n$ first-order differential equations for the $2n$ variables $\xi_1, \ldots, \xi_n,$ $\pi_1, \ldots, \pi_n,$ have the same physical content as the n second-order Lagrange equations. The equations (9–5) are called *Hamilton's canonical equations* of motion, and the function $\mathcal{H}(\xi,\pi)$ is called the *hamiltonian*.

The structure of the hamiltonian function itself can be further delineated if we put $\mathcal{L} = T^* - V$ in (9–4). Then we find

$$\mathcal{H}(\xi,\pi) = \sum_j \dot{\xi}_j \pi_j - T^* + V$$

$$= T + V \tag{9-6}$$

on using the Legendre transformation (2–199) between T and T^*, which is valid when time-varying constraints are absent. Thus, for conservative holonomic systems, the hamiltonian function is the total energy expressed in terms of generalized coordinates and generalized momenta. Because of (2–244) we know that the magnitude E of the total energy remains constant in a conservative system. Although we restrict our attention in this chapter to conservative systems, it is worthwhile to note that the hamiltonian (9–4) and the canonical equations (9–5) have applications to (nonconservative) systems with time-varying constraints.

In these cases, however, the hamiltonian function is *not* in general equal to the total energy.[1]

When Hamilton's equations (9–5) are used to study the motion of a system, the $2n$ variables $\xi_1, \ldots, \xi_n$ and $\pi_1, \ldots, \pi_n$ are treated as independent coordinates. It is convenient to consider a $2n$-dimensional cartesian space in which these variables are measured along the $2n$ orthogonal coordinate axes. Such a hyperspace is usually called a *phase space*. A single point in phase space represents the instantaneous values of the n generalized coordinates (fixing the entire configuration of the holonomic system in physical space) *and* the instantaneous values of the n generalized momenta (fixing the instantaneous states of motion through-

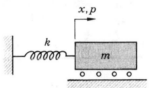

Fig. 9–1.   Example 9–1.   Conservative spring-mass system
with a single degree of freedom.

out the system in physical space). In phase space Hamilton's canonical equations give the time rates of change of the $2n$ coordinates in phase space as a function of the present values of these coordinates. It is helpful to think of the $\dot{\xi}_j$ and $\dot{\pi}_j$ as the $2n$ components of the velocity vector of the representative point in phase space. The motion of a single point (albeit in a hyperspace of $2n$ dimensions) governed by Hamilton's canonical equations thus represents[2] the motion of a holonomic system of n degrees of freedom governed by Lagrange's equations.

Example 9–1. As a simple illustration we consider the linear system of Fig. 9–1. Taking the displacement x as the single generalized coordinate, the lagrangian $T^* - V$ is

$$\mathcal{L}(x,\dot{x}) = \tfrac{1}{2}m\dot{x}^2 - \tfrac{1}{2}kx^2. \tag{9–7}$$

Next, introducing the momentum,

$$p = m\dot{x}, \tag{9–8}$$

we construct the hamiltonian (9–4),

$$\mathcal{H}(x,p) = p\dot{x} - \mathcal{L}(x,\dot{x}) = p\dot{x} - \tfrac{1}{2}m\dot{x}^2 + \tfrac{1}{2}kx^2 = \frac{p^2}{2m} + \tfrac{1}{2}kx^2, \tag{9–9}$$

---

[1] See Prob. 9–4.

[2] When this same kind of representation is used to describe the behavior of general dynamic systems in automatic control theory the hyperspace is usually called a *state space*.

which we recognize as $T + V$. Several contours of constant $\mathcal{3C}$ are sketched in the phase space in Fig. 9–2. Here, since $n = 1$, the phase space is only two-dimensional, and can be drawn in a plane.

Hamilton's equations (9–5), which dictate the motion of the representative point P in Fig. 9–2, are

$$\dot{p} = -\frac{\partial \mathcal{3C}}{\partial x} = -kx$$

$$\dot{x} = \frac{\partial \mathcal{3C}}{\partial p} = \frac{p}{m}. \tag{9–10}$$

These equations can be integrated to provide the time history of the rep-

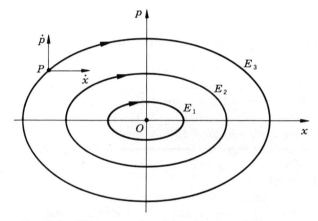

FIG. 9–2. Phase space for physical system of Fig. 9–1.

resentative point P as it moves around the path sketched in Fig. 9–2. If at $t = 0$ the system of Fig. 9–1 is started with initial values $x = x_0$ and $p = p_0$, then the integrals of (9–10) are, using the notation ω_n for $\sqrt{k/m}$,

$$x = x_0 \cos \omega_n t + \frac{\omega_n p_0}{k} \sin \omega_n t \tag{9–11}$$

$$p = -m\omega_n x_0 \sin \omega_n t + p_0 \cos \omega_n t.$$

These are parametric equations for an ellipse. The constant total energy E associated with the ellipse is determined by the initial conditions; i.e.,

$$E = \frac{p_0^2}{2m} + \tfrac{1}{2}kx_0^2. \tag{9–12}$$

In terms of the total energy E, the semiaxis of the ellipse along the x axis is $\sqrt{2E/k}$ and the semiaxis of the ellipse along the p axis is $\sqrt{2mE}$. An alternative form for the parametric equations (9–11) is

$$x = \sqrt{\frac{2E}{k}} \sin (\omega_n t + \varphi_0)$$

$$p = \sqrt{2mE} \cos (\omega_n t + \varphi_0),$$

(9–13)

where the *initial phase angle* φ_0 is given by

$$\tan \varphi_0 = \frac{m\omega_n x_0}{p_0}.$$

(9–14)

In (9–13) the initial conditions enter through the energy level E and the starting phase φ_0. In applications to microscopic models, energy levels and starting phases are often more convenient than initial displacements and initial momenta.

9–2 A microscopic mechanical model

The concept of representing the dynamic state of a system by a single point in phase space is especially attractive when considering a microscopic mechanical model for a macroscopic sample of material. The number of molecules (mass particles) in the system is then of the order of Avogadro's number, 6.0225×10^{26} molecules per kilogram mole. For the sake of definiteness, we consider the following particular model, which is widely used (although it is by no means the only one discussed) in statistical mechanics.

We consider, within the framework of classical newtonian dynamics, a system consisting of a large number N of identical mass particles, each with mass m. This system moves in a three-dimensional physical space with no geometric constraints. The particles interact with each other according to conservative force mechanisms, so that all force interactions are derivable from a total-system potential-energy function V, which depends on the positions of all the particles in the system.

In physical space each mass particle has three degrees of freedom, so the system has $n = 3N$ degrees of freedom. The corresponding phase space has $6N$ dimensions. We shall adopt the usual notation of statistical mechanics and denote the generalized coordinates of the system by the symbols $q_i(i = 1, \ldots, 3N)$. The coordinates x_1, y_1, and z_1 of the first mass particle with respect to an inertial frame are labeled q_1, q_2, and q_3. The coordinates x_2, y_2, and z_2 of the second mass particle are labeled q_4, q_5, and q_6, and so on up to q_{3N}. The generalized momenta in this case are the actual momentum components. The momentum components $m\dot{x}_1$, $m\dot{y}_1$, and $m\dot{z}_1$ of the first particle are labeled p_1, p_2, and p_3. The momentum components $m\dot{x}_2$, $m\dot{y}_2$, and $m\dot{z}_2$ of the second particle are labeled p_4, p_5, and p_6, and so on up to p_{3N}.

For this system[1] the lagrangian is

$$\mathcal{L}(q,\dot{q}) = \sum_{i=1}^{3N} \tfrac{1}{2}m\dot{q}_i^2 - V(q), \tag{9–15}$$

and the hamiltonian is

$$\mathcal{H}(q,p) = \sum_{i=1}^{3N} \frac{p_i^2}{2m} + V(q). \tag{9–16}$$

The phase space is a $6N$-dimensional space with $3N$ *configurational coordinates* q_i and $3N$ *momentum coordinates* p_i. Because the system is conservative, the hamiltonian (9–16) is a constant along the trajectory of the representative point in phase space. The rate at which this trajectory is traced out is given by Hamilton's canonical equations (9–5), which, for this case, take the form

$$\dot{p}_i = -\frac{\partial \mathcal{H}}{\partial q_i} \qquad \dot{q}_i = \frac{\partial \mathcal{H}}{\partial p_i} \qquad i = 1, \ldots, 3N. \tag{9–17}$$

9–3 The distribution function

If, conceptually, we start the mechanical model of the preceding section from given initial values of the $3N$ coordinates and $3N$ momenta, the representative point in phase space will trace out, according to (9–15), a trajectory in phase space. This trajectory will always lie on the hypersurface $\mathcal{H}(q,p) = E$, where the initial conditions fix the value of the constant total energy E.

At any time t the position of the representative point in phase space describes the complete dynamic state of the microscopic model. This is called the *microstate* of the system. Now the problem of statistical mechanics is to develop a useful correspondence between the dynamics of such microscopic models and the constitutive behavior of macroscopic objects. For example, statistical mechanics can be used to derive, from a microscopic model, a macroscopic relation between the internal energy, entropy, and volume for a gas. The fundamental hypothesis of statistical mechanics is that the observed macroscopic properties correspond not to a *single* microstate, but rather to a *distribution* of microstates. A macroscopic property in statistical mechanics is represented by a statistical average over the microstates accessible to the microscopic model.

In order to deal with distributions of microstates, we introduce a major conceptual extension of the microscopic model. Instead of a single model

---

[1] In (9–15) and (9–16) we use the abbreviations $V(q)$ for $V(q_1, q_2, \ldots, q_{3N})$, $\mathcal{L}(q,\dot{q})$ for $\mathcal{L}(q_1, q_2, \ldots, q_{3N}, \dot{q}_1, \dot{q}_2, \ldots, \dot{q}_{3N})$, and $\mathcal{H}(q,p)$ for $\mathcal{H}(q_1, q_2, \ldots, q_{3N}, p_1, p_2, \ldots, p_{3N})$.

system represented by a single point in phase space, we consider a very large number of identical microscopic systems. This set of systems is called an *ensemble*, and the individual members of the ensemble are referred to as *sample* systems. The sample systems are taken to be identical, but the microstates of the samples are generally different. The distribution of microstates among the samples can be considered to be the result of starting off the systems from different initial conditions. As time goes on the representative point in phase space for any one sample system traces out its trajectory according to the canonical equations (9–17). If two sample systems start from different initial conditions, their representative points will in general follow different trajectories which remain distinct. If a distribution of initial conditions were given, it would, in principle, be possible to integrate the canonical equations along each sample trajectory and establish the distribution of representative points in phase space at any subsequent time t.

In terms of the conceptual framework just presented, we can give a brief preview of the discipline of statistical mechanics. There are essentially three problems: The first is the problem of assigning initial distributions of microstates so as to provide useful representations of prescribed macroscopic conditions. The second is the problem of predicting the subsequent evolution of the distribution as the individual representative points generate trajectories according to the canonical equations. The third problem is that of developing appropriate statistical averages over the ensemble of microstates which can be used to predict macroscopic properties. In this prelude to statistical mechanics we do not enter into the first or third problem, but we do develop a powerful tool for studying the purely dynamical problem of predicting the evolution of an ensemble distribution.

Example 9–2. To illustrate the concept of an ensemble, we consider again the single-degree-of-freedom system of Example 9–1, for which the phase space reduces to a plane. For a single-sample system like that of Fig. 9–1, there is a single representative point P in the phase plane, as sketched in Fig. 9–3a. The elliptical trajectory (9–13) traversed by P can be fixed by assigning the total energy E and the starting phase.

Now, suppose we consider a very large number of sample systems, all identical with Fig. 9–1. If we start them all off with the *same* total energy E but with a random distribution of starting phases, the distribution of representative points might appear as sketched in Fig. 9–3b. As time goes by the sample points will move in single file around the same ellipse. The distribution of microstates at each instant will vary, but this distribution can be traced back to the initial distribution.

In Fig. 9–3c a more complex ensemble of microstates is sketched. Here there is a random distribution of total energy, as well as a random

distribution of phase.   As time goes by each sample point traces out the
ellipse (9–13) appropriate to its own initial conditions.   The ensembles
sketched in Fig. 9–3*b* and *c* give a suggestion of the sort of distributions
that have been found useful in statistical mechanics.   As previously

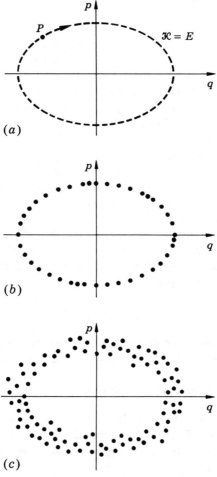

FIG. 9–3.   (*a*) Single microstate.   (*b*) Ensemble of microstates with equal energies.
(*c*) Ensemble with a range of energies.

stated, an important part of the discipline of statistical mechanics involves
working out the consequences of various hypotheses concerning the initial
distribution of microstates.   We do not, however, pursue this question
any further here.

We return to the general microscopic model with an ensemble of systems, each consisting of N identical mass particles. The microstate of each sample system is represented by a single point in a $6N$-dimensional phase space. As time goes on, each representative point moves in the phase space according to the canonical equations (9–17). As a result of these individual motions, the distribution of the ensemble generally changes with time.

We now introduce a distribution function f as a quantitative description of the distribution of sample points in the phase space, and in the next section we study its dynamic evolution. The *distribution function f* is conceived as a function which at any instant is defined throughout the phase space and which has the property that the integral of the distribution function over any subvolume of the phase space is equal to the fraction of sample points of the ensemble which are inside that particular subvolume at that particular instant.

To give an analytical representation of the preceding statement, it is convenient to consider that the number of samples in the ensemble is so large that we can pass from a discrete to a continuous description. Let a point P in $6N$-dimensional phase space have coordinates $q_1, \ldots,$ $q_{3N}, p_1, \ldots, p_{3N}$, and let $d\mathcal{V} = dq_1 \cdots dq_{3N} \, dp_1 \cdots dp_{3N}$, be the $6N$-dimensional differential volume element centered on P. Then the distribution function $f(P,t)$ is a function of position P in phase space and time t, which has the property that at any instant the integral

$$\int_{\mathcal{V}} f(P,t) \, d\mathcal{V}, \qquad (9\text{--}18)$$

over any subvolume $\mathcal{V}$ of phase space, represents the fraction of the ensemble which lies inside $\mathcal{V}$ at that instant. As an extreme case of (9–18),

$$f(P,t) \, dq_1 \cdots dq_{3N} \, dp_1 \cdots dp_{3N} \qquad (9\text{--}19)$$

represents the fraction of ensemble members which lie within the infinitesimal volume element $d\mathcal{V}$. At time t the distribution function $f(P,t)$ thus represents the *fraction of ensemble members per unit volume* at the position P at time t.

When we integrate (9–19) over *all* phase space, we get

$$\int_{-\infty}^{\infty} \cdots \int_{-\infty}^{\infty} f(P,t) \, dq_1 \cdots dq_{3N} \, dp_1 \cdots dp_{3N} = 1, \quad (9\text{--}20)$$

at any time, because *every* sample within the ensemble lies *somewhere* in the phase space. The distribution function $f(P,t)$ has the properties of a *probability-density* function; i.e., it is a nonnegative function of position in the space, so normalized that its integral over the entire space is unity.

9–4   The phase fluid

Our final extension of concept is to consider that the ensemble of sample points in the phase space can be thought of as a *fluid continuum*. We imagine that this fluid flows in the $6N$-dimensional space as the individual points move according to the canonical equations. If we take the total "mass" of the fluid flowing to be *unity*, we can use the distribution func-- tion $f(P,t)$ to represent the fluid *density* at the location P at time t.

We can now show how the dynamic requirements of the original mechanical model imply a simple restriction on the flow of the phase fluid. We begin with the continuity equation (8–71) for a compressible fluid in three dimensions,

$$\frac{D\rho}{Dt} + \rho \mathbf{\nabla} \cdot \mathbf{V} = 0. \tag{9-21}$$

To extend this to phase space, we let $f(p,t)$ play the role of the density ρ and take for the velocity vector the $6N$-dimensional vector with compo- nents $\dot{q}_1, \ldots, \dot{q}_{3N}, \dot{p}_1, \ldots, \dot{p}_{3N}$. The vector differential operator $\mathbf{\nabla}$ in phase space has the $6N$ components

$$\frac{\partial}{\partial q_1}, \frac{\partial}{\partial q_2}, \cdots, \frac{\partial}{\partial q_{3N}}, \frac{\partial}{\partial p_1}, \frac{\partial}{\partial p_2}, \cdots, \frac{\partial}{\partial p_{3N}}. \tag{9-22}$$

The continuity equation in phase space thus takes the form

$$\frac{Df}{Dt} + f \sum_{i=1}^{3N} \left(\frac{\partial \dot{q}_i}{\partial q_i} + \frac{\partial \dot{p}_i}{\partial p_i} \right) = 0. \tag{9-23}$$

This continuity equation states that the "mass" of the phase fluid must be conserved during the flow.

Next we use the dynamic requirements which dictate the motion of the representative points to evaluate the summation in (9–23). According to the canonical equations (9–17), the expression in the parentheses of (9–23) is

$$\frac{\partial \dot{q}_i}{\partial q_i} + \frac{\partial \dot{p}_i}{\partial p_i} = \frac{\partial}{\partial q_i}\left(\frac{\partial \mathcal{K}}{\partial p_i}\right) + \frac{\partial}{\partial p_i}\left(-\frac{\partial \mathcal{K}}{\partial q_i}\right) = 0, \tag{9-24}$$

which means that (9–23) reduces to

$$\frac{Df}{Dt} = 0. \tag{9-25}$$

Equation (9–25) is known as Liouville's theorem. It states that, to an observer moving with the phase fluid, the density $f(p,t)$ appears to remain constant. When the material derivative in (9–25) is expanded, we obtain

$$\frac{\partial f}{\partial t} + \sum_{i=1}^{3N} \left(\dot{q}_i \frac{\partial f}{\partial q_i} + \dot{p}_i \frac{\partial f}{\partial p_i} \right) = 0 \tag{9-26}$$

as an alternative statement of Liouville's theorem.   It is important to remember that the only essential property of the microscopic model used in obtaining this result is the assumption that the system is *conservative*. Thus Liouville's theorem is the starting point for the discussion of any conservative microscopic model.

At this point we have gone about as far as possible into statistical mechanics without actually introducing statistical concepts.   The major steps remaining involve establishing the ensemble (i.e., assigning particular statistical distributions of initial conditions) and then developing procedures for associating statistical averages across the ensemble with macroscopic properties.

Originally, statistical mechanics was developed in order to provide a mechanical model for the theory of thermodynamics.   In addition there has been considerable success in using statistical mechanics to predict and explain phenomena that lie beyond the domain of macroscopic continuum theories.   The union of statistical mechanics with quantum mechanics has been fruitful in providing more precise quantitative thermodynamic relations and in explaining phenomena at very low temperatures.   Currently, there is much active research in the area of nonequilibrium statistical mechanics which leads to the prediction of transport properties such as viscosity and heat conductivity.

PROBLEMS

9–1. Consider the simple pendulum with the single generalized coordinate $\xi = \theta$.

(a)   Construct the hamiltonian $\mathcal{3C}(\xi,\pi)$.

(b)   Obtain the canonical equations.   Eliminate time between these to get a single

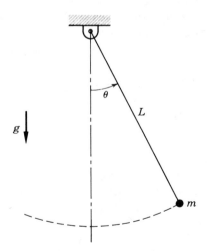

differential equation connecting π and ξ. Integrate to get the phase-plane trajectories $\pi = \pi(\xi)$.

(c) Sketch the phase-plane trajectories for various values of the total energy E. Note the special character of the trajectory with $E = 2mgL$. (Take $E = 0$ when $\xi = \pi = 0$.)

9-2. Consider a single particle with mass m in a plane with the generalized coordinates $\xi_1 = r$ and $\xi_2 = \theta$. Let the particle be attracted to the origin by a central force field whose potential energy is $V(r)$.

(a) Construct the hamiltonian $\mathcal{3C}(\xi_1, \xi_2, \pi_1, \pi_2)$.
(b) Obtain the canonical equations and discuss their significance.

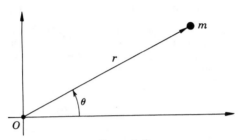

PROB. 9-2

9-3. The rigid block has mass M and centroidal moment of inertia I and is supported by equal springs. Consider *small* vertical and rocking motions only, with the generalized coordinates $\xi_1 = y$ and $\xi_2 = \theta$.

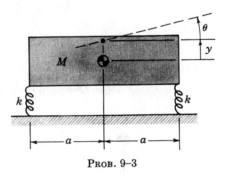

PROB. 9-3

(a) Establish the hamiltonian function $\mathcal{3C}(\xi_1, \xi_2, \pi_1, \pi_2)$.
(b) Obtain the canonical equations and discuss their significance.

9-4. Consider the system of the mass m_2 and the spring in Fig. 2-58b with $\xi = x_2 - x_1$ as the generalized coordinate and $x_1 = g(t)$ as a prescribed time-varying constraint.

(a) Establish the hamiltonian $\mathcal{3C}(\xi, \pi)$ as defined by (9-4).
(b) What is the total energy $T + V$?

9-5. The mass shown has a free-travel length of L, at the ends of which are ideal linear springs. Sketch the phase-plane trajectory if the mass has a total energy E.

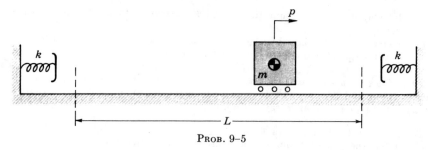

PROB. 9-5

9-6. Consider the two identical masses with ideal spring bumpers.

(a) Sketch separate phase planes p_1 versus q_1 and p_2 versus q_2 for the two masses. Indicate the nature of the trajectories when there is a collision between one mass and a wall and when there is a collision between the two masses.

(b) Show that when the two masses collide (away from a wall) they *exchange* momenta and energies.

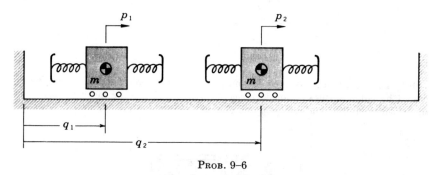

PROB. 9-6

9-7. The system shown consists of an inverted simple pendulum stabilized by a linear spring which exerts a restoring torque $kL^2\theta$ when the pendulum-displacement

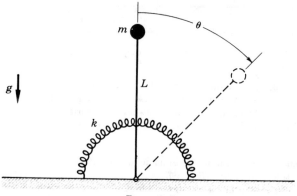

PROB. 9-7

angle is θ. Consider the generalized coordinate $\xi = \theta$, and take the datum position for the potential energy to be $\theta = 0$.

(a)   Establish the hamiltonian $\mathcal{H}(\xi,\pi)$.

(b)   Obtain the canonical equations.

(c)   Assuming $kL < mg$, determine the values of $\xi = \theta$ for which $\dot\pi = 0$.

(d)   Possible equilibrium configurations of the system correspond to points on a phase-plane trajectory, where $\dot\xi$ and $\dot\pi$ simultaneously equal zero. What are the values of total energy E for these equilibrium configurations?

Appendix A

Variational calculus

The variational calculus is a mathematical tool which facilitates the comparison of neighboring states or neighboring motions. Consider a time history of motion $\xi(t)$ and neighboring motions of the form

$$\xi(t) + \epsilon u(t), \tag{A-1}$$

as sketched in Fig. A–1. For a fixed function $u(t)$ the parameter ϵ governs the magnitude of the difference between (A–1) and $\xi(t)$. When $\epsilon = 0$, (A–1) reduces to $\xi(t)$. For fixed ϵ the time history of the difference between (A–1) and $\xi(t)$ is governed by the behavior of the variational shape function $u(t)$. In some applications $u(t)$ can be an arbitrary (sufficiently smooth) function. In other cases admissibility requirements place certain restrictions on $u(t)$. For example, it might be required that all neighboring motions coincide with $\xi(t)$ at $t = 0$. In Fig. A–1 we see that all motions of the form $\xi + \epsilon u$ will meet this requirement if $u(0) = 0$.

When the parameter ϵ in (A–1) is considered to be a first-order differential quantity, the difference $\epsilon u(t)$ between (A–1) and $\xi(t)$ is called a *variation*, denoted by the symbol

$$\delta \xi = \epsilon u. \tag{A-2}$$

A neighboring motion (A–1) is then called a *varied* motion $\xi + \delta \xi$. Consider next a function $f(\xi)$. When its argument $\xi(t)$ is replaced by (A–1), the function is replaced by

$$f(\xi + \epsilon u) = f(\xi) + \epsilon u \frac{df}{d\xi} + \tfrac{1}{2}\epsilon^2 u^2 \frac{d^2 f}{d\xi^2} + \cdots, \tag{A-3}$$

where the right-hand side is a Taylor's series expansion. The variation

417

in the function is *defined* to be the *first-order* term in ϵ in (A–3).

$$\delta f = \epsilon u \frac{df}{d\xi} = \frac{df}{d\xi} \delta\xi. \tag{A-4}$$

An equivalent alternative definition of the *variational operation* is

$$\delta f = \epsilon \left\{ \left[\frac{\partial f(\xi + \epsilon u)}{\partial \epsilon} \right]_{\epsilon=0} \right\}. \tag{A-5}$$

The latter form emphasizes that the variation operation is performed with t fixed. The variational operation can be extended to derivatives

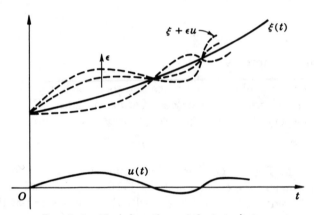

Fig. A–1.   Varied motions of the form $\xi + \epsilon u$.

and integrals of $\xi(t)$. Thus, if $\dot{\xi} = d\xi/dt$, the extension of (A–5) is

$$\delta\dot{\xi} = \epsilon \left\{ \left[\frac{\partial}{\partial\epsilon} (\dot{\xi} + \epsilon\dot{u}) \right]_{\epsilon=0} \right\}$$

$$= \epsilon\dot{u}. \tag{A-6}$$

Note that this is also the time derivative of (A–2); i.e.,

$$\delta\dot{\xi} = \frac{d}{dt} (\delta\xi); \tag{A-7}$$

the operations of variation and time differentiation are *commutative*.

Similarly, if $I(\xi)$ denotes the integral,

$$I(\xi) = \int_0^\xi f(\theta)\, d\theta, \tag{A-8}$$

the extension of (A–5) is

$$\delta I = \epsilon \left\{ \left[\frac{\partial}{\partial\epsilon} \int_0^{\xi+\epsilon u} f(\theta)\, d\theta \right]_{\epsilon=0} \right\}$$

$$= \epsilon u f(\xi) = f(\xi)\, \delta\xi. \tag{A-9}$$

The variational operation can also be extended to motions which have both time *and space* dependence. If $\xi(x,t)$ is such a motion, we consider neighboring motions of the form

$$\xi(x,t) + \epsilon u(x,t). \tag{A-10}$$

When the parameter ϵ is taken to be a first-order differential, the difference between (A–10) and $\xi(x,t)$ is called the variation $\delta\xi$. Variations of functions, derivatives, and integrals can be evaluated from extensions of (A–5). Thus

$$\delta\left(\frac{\partial\xi}{\partial x}\right) = \epsilon\left\{\left[\frac{\partial}{\partial\epsilon}\left(\frac{\partial\xi}{\partial x} + \epsilon\frac{\partial u}{\partial x}\right)\right]_{\epsilon=0}\right\}$$

$$= \epsilon\frac{\partial u}{\partial x} = \frac{\partial}{\partial x}(\epsilon u) = \frac{\partial}{\partial x}(\delta\xi), \tag{A-11}$$

which indicates that space differentiation is also commutative with the variational operation.

NECESSARY CONDITIONS FOR A VARIATIONAL INDICATOR TO VANISH

The outputs of the variational principles discussed in this book are always obtained as necessary conditions for the vanishing of a variational indicator under admissible variations. Each variational principle includes the definition of the appropriate indicator and the limitations on admissible variations. The process of drawing out the necessary conditions for the vanishing of the indicator is an important application of the calculus of variations. This process is outlined in several places in the text. Here we give a somewhat more detailed discussion.

Consider a variational indicator of the form

$$\text{V.I.} = \delta\int_{t_1}^{t_2} F(\xi,\dot{\xi},t)\,dt, \tag{A-12}$$

where F is a known function of its arguments, and $\xi(t)$ is the (unknown) natural motion which is to be determined from the condition that (A–12) vanish for arbitrary admissible variations. For simplicity we consider that the only restrictions on $\delta\xi(t)$ (beyond differentiability) are that $\delta\xi$ should vanish at $t = t_1$ and $t = t_2$.

We take the varied motions to have the form (A–1) and use the following extension of (A–5) to evaluate the variation operation in (A–12):

$$\text{V.I.} = \epsilon\left\{\left[\frac{\partial}{\partial\epsilon}\int_{t_1}^{t_2} F(\xi + \epsilon u, \dot{\xi} + \epsilon\dot{u}, t)\,dt\right]_{\epsilon=0}\right\}$$

$$= \epsilon\int_{t_1}^{t_2}\left(\frac{\partial F}{\partial\xi}u + \frac{\partial F}{\partial\dot{\xi}}\dot{u}\right)dt. \tag{A-13}$$

In order to make the arbitrary shape function $u(t)$ a common factor, we integrate the second term in the integrand, by parts.

$$\text{V.I.} = \epsilon \int_{t_1}^{t_2} \left[\frac{\partial F}{\partial \xi} - \frac{d}{dt} \left(\frac{\partial F}{\partial \dot{\xi}} \right) \right] u \, dt + \epsilon \left. \frac{\partial F}{\partial \dot{\xi}} u \right|_{t_1}^{t_2}. \qquad (A\text{--}14)$$

If variations in ξ are required to vanish at $t = t_1$ and $t = t_2$, then $u(t)$ must vanish at these times, and the expression between limits on the right of (A–14) is zero. It then remains to draw out the equation for the natural motion $\xi(t)$ from the requirement that the variational indicator vanish for *arbitrary* variations within the interval from t_1 to t_2. If the shape function $u(t)$ can be arbitrary, we conclude that a necessary condition for (A–14) to vanish is that

$$\frac{d}{dt} \left(\frac{\partial F}{\partial \dot{\xi}} \right) - \frac{\partial F}{\partial \xi} = 0 \qquad t_1 < t < t_2. \qquad (A\text{--}15)$$

Proof of this statement follows from a *reductio ad absurdum* argument. We suppose that (A–15) does *not* hold at some time T, where $t_1 < T < t_2$. We then show that the indicator (A–14) *cannot* vanish for arbitrary $u(t)$. If (A–15) does not hold at $t = T$, the contents of the square brackets in (A–14) are nonzero at $t = T$. Without loss of generality we may assume the contents of the brackets to be positive. Because of continuity there will be some (possibly small) range of t, say, from t_3 to t_4, for which the contents of the brackets remain positive, as indicated in Fig. A–2a. But if this were the case, the indicator (A–14) could not vanish for arbitrary variations. Figure A–2b shows an admissible shape function $u(t)$ which, when multiplied by the contents of the square brackets, would give a nonzero value to the integral in (A–14). Therefore we are forced to conclude that (A–15) must hold at $t = T$. The same argument applies for any T between t_1 and t_2. Thus the satisfaction of (A–15) throughout the interval is a necessary condition for the indicator to vanish for arbitrary variations.

In the calculus of variations, (A–15) is called *Euler's equation*. Solutions $\xi(t)$ to (A–15) are called *extremals* for the integral in (A–12). When the function F in (A–12) is the lagrangian of a system (with no nonconservative forces), then (A–15) is Lagrange's equation (2–238). The derivation just given can be recast, using variational notation. Thus, with $\delta\xi$ given by (A–2) and $\delta\dot{\xi}$ given by (A–6) and (A–7), we can write (A–13) as

$$\text{V.I.} = \int_{t_1}^{t_2} \left[\frac{\partial F}{\partial \xi} \delta\xi + \frac{\partial F}{\partial \dot{\xi}} \frac{d}{dt} (\delta\xi) \right] dt, \qquad (A\text{--}16)$$

and after integration-by-parts, we have

$$\text{V.I.} = \int_{t_1}^{t_2} \left[\frac{\partial F}{\partial \xi} - \frac{d}{dt} \left(\frac{\partial F}{\partial \dot{\xi}} \right) \right] \delta \xi \, dt + \frac{\partial F}{\partial \dot{\xi}} \delta \xi \Big|_{t_1}^{t_2}, \tag{A-17}$$

in place of (A–14). In the text these manipulations are always described in variational notation.

When a variational indicator includes both time *and* space integrals, an extension of the preceding derivation can be used to show that the

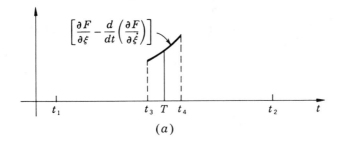

(a)

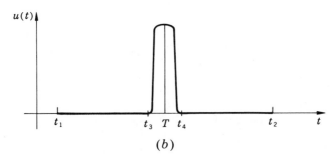

(b)

FIG. A–2. Variation which contradicts assumption that (A–15) does not hold at $t = T$.

necessary conditions for the indicator to vanish for arbitrary admissible variations generally involve *partial differential equations* in space and time and *natural boundary conditions*. To give one simple example, we consider the variational indicator (7–17), which we abbreviate as follows:

$$\text{V.I.} = \int_{t_1}^{t_2} dt \left\{ \int_0^L D[\xi(x,t)] \, \delta\xi(x,t) \, dx - B[\xi(L,t)] \, \delta\xi(L,t) \right\}. \tag{A-18}$$

To reach this stage we have already taken account of the requirements that $\delta\xi(x,t)$ vanish at $t = t_1$ and $t = t_2$ for all x and at $x = 0$ for all t. Beyond this, admissible variations can be arbitrary. We therefore conclude that the necessary conditions for (A–18) to vanish for arbitrary admissible variations are the partial differential equation

$$D[\xi(x,t)] = 0 \qquad \begin{cases} 0 < x < L \\ t_1 < t < t_2 \end{cases} \qquad \text{(A–19)}$$

and the natural boundary condition

$$B[\xi(L,t)] = 0 \qquad t_1 < t < t_2. \qquad \text{(A–20)}$$

The proof of this statement follows the same *reductio ad absurdum* approach. If it is assumed that (A–19) does *not* hold for some X and T, where $0 < X < L$ and $t_1 < T < t_2$, we have only to construct an admissible variation $\delta(x,t)$ which vanishes almost everywhere except for a large

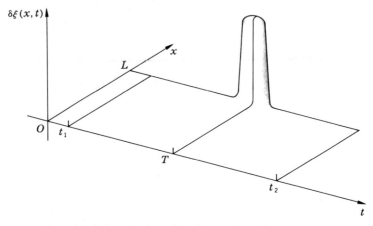

Fɪɢ. A–3.   Variation which contradicts assumption that (A–20) does not hold at $t = T$.

local bump in the neighborhood of (X,T), in order to reach a contradiction. Similarly, if it is assumed that (A–20) does *not* hold at some instant T, where $t_1 < T < t_2$, we are led to a contradiction when we consider an admissible variation of the shape sketched in Fig. A–3. This variation $\delta\xi(x,t)$ is zero over almost all the domain of the double integral in (A–18), but is large in the neighborhood of $t = T$ on $x = L$.

Lᴀɢʀᴀɴɢᴇ ᴍᴜʟᴛɪᴘʟɪᴇʀꜱ

Consider an algebraic variational indicator of the form

$$\text{V.I.} = \delta F(\xi_1,\xi_2) = \frac{\partial F}{\partial \xi_1}\,\delta\xi_1 + \frac{\partial F}{\partial \xi_2}\,\delta\xi_2. \qquad \text{(A–21)}$$

If the variations $\delta\xi_1$ and $\delta\xi_2$ are independent, the necessary conditions for (A–21) to vanish for arbitrary variations are, simply,

$$\frac{\partial F}{\partial \xi_1} = 0 \qquad \frac{\partial F}{\partial \xi_2} = 0. \tag{A–22}$$

The natural state (ξ_1, ξ_2) is obtained by solving the two equations of (A–22) simultaneously.

If, on the other hand, the variations $\delta\xi_1$ and $\delta\xi_2$ are not independent, it is necessary to introduce the constraint between $\delta\xi_1$ and $\delta\xi_2$ into (A–21) before drawing out the necessary conditions for the indicator to vanish. For example, suppose that ξ_1 and ξ_2 are related by the (holonomic) constraint

$$h(\xi_1, \xi_2) = 0; \tag{A–23}$$

then the variations $\delta\xi_1$ and $\delta\xi_2$ must satisfy

$$\delta h = \frac{\partial h}{\partial \xi_1} \delta\xi_1 + \frac{\partial h}{\partial \xi_2} \delta\xi_2 = 0. \tag{A–24}$$

If we solve (A–24) for $\delta\xi_2$ and insert in (A–21), the indicator becomes

$$\text{V.I.} = \left(\frac{\partial F}{\partial \xi_1} - \frac{\partial F}{\partial \xi_2} \frac{\partial h/\partial \xi_1}{\partial h/\partial \xi_2} \right) \delta\xi_1. \tag{A–25}$$

The necessary condition, then, for the indicator (A–21) to vanish for arbitrary variations *which satisfy the constraint* (A–24) is

$$\frac{\partial F}{\partial \xi_1} - \frac{\partial F}{\partial \xi_2} \frac{\partial h/\partial \xi_1}{\partial h/\partial \xi_2} = 0. \tag{A–26}$$

The natural state (ξ_1, ξ_2) is obtained by solving (A–26) and (A–23) simultaneously.

A related variational problem, which has the *same solution*, is that of determining the necessary conditions for the *extended indicator*,

$$\text{V.I.} = \delta F(\xi_1, \xi_2) + \lambda \, \delta h(\xi_1, \xi_2), \tag{A–27}$$

to vanish for arbitrary *independent* variations $\delta\xi_1$ and $\delta\xi_2$. The parameter λ in (A–27) is called a *Lagrange multiplier*. Note that the extended indicator (A–27) consists of the original indicator (A–21) plus the product of the Lagrange multiplier times the constraint (A–24).

To verify that the solution of the related problem is actually the same as (A–26), we expand (A–27) as follows:

$$\text{V.I.} = \left(\frac{\partial F}{\partial \xi_1} + \lambda \frac{\partial h}{\partial \xi_1} \right) \delta\xi_1 + \left(\frac{\partial F}{\partial \xi_2} + \lambda \frac{\partial h}{\partial \xi_2} \right) \delta\xi_2. \tag{A–28}$$

The conditions for (A–28) to vanish for arbitrary independent variations

$\delta\xi_1$ and $\delta\xi_2$ are

$$\frac{\partial F}{\partial \xi_1} + \lambda \frac{\partial h}{\partial \xi_1} = 0$$

$$\frac{\partial F}{\partial \xi_2} + \lambda \frac{\partial h}{\partial \xi_2} = 0. \tag{A-29}$$

These two equations are equivalent to the single relation (A-26), as can be seen when λ is eliminated between the two equations. We have thus verified that an algebraic variational problem with *constrained* variations can be replaced by a related variational problem with independent variations. In some applications the algebraic manipulations are simpler for the latter problem.

The preceding result can be generalized in several directions. For example, the constrained variational problem with the indicator

$$\delta F(\xi_1, \xi_2, \ldots, \xi_m) \tag{A-30}$$

and the p constraints ($p < m$)

$$\delta h_j(\xi_1, \xi_2, \ldots, \xi_m) = 0 \qquad j = 1, \ldots, p \tag{A-31}$$

can be shown to have the same solution as the unconstrained variational problem with the indicator

$$\delta F + \lambda_1 \, \delta h_1 + \lambda_2 \, \delta h_2 + \cdots + \lambda_p \, \delta h_p, \tag{A-32}$$

where $\lambda_1, \lambda_2, \ldots, \lambda_p$ are Lagrange multipliers.

As another example, consider the variational problem with a variational indicator in integral form,

$$\text{V.I.} = \int_{t_1}^{t_2} \delta F(\xi_1, \dot{\xi}_1, \xi_2, \dot{\xi}_2, t) \, dt, \tag{A-33}$$

where the variations $\delta\xi_1$ and $\delta\xi_2$ must satisfy a constraint of the form

$$\delta h(\xi_1, \xi_2) = \frac{\partial h}{\partial \xi_1} \, \delta\xi_1 + \frac{\partial h}{\partial \xi_2} \, \delta\xi_2 = 0. \tag{A-34}$$

It can be verified that a related variational problem (having the same solution) with independent variations has the extended indicator

$$\text{V.I.} = \int_{t_1}^{t_2} (\delta F + \lambda \, \delta h) \, dt. \tag{A-35}$$

In this case the Lagrange multiplier λ is generally a function of t. If the variations $\delta\xi_1$ and $\delta\xi_2$ in (A-33) are constrained, not by the holonomic constraint (A-34) but by the nonholonomic constraint

$$g_1(\xi_1, \xi_2) \, \delta\xi_1 + g_2(\xi_1, \xi_2) \, \delta\xi_2 = 0, \tag{A-36}$$

then the indicator for the related unconstrained problem would have the form

$$\text{V.I.} = \int_{t_1}^{t_2} [\delta F + \lambda(g_1 \, \delta\xi_1 + g_2 \, \delta\xi_2)] \, dt. \tag{A-37}$$

Appendix B

Matrix algebra

Matrix notation provides a convenient abbreviation for a set of simultaneous linear algebraic equations. Operations on such sets of equations can be described concisely by using matrix algebra.

A *matrix* is an array of elements arranged in rows and columns. For example, a rectangular matrix of elements e_{ij} with m rows and n columns has the form

$$\begin{bmatrix} e_{11} & e_{12} & \cdots & e_{1n} \\ e_{21} & e_{22} & \cdots & e_{2n} \\ \cdots & \cdots & \cdots & \cdots \\ e_{m1} & e_{m2} & \cdots & e_{mn} \end{bmatrix}. \tag{B-1}$$

For some purposes it is convenient to treat such an array as a single entity and to represent it by a single symbol such as $[E]$. Of particular interest are *square* matrices of order n,

$$[A] = \begin{bmatrix} a_{11} & a_{12} & \cdots & a_{1n} \\ a_{21} & a_{22} & \cdots & a_{2n} \\ \cdots & \cdots & \cdots & \cdots \\ a_{n1} & a_{n2} & \cdots & a_{nn} \end{bmatrix}, \tag{B-2}$$

column matrices of order n,

$$\{x\} = \begin{Bmatrix} x_1 \\ x_2 \\ \cdot \\ \cdot \\ \cdot \\ x_n \end{Bmatrix}, \tag{B-3}$$

and *row* matrices of order n,

$$\{w\} = \{w_1\, w_2\, \cdots\, w_n\}. \tag{B-4}$$

A square matrix (B–2) is said to be *symmetric* if, for every i and j, $a_{ij} = a_{ji}$. A square matrix (B–2) is said to be *diagonal* if $a_{ij} = 0$ whenever i and j are different; i.e., in a diagonal matrix the only nonzero elements lie along the *main diagonal*, where $i = j$. An important diagonal matrix is the *identity* matrix, or *unit* matrix,

$$[1] = \begin{bmatrix} 1 & 0 & \cdots & 0 \\ 0 & 1 & \cdots & 0 \\ & & \cdots & \\ 0 & 0 & \cdots & 1 \end{bmatrix}. \tag{B–5}$$

The *transpose* $[E]^t$ of a rectangular matrix $[E]$ with m rows and n columns is a rectangular matrix with n rows and m columns such that the rows of $[E]^t$ are the same as the columns of $[E]$. The transpose of a square matrix may be visualized as the matrix obtained by flipping the given matrix over about its main diagonal. Note that a symmetric matrix is its own transpose. The transpose of a column matrix is a row matrix, and the transpose of a row matrix is a column matrix.

The rules for operating with matrices have been developed so as to facilitate the manipulation of sets of simultaneous algebraic equations. Let $[A]$ and $[B]$ be two rectangular matrices with the same number of elements arranged in the same number of rows and columns. The two matrices are said to be *equal* if, for every i and j, $a_{ij} = b_{ij}$. The matrix *sum* $[A] + [B]$ is a matrix of the same format whose elements, for every i and j, are given by $a_{ij} + b_{ij}$. The matrix difference $[A] - [B]$ is a similar matrix, with elements $a_{ij} - b_{ij}$. The matrix $[B]$ is said to be the *product* of a *scalar* c and the *matrix* $[A]$ if, for every i and j, $b_{ij} = ca_{ij}$. The following numerical example illustrates these rules:

$$\begin{bmatrix} 1 & 2 & 3 \\ 2 & 0 & 3 \\ -1 & 1 & 2 \end{bmatrix} + 2 \begin{bmatrix} 2 & 1 & -1 \\ 0 & 3 & 1 \\ -1 & 0 & -2 \end{bmatrix} = \begin{bmatrix} 5 & 4 & 1 \\ 2 & 6 & 5 \\ -3 & 1 & -2 \end{bmatrix}. \tag{B–6}$$

The operation which is the essential ingredient of matrix algebra is the *matrix product* $[A][B]$. This operation is defined only when the number of columns in $[A]$ is the same as the number of rows in $[B]$. Let $[A]$ be a rectangular matrix of elements a_{ij}, with m rows and p columns, and let $[B]$ be a rectangular matrix of elements b_{jk}, with p rows and n columns. The matrix product $[C] = [A][B]$ is a rectangular matrix of elements c_{ik}, with m rows and n columns, where for every i and k,

$$c_{ik} = \sum_{j=1}^{p} a_{ij}b_{jk}. \tag{B–7}$$

The elements entering into (B–7) are indicated in Fig. B–1.

The matrix-multiplication rule permits us to represent the set of simultaneous equations

$$a_{11}x_1 + a_{12}x_2 + \cdots + a_{1n}x_n = y_1$$
$$a_{21}x_1 + a_{22}x_2 + \cdots + a_{2n}x_n = y_2 \qquad \text{(B-8)}$$
$$\cdots \cdots \cdots \cdots \cdots \cdots \cdots \cdots$$
$$a_{n1}x_1 + a_{n2}x_2 + \cdots + a_{nn}x_n = y_n$$

by the matrix product

$$[A]\{x\} = \{y\}, \qquad \text{(B-9)}$$

where $[A]$ is the square matrix (B-2) of the coefficients in (B-8), $\{x\}$ is

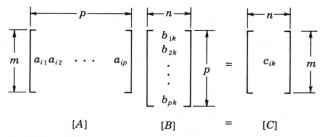

$$[A] \qquad\qquad [B] \qquad = \qquad [C]$$

Fig. B-1. Schematic representation of matrix multiplication:
$c_{ik} = a_{i1}b_{1k} + a_{i2}b_{2k} + \cdots + a_{ip}b_{pk}.$

the column matrix (B-3) of the variables in (B-8), and $\{y\}$ is a similar column matrix of the right-hand sides of (B-8). Other examples of matrix multiplication are furnished by the following numerical examples:

$$\{1 \quad 2 \quad -1\} \begin{bmatrix} 3 & -1 & 1 \\ 0 & 2 & 2 \\ 1 & 0 & -1 \end{bmatrix} = \{2 \quad 3 \quad 6\}$$

$$\{1 \quad 3 \quad 5\} \begin{Bmatrix} 2 \\ -1 \\ 1 \end{Bmatrix} = 4 \qquad \text{(B-10)}$$

$$\begin{Bmatrix} 2 \\ -1 \\ 1 \end{Bmatrix} \{1 \quad 3 \quad 5\} = \begin{bmatrix} 2 & 6 & 10 \\ -1 & -3 & -5 \\ 1 & 3 & 5 \end{bmatrix}.$$

Note that the product of a column into a row is a square matrix, while the product of a row into a column is a matrix with one row and one column, which can be taken as a scalar. This latter kind of product is related to the *dot*, or *inner*, product of vectors. Thus, if $\{v\}$ is a column of the rectangular components of a vector $\mathbf{V}$ and if $\{u\}$ is a similar column of the components of $\mathbf{U}$, their dot product is

$$\mathbf{V} \cdot \mathbf{U} = \{v\}^t\{u\} = \{u\}^t\{v\}. \qquad \text{(B-11)}$$

Matrix multiplication is generally *not* commutative. The only case where $[A][B]$ and $[B][A]$ are both defined and have the same format is when both $[A]$ and $[B]$ are square. Even here, $[A][B]$ is generally different from $[B][A]$; e.g.,

$$\begin{bmatrix} 1 & 2 \\ 3 & 4 \end{bmatrix} \begin{bmatrix} 1 & -1 \\ 0 & 2 \end{bmatrix} = \begin{bmatrix} 1 & 3 \\ 3 & 5 \end{bmatrix} \tag{B-12}$$

$$\begin{bmatrix} 1 & -1 \\ 0 & 2 \end{bmatrix} \begin{bmatrix} 1 & 2 \\ 3 & 4 \end{bmatrix} = \begin{bmatrix} -2 & -2 \\ 6 & 8 \end{bmatrix}.$$

It can be shown that matrix multiplication *is* associative; i.e., if the continued matrix product $[A][B][C]$ is definable, it satisfies

$$([A][B])[C] = [A]([B][C]). \tag{B-13}$$

An important triple-matrix product is the *scalar product* $\{x\}'[A]\{y\}$, where $[A]$ is a square matrix and $\{x\}$ and $\{y\}$ are columns of the same order; e.g.,

$$\{1 \quad 2 \quad 3\} \begin{bmatrix} 1 & 0 & -1 \\ 0 & 2 & 1 \\ -1 & 1 & 1 \end{bmatrix} \begin{Bmatrix} 1 \\ 0 \\ -1 \end{Bmatrix} = \{1 \quad 2 \quad 3\} \begin{Bmatrix} 2 \\ -1 \\ -2 \end{Bmatrix} = -6$$

$$= \{-2 \quad 7 \quad 4\} \begin{Bmatrix} 1 \\ 0 \\ -1 \end{Bmatrix} = -6. \tag{B-14}$$

If, for a given square matrix $[A]$, a square matrix $[B]$ can be found such that

$$[A][B] = [1], \tag{B-15}$$

the matrix $[B]$ is called the *inverse* of $[A]$ and denoted by $[A]^{-1}$. It can be shown[1] that no such matrix exists if the determinant of the elements of $[A]$ vanishes. The matrix $[A]$ is then said to be *singular*. It can also be shown that every nonsingular matrix has a unique nonsingular inverse. Furthermore, it can be shown that the inverse of $[A]^{-1}$ is $[A]$ itself; i.e.,

$$[A][A]^{-1} = [A]^{-1}[A] = [1]. \tag{B-16}$$

The solution of the simultaneous equations (B-8) can be indicated neatly by using the inverse of the coefficient matrix. Thus, if we multiply both sides of (B-9) by $[A]^{-1}$ and use (B-16), we obtain

$$\{x\} = [A]^{-1}\{y\}, \tag{B-17}$$

whenever $[A]$ is nonsingular. It should be pointed out that although the

[1] See, for example, F. B. Hildebrand, "Methods of Applied Mathematics," Prentice-Hall, Inc., Englewood Cliffs, N.J., 1952, p. 15.

inverse notation *indicates* the solution of a set of simultaneous equations, it does not facilitate the *actual* numerical solution of a particular system. More labor is required[1] to compute the inverse of $[A]$ than is required to solve (B-8).

Suppose that, in addition to (B-9), we know that

$$\{z\} = [B]\{y\}, \tag{B-18}$$

where $[B]$ is a square matrix.   Then

$$\{z\} = [B][A]\{x\}, \tag{B-19}$$

or $\qquad\qquad \{z\} = [C]\{x\}, \tag{B-20}$

if we define $[C]$ to be the matrix product $[B][A]$.   In the nonsingular case the inverses of (B-9), (B-18), and (B-20) are

$$\{x\} = [A]^{-1}\{y\}$$
$$\{y\} = [B]^{-1}\{z\} \tag{B-21}$$
$$\{x\} = [C]^{-1}\{z\}.$$

If we substitute the second of these in the first and compare with the third, we find

$$([B][A])^{-1} = [A]^{-1}[B]^{-1}; \tag{B-22}$$

i.e., the inverse of a product of square matrices is the product of the inverses *taken in reverse order*.

A parallel demonstration can be given to show that the *transpose* of a matrix product is the product of the transposes *taken in reverse order;* i.e.,

$$([B][A])^t = [A]^t[B]^t. \tag{B-23}$$

When this rule is applied to the scalar product $\{x\}^t[A]\{y\}$, we have

$$(\{x\}^t[A]\{y\})^t = \{y\}^t[A]^t\{x\}. \tag{B-24}$$

Since the scalar product is a scalar (a square matrix with a single row and a single column), it is its own inverse.   Therefore

$$\{x\}^t[A]\{y\} = \{y\}^t[A]^t\{x\} \tag{B-25}$$

for any square matrix $[A]$.   If $[A]$ is symmetric, $[A] = [A]^t$ and

$$\{x\}^t[A]\{y\} = \{y\}^t[A]\{x\} \tag{B-26}$$

for any pair of columns $\{x\}$ and $\{y\}$.   In the opposite direction it can also be shown that if (B-26) is true *for every pair of columns* $\{x\}$ and $\{y\}$, then the square matrix $[A]$ must be symmetric.

[1] See, for example, S. H. Crandall, "Engineering Analysis," McGraw-Hill Book Company, New York, 1956, p. 35.

Appendix C

Vector analysis

In this appendix we list the more common algebraic identities, differentiation rules, and integral theorems of vector analysis. Let $\mathbf{f}$ and $\mathbf{g}$ be vectors with rectangular components f_1, f_2, f_3 and g_1, g_2, g_3, respectively. The *dot product*, or *inner product*, of $\mathbf{f}$ and $\mathbf{g}$ is the scalar

$$\mathbf{f} \cdot \mathbf{g} = \mathbf{g} \cdot \mathbf{f} = f_1 g_1 + f_2 g_2 + f_3 g_3. \tag{C-1}$$

The *cross product*, or *vector product*, of $\mathbf{f}$ into $\mathbf{g}$ is

$$\mathbf{f} \times \mathbf{g} = -\mathbf{g} \times \mathbf{f} = \det \begin{vmatrix} \mathbf{u}_1 & \mathbf{u}_2 & \mathbf{u}_3 \\ f_1 & f_2 & f_3 \\ g_1 & g_2 & g_3 \end{vmatrix}. \tag{C-2}$$

If $\mathbf{f}$, $\mathbf{g}$, and $\mathbf{h}$ are three vectors, the *triple scalar product*

$$\mathbf{f} \times \mathbf{g} \cdot \mathbf{h} = \mathbf{f} \cdot \mathbf{g} \times \mathbf{h} = \mathbf{g} \times \mathbf{h} \cdot \mathbf{f} = \mathbf{h} \times \mathbf{f} \cdot \mathbf{g}$$

$$= \det \begin{vmatrix} f_1 & f_2 & f_3 \\ g_1 & g_2 & g_3 \\ h_1 & h_2 & h_3 \end{vmatrix} \tag{C-3}$$

is the volume of the parallelepiped, which has $\mathbf{f}$, $\mathbf{g}$, and $\mathbf{h}$ as coterminous edges. Note that the dot and cross can be interchanged and the vectors can be cyclically permuted without altering the result. The *vector triple products* satisfy the following identities:

$$\mathbf{f} \times (\mathbf{g} \times \mathbf{h}) = (\mathbf{f} \cdot \mathbf{h})\mathbf{g} - (\mathbf{f} \cdot \mathbf{g})\mathbf{h}$$

$$(\mathbf{f} \times \mathbf{g}) \times \mathbf{h} = (\mathbf{f} \cdot \mathbf{h})\mathbf{g} - (\mathbf{g} \cdot \mathbf{h})\mathbf{f}. \tag{C-4}$$

In *rectangular* coordinates the *gradient* of a scalar field $\varphi(x_1,x_2,x_3)$ is

$$\text{grad } \varphi \equiv \nabla\varphi = \frac{\partial\varphi}{\partial x_1}\,\mathbf{u}_1 + \frac{\partial\varphi}{\partial x_2}\,\mathbf{u}_2 + \frac{\partial\varphi}{\partial x_3}\,\mathbf{u}_3. \qquad \text{(C–5)}$$

The *divergence* of a vector field $\mathbf{f}(x_1,x_2,x_3)$ is

$$\text{div } \mathbf{f} \equiv \nabla\cdot\mathbf{f} = \frac{\partial f_1}{\partial x_1} + \frac{\partial f_2}{\partial x_2} + \frac{\partial f_3}{\partial x_3}, \qquad \text{(C–6)}$$

and the *curl* of $\mathbf{f}$ is

$$\text{curl } \mathbf{f} \equiv \nabla\times\mathbf{f} = \det \begin{vmatrix} \mathbf{u}_1 & \mathbf{u}_2 & \mathbf{u}_3 \\ \dfrac{\partial}{\partial x_1} & \dfrac{\partial}{\partial x_2} & \dfrac{\partial}{\partial x_3} \\ f_1 & f_2 & f_3 \end{vmatrix}. \qquad \text{(C–7)}$$

The notation $\mathbf{f}\cdot\nabla$ is used for the convective differential operator,

$$\mathbf{f}\cdot\nabla = f_1\frac{\partial}{\partial x_1} + f_2\frac{\partial}{\partial x_2} + f_3\frac{\partial}{\partial x_3}. \qquad \text{(C–8)}$$

The following identities are valid for sufficiently differentiable vector fields $\mathbf{f}$ and $\mathbf{g}$ and scalar fields φ and ψ:

$$\nabla\cdot(\varphi\mathbf{f}) = \varphi(\nabla\cdot\mathbf{f}) + \mathbf{f}\cdot(\nabla\varphi). \qquad \text{(C–9)}$$

$$\nabla\times(\varphi\mathbf{f}) = \varphi(\nabla\times\mathbf{f}) + (\nabla\varphi)\times\mathbf{f}. \qquad \text{(C–10)}$$

$$\nabla\cdot(\mathbf{f}\times\mathbf{g}) = \mathbf{g}\cdot(\nabla\times\mathbf{f}) - \mathbf{f}\cdot(\nabla\times\mathbf{g}). \qquad \text{(C–11)}$$

$$\nabla\times(\mathbf{f}\times\mathbf{g}) = (\mathbf{g}\cdot\nabla)\mathbf{f} - (\mathbf{f}\cdot\nabla)\mathbf{g} + \mathbf{f}(\nabla\cdot\mathbf{g}) - \mathbf{g}(\nabla\cdot\mathbf{f}). \qquad \text{(C–12)}$$

$$\nabla(\mathbf{f}\cdot\mathbf{g}) = (\mathbf{f}\cdot\nabla)\mathbf{g} + (\mathbf{g}\cdot\nabla)\mathbf{f} + \mathbf{f}\times(\nabla\times\mathbf{g}) + \mathbf{g}\times(\nabla\times\mathbf{f}). \qquad \text{(C–13)}$$

$$\nabla\cdot(\nabla\varphi\times\nabla\psi) = 0. \qquad \text{(C–14)}$$

$$\nabla\times(\nabla\varphi) = 0. \qquad \text{(C–15)}$$

$$\nabla\cdot(\nabla\times\mathbf{f}) = 0. \qquad \text{(C–16)}$$

$$\nabla\times(\nabla\times\mathbf{f}) = \nabla(\nabla\cdot\mathbf{f}) - (\nabla\cdot\nabla)\mathbf{f}. \qquad \text{(C–17)}$$

In *spherical* coordinates (see Fig. 2–41) the *gradient* of a scalar field $\psi(r,\theta,\varphi)$ is

$$\text{grad } \psi \equiv \nabla\psi = \frac{\partial\psi}{\partial r}\,\mathbf{u}_r + \frac{1}{r}\frac{\partial\psi}{\partial\theta}\,\mathbf{u}_\theta + \frac{1}{r\sin\theta}\frac{\partial\psi}{\partial\varphi}\,\mathbf{u}_\varphi. \qquad \text{(C–18)}$$

The *divergence* of a vector field $\mathbf{f}(r,\theta,\varphi)$ is

$$\text{div } \mathbf{f} \equiv \nabla\cdot\mathbf{f} = \frac{1}{r^2}\frac{\partial}{\partial r}(r^2 f_r) + \frac{1}{r\sin\theta}\frac{\partial}{\partial\theta}(f_\theta\sin\theta) + \frac{1}{r\sin\theta}\frac{\partial f_\varphi}{\partial\varphi}, \qquad \text{(C–19)}$$

and the *curl* of f is

$$\text{curl } \mathbf{f} \equiv \nabla \times \mathbf{f} = \frac{1}{r^2 \sin \theta} \det \begin{vmatrix} \mathbf{u}_r & r\mathbf{u}_\theta & r \sin \theta \, \mathbf{u}_\varphi \\ \dfrac{\partial}{\partial r} & \dfrac{\partial}{\partial \theta} & \dfrac{\partial}{\partial \varphi} \\ f_r & rf_\theta & r \sin \theta f_\varphi \end{vmatrix}. \qquad (\text{C-20})$$

Consider a closed region of space with volume V and bounding surface S, and let the vector field $\mathbf{f}$ be continuously differentiable in this region. The *divergence theorem* states that the volume integral of the divergence of $\mathbf{f}$ equals the net flux of $\mathbf{f}$ through S; i.e.,

$$\int_V \nabla \cdot \mathbf{f} \, dV = \oint_S \mathbf{f} \cdot \mathbf{u}_n \, dS, \qquad (\text{C-21})$$

where $\mathbf{u}_n$ is the unit vector normal to S which points *out* from the volume enclosed by S.

Consider a portion of a surface S bounded by a simple closed curve C in a region where the vector field $\mathbf{f}$ has continuous second derivatives. *Stokes' theorem* states that the net flux of the curl of $\mathbf{f}$ through S equals the circulation of $\mathbf{f}$ around C; i.e.,

$$\int_S (\nabla \times \mathbf{f}) \cdot \mathbf{u}_n \, dS = \oint_C \mathbf{f} \cdot d\mathbf{R}, \qquad (\text{C-22})$$

where $\mathbf{u}_n$ is the unit vector normal to the surface S, and $d\mathbf{R}$ is the element of arc along the periphery of C so directed that $d\mathbf{R} \times \mathbf{u}_n$ points *out* from the area enclosed by C.

Appendix D

The complementary mechanical variational principle

In Chap. 1 complementary variational principles for statics are discussed, and in Chaps. 5 and 6 complementary forms of Hamilton's principle for electric circuits are employed. For dynamic mechanical systems, however, the text discussion is limited to the form of Hamilton's principle given in connection with (1–98) and (1–99): the lagrangian is $T^* - V$, and the admissibility conditions are geometric. In this appendix we point out the existence[1] of a complementary form of Hamilton's principle in which the lagrangian is $V^* - T$ and the admissibility conditions are restrictions on forces and momenta.

In the complementary formulation, the motion of a system is described in terms of momentum and force variables. The variational indicator has the form

$$\text{V.I.} = \int_{t_1}^{t_2} \left[\delta(V^* - T) + \sum_i v_i \, \delta p_i \right] dt, \tag{D–1}$$

where the potential coenergy V^* is a function of force variables, and the kinetic energy is a function of momentum variables. The summation in (D–1) is a work expression which includes the work done by all elements not already accounted for in the lagrangian $V^* - T$. To see that $v_i \, \delta p_i$ represents a work increment, consider the element shown in Fig. D–1 which exerts force f_i against its terminals when the relative velocity of separation of the terminals is v_i. The work done on the system which forms the environment for such an element during an interval dt is

$$f_i v_i \, dt = v_i \, dp_i, \tag{D–2}$$

---

[1] The principle was enunciated by R. A. Toupin, A Variational Principle for the Mesh-type Analysis of a Mechanical System, *J. Appl. Mech.*, **19**:151–152 (1952).

where p_i is a momentum coordinate defined by the relation

$$\frac{dp_i}{dt} = f_i. \tag{D-3}$$

The work (D–2) is an increment due to a differential increase in p_i. Under a differential variation δp_i, the increment of work done by the element is

$$v_i\, \delta p_i. \tag{D-4}$$

For forces and momenta to be admissible in the complementary variational principle it is necessary that forces balance at every mass-free junction and that corresponding momentum and force variables satisfy relations of the form of (D–3). For an inertia element, (D–3) represents a momentum principle and p_i is a physical momentum. In other

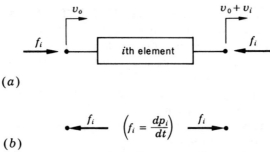

(a)

(b)

FIG. D–1. Power delivered by element is $f_i v_i$, where v_i is rate of separation of terminals and f_i is compressive force. (a) Forces acting *on element;* (b) forces acting *on environment.*

cases (D–3) is simply the *definition* of a momentum coordinate p_i. A momentum coordinate p_i *defined* by (D–3) is sometimes called the *impulse* of the force f_i.

The complementary form of Hamilton's principle states that *an admissible motion of a mechanical system from a fixed configuration of momenta at time t_1 to another fixed configuration of momenta at time t_2 is a natural motion if, and only if, the variational indicator* (D–1) *vanishes for admissible variations of momenta and forces.* The inputs to this principle consist of the force and momentum admissibility conditions, the constitutive relations for deformable elements (e.g., springs, dashpots, and motion sources), and the velocity-momentum relations for the inertia elements. The output from the principle (i.e., the necessary conditions for the indicator to vanish) can be interpreted as geometric-compatibility requirements stated in terms of momentum coordinates.

Example D–1. To illustrate the application of the complementary variational principle we consider the system shown in Fig. D–2. Examining forces and momenta, we note that the force acting to the right on the mass must be the same as the tensile force in the spring, which in turn must be the same as the tensile force in the dashpot. If we call this force f, the linear-momentum principle requires that

$$f = \frac{dp}{dt} = \dot{p}, \qquad (D\text{–}5)$$

where p is the momentum of the mass. The momentum and force admissibility conditions will automatically be satisfied if p is taken as the single independent momentum coordinate and (D–5) is used to repre-

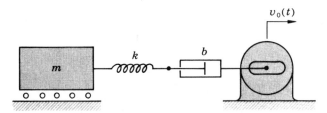

Fig. D–2. Example D–1. Mass-spring-dashpot system excited by velocity source.

sent the force. The lagrangian (for a linear spring and a newtonian mass particle) is

$$\mathcal{L} = V^* - T$$
$$= \frac{\dot{p}^2}{2k} - \frac{p^2}{2m}, \qquad (D\text{–}6)$$

and the work expression (for a linear dashpot and a prescribed velocity source) is

$$\sum_i v_i \, \delta p_i = \left(-\frac{\dot{p}}{b} + v_0 \right) \delta p. \qquad (D\text{–}7)$$

The equations of motion can then be obtained by substituting (D–6) and (D–7) into the variational indicator (D–1) and applying the calculus of variations to obtain the necessary condition for the indicator to vanish for arbitrary δp. An alternative method is to apply Lagrange's equation (2–228), with p playing the role of the generalized coordinate. In either case we obtain

$$\frac{\ddot{p}}{k} + \frac{\dot{p}}{b} + \frac{p}{m} = v_0(t) \qquad (D\text{–}8)$$

as the equation of motion in terms of the momentum coordinate p. This equation can be interpreted as a statement of geometric compatibility. The first term on the left represents the rate of extension of the spring; the second term represents the rate of extension of the dashpot; and the final term on the left represents the velocity of the mass. Geometric compatibility requires that the sum of these should equal the velocity of the source.

The parallelism between the above example and Example 2–9 should be noted. These two examples constitute a case of the mechanical dual analogy mentioned in Sec. 5–5.

The complementary variational principle should provide an attractive approach to the analysis of any system in which the force and momentum relations are simple and direct while the geometric requirements are

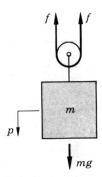

FIG. D–3. Forces acting on mass in system of Fig. 2–53.

complex or obscure. As an indication of this possibility we reconsider Example 2–11.

Example D–2. In the pulley system of Fig. 2–53 it is possible that some analysts would consider the geometric constraint obscure, but would see at once that the magnitude of the force in the string was the same everywhere. If this force is called f, the forces acting on the mass are as shown in Fig. D–3. The momentum principle applied to the mass yields

$$mg - 2f = \frac{dp}{dt} = \dot{p}, \qquad \text{(D–9)}$$

so that momentum and force admissibility will be automatically ensured if p is taken as the independent momentum coordinate and the force f is expressed as

$$f = \tfrac{1}{2}(mg - \dot{p}). \qquad \text{(D–10)}$$

The lagrangian for the spring and the mass is

$$\mathcal{L} = V^* - T$$
$$= \frac{(mg - \dot{p})^2}{8k} - \frac{p^2}{2m}, \tag{D-11}$$

and there is no work expression. Insertion of (D–11) into the indicator (D–1) or direct application of Lagrange's equation,

$$\frac{d}{dt}\left(\frac{\partial\mathcal{L}}{\partial\dot{p}}\right) - \frac{\partial\mathcal{L}}{\partial p} = 0 \tag{D-12}$$

yields

$$\frac{\ddot{p}}{4k} + \frac{p}{m} = 0 \tag{D-13}$$

as the equation of motion for the system of Example 2–11 in terms of the momentum coordinate p. This result should be compared with (2–165). Note that, by working back through (D–10), it is possible to interpret (D–13) as a statement that the velocity of the mass is half the rate of extension of the spring.

LIMITATION OF THE COMPLEMENTARY VARIATIONAL PRINCIPLE

The principle just described can be applied in the manner illustrated to any mechanical system in which the force and momentum admissibility conditions are independent of the system's geometric displacements. Systems of this type include uniaxial translational systems (in which all dynamic forces and all displacements are parallel to a fixed direction) and fixed-axis rotation systems (in which all dynamic torques and all angular displacements are about a fixed axis of rotation).   There are also many systems which are not of this type for large geometric motions but which can be considered to be so when "small" motions are analyzed.

For large motions of general mechanical systems, however, it is the rule that force and momentum requirements depend on the instantaneous configuration of the system.   For example, the linear-momentum principle embodied in (2–151) for the flyball governor cannot be stated without reference to the system's position angle θ.   In these systems it is not possible to represent an admissible force-and-momentum state in terms of momentum coordinates alone, and the complementary variational principle cannot be applied in the manner described above.   In this respect the analogy between mechanical systems and electrical networks breaks down.   In electrical networks the two Kirchhoff laws are always uncoupled.   In mechanical systems geometric compatibility can always be expressed in terms of geometrical entities alone, but the force and momentum requirements cannot generally be expressed solely in terms of forces and momenta.

Example D-3.   To illustrate a simple case where the force and momentum requirements are inherently dependent on the geometric configuration, we consider the system of Fig. D–4*a*.   When $x = 0$ the spring is unstretched and the strut is vertical.   The force f_0 is assumed to be constant, and the block and strut which transmit the effect of f_0 to the mass m are assumed to be massless.   The balance of forces acting on the massless strut is indicated in Fig. D–4*b*.   Note that the magnitude of the horizontal component *depends on the displacement x.*

The force and momentum admissibility condition for the system, in

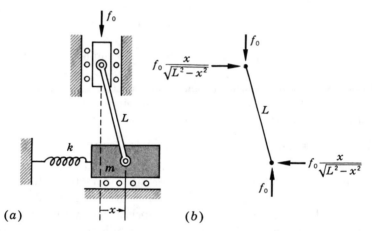

FIG. D–4.   Example D–3.   Force f_0 acts on mass m through massless strut of length L.

terms of the spring tension f and the linear momentum p of the mass, is

$$f_0 \frac{x}{\sqrt{L^2 - x^2}} - f = \dot{p}. \qquad (D\text{--}14)$$

The presence of the geometric displacement x in this statement makes it impossible to describe an admissible state solely in terms of momentum coordinates, as was the case in the preceding examples.   As a consequence, the complementary Hamilton's principle for mechanical dynamics cannot be applied directly to this system

———◆◆———

The complementary Hamilton's principle does, however, apply to the fairly wide class of systems in which momentum and force admissibility conditions are independent of the geometric motions.   The complementary approach may even be more convenient than the usual one when dealing with systems where there are many independent geometrical coordinates but only a few independent momentum coordinates. This is often the case in redundant structures. There are also applications where the required dynamic-response quantities are forces or momenta. In these cases the natural analysis procedure[1] may entail using the complementary approach.

Although the complementary Hamilton's principle cannot be applied directly to a system in which the momentum and force conditions are coupled with the geometry, it is sometimes possible to modify the original system in such a way that the complementary approach becomes valid and the results from the modified system can be directly interpreted in terms of the original system. The modification consists of introducing fictitious deformable force elements into the system in such a way that

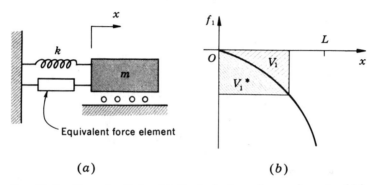

(a) (b)

Fig. D-5. Example D-4. (a) Equivalent nonlinear element which represents effect of strut. (b) Constitutive relation for nonlinear element.

the coupling between forces and geometric motions is transferred from the momentum and force admissibility conditions to the constitutive relations of the fictitious force elements. To illustrate this device we reconsider the system of Example D-3.

Example D-4. In Fig. D-5 we show a modified system which is "equivalent" to the system of Fig. D-4 for horizontal motions. The nonlinear element indicated is supposed to supply the horizontal force component of the strut in Fig. D-4; i.e., the tension f_1 in the nonlinear element is

$$f_1 = -f_0 \frac{x}{\sqrt{L^2 - x^2}}. \tag{D-15}$$

The nonlinear element exerts a repulsive force which depends (nonlinearly) on the displacement x. The constitutive equation (D-15) is sketched in Fig. D-5(b). This relation can also be described in terms

[1] See D. C. Karnopp, Coupled Vibratory-system Analysis, Using the Dual Formulation, *J. Acoust. Soc. Am.*, **40**:380-384 (1966).

of a potential *energy* function,

$$V_1(x) = \int_0^x f_1\,dx = f_0L\left(\sqrt{1 - \left(\frac{x}{L}\right)^2} - 1\right),\qquad \text{(D–16)}$$

or a potential *coenergy* function,

$$V_1^*(f_1) = xf_1 - V_1 = -f_0L\left(\sqrt{1 + \left(\frac{f_1}{f_0}\right)^2} - 1\right).\qquad \text{(D–17)}$$

The complementary variational principle can now be applied to the equivalent system of Fig. D–5a. The momentum and force admissibility condition (D–14) becomes

$$-f_1 - f = \dot{p}.\qquad \text{(D–18)}$$

If we introduce the momentum coordinate p_1 *defined* by

$$\dot{p}_1 = f_1,\qquad \text{(D–19)}$$

the force and momentum admissibility conditions will be automatically satisfied when p and p_1 are taken as independent momentum coordinates and the spring force f is taken as

$$f = -\dot{p}_1 - \dot{p}.\qquad \text{(D–20)}$$

The lagrangian is

$$\begin{aligned}
\mathcal{L} &= V^* - T \\
&= -f_0L\left(\sqrt{1 + \left(\frac{\dot{p}_1}{f_0}\right)^2} - 1\right) + \frac{(\dot{p}_1 + \dot{p})^2}{2k} - \frac{p^2}{2m},
\end{aligned}\qquad \text{(D–21)}$$

and there is no work expression. The equations of motion are obtained by inserting (D–21) into the variational indicator (D–1) or by applying Lagrange's equations. In either case we obtain

$$\frac{d}{dt}\left(\frac{\dot{p}_1 + \dot{p}}{k}\right) + \frac{p}{m} = 0$$
$$\frac{d}{dt}\left[\frac{-L(\dot{p}_1/f_0)}{\sqrt{1 + (\dot{p}_1/f_0)^2}} + \frac{\dot{p}_1 + \dot{p}}{k}\right] = 0\qquad \text{(D–22)}$$

as the equations of motion in terms of the momentum coordinates p and p_1. The first of (D–22) states that the velocity of the spring terminal must be the same as the velocity of the mass. The second states that the velocity of the terminal of the nonlinear element must be equal to the velocity of the spring terminal.

This example indicates how the presence of geometric factors in the momentum and force admissibility conditions can be transferred to the constitutive relations of equivalent (usually nonlinear) force elements.

Appendix E

Natural dependence of state-functions

Energy and coenergy functions of several types (e.g., potential, kinetic, electrical, and magnetic) are employed throughout the text. In each case, the energy function (unstarred) is defined by integrating a work increment, and the coenergy (starred) is defined by a Legendre transformation. When the constitutive relation is nonlinear, there is a clear distinction between the starred and the unstarred functions. When the constitutive relation is linear, however, the two state-functions have equal magnitudes, and they are distinguished only by the fact that they depend upon complementary independent variables. It is the purpose of this appendix to show that there is a criterion for determining which independent variable (or variables) should be used in each case, and that this criterion applies even in the case of linear constitutive relations. We say that a particular state-function depends *naturally* on a certain independent variable if the underlying constitutive relation can be *completely* derived from the state-function. We shall show that whenever a *natural* independent variable is eliminated by introducing the corresponding *complementary* variable, a certain loss of information occurs which makes it impossible to determine the constitutive relation completely.

POTENTIAL ENERGY AND COENERGY

To present the argument[1] in its simplest form, we reconsider the ideal two-force element of Fig. 2–44b. When the element is elastic, with a single-valued constitutive relation between the force f and the elongation e, the potential energy $V(e)$ is obtained by integrating the work

---

[1] This argument is essentially that given by H. B. Callen, "Thermodynamics," John Wiley & Sons, Inc., New York, 1960, pp. 91–92.

increment $f\,de$ from the datum configuration, with no elongation, to the present configuration with elongation e.

$$V(e) = \int_0^e f\,de. \tag{E-1}$$

The potential energy $V(e)$ has the property that it implicitly contains the underlying constitutive relation between f and e. If $V(e)$ is given, the constitutive relation can be obtained by differentiation,

$$f = \frac{dV}{de}. \tag{E-2}$$

The right-hand side of (E-2) is a function of e; let us call it $\varphi(e)$. Then, writing (E-2) in the form

$$f = \varphi(e) \tag{E-3}$$

emphasizes the fact that it *is* the constitutive relation between force and elongation.

Now suppose we wish to change our independent variable from e to f and obtain a state-function which depends on f. We shall examine two procedures: The first is the familiar Legendre transformation which leads to the potential coenergy $V^*(f)$; the second is a simple transformation of the independent variable in $V(e)$ to obtain a function of f (which still has the magnitude of the potential energy). We shall see that, although both procedures are correct and have significance, there is a certain *loss of information* involved in the latter procedure. We therefore say that V depends *naturally* on e but does *not* depend naturally on f. A similar argument can be used to say that V^* depends naturally on f but does not depend naturally on e.

Let us assume that we know the general properties (E-1) and (E-2) of potential energy and that we are given the potential energy $V(e)$ of a particular elastic element. The detailed procedure by which we obtain the complementary state-function $V^*(f)$ is as follows: We first differentiate the given $V(e)$ according to (E-2) to obtain (E-3). Then we invert (E-3) to obtain

$$e = \varphi^{-1}(f). \tag{E-4}$$

Finally, we introduce the Legendre transformation

$$V^* = ef - V \tag{E-5}$$

and use (E-4) to eliminate e in favor of f to obtain $V^*(f)$. The potential coenergy $V^*(f)$ so obtained is a state-function which also implicitly contains the underlying constitutive relation between f and e. Given a particular coenergy function $V^*(f)$ and knowing only the fundamental relations (E–1) and (E–2), together with the theoretical structure (E–5) of the Legendre transformation, we can recover the constitutive relation. The total differential of (E–5) is

$$dV^* = e\,df + f\,de - \frac{dV}{de}\,de \qquad (\text{E–6})$$
$$= e\,df,$$

on using (E–2), and thus the derivative of $V^*(f)$ is

$$\frac{dV^*}{df} = e, \qquad (\text{E–7})$$

which *is* the constitutive relative in the form (E–4).

Now let us consider the second procedure. Starting with the same information, we proceed as before up to (E–4). Then, instead of introducing the Legendre transformation, we simply use (E–4) to eliminate the elongation e in $V(e)$, thereby obtaining $V[\varphi^{-1}(f)]$. To avoid confusion, let us introduce the notation $\bar{V}(f)$ for this function; i.e.,

$$\bar{V}(f) = V[\varphi^{-1}(f)]. \qquad (\text{E–8})$$

The function $\bar{V}(f)$ is a state-function which has the magnitude of the potential energy but which is a function of the force f. We now show that $\bar{V}(f)$ does not contain as much information as $V^*(f)$. We cannot entirely recover the constitutive relation between f and e, from a given $\bar{V}(f)$, if we know only the fundamental relations (E–1) and (E–2) and the structural form of (E–8). Let us see how far we can get.

We begin by writing (E–2) as follows:

$$f = \frac{dV}{de} = \frac{dV}{df}\frac{df}{de}$$
$$= \frac{d\bar{V}}{df}\frac{df}{de}. \qquad (\text{E–9})$$

Then, inserting our given $\bar{V}(f)$ in (E–9), we rearrange and integrate to obtain

$$\frac{d\bar{V}}{df}\frac{df}{f} = de$$
$$\psi(f) = e + C, \qquad (\text{E–10})$$

where $\psi(f)$ is an indefinite integral, and C is a constant of integration. At this point we have exhausted our given information, but we still have an undetermined constant of integration to evaluate before we can recover the explicit constitutive relation (E–4). Whereas $V(e)$ and $V^*(f)$ both imply the complete constitutive relation, the function $\bar{V}(f)$ fixes the constitutive relation only to within an undetermined constant of integration.

Example E–1. Consider a spring, characterized by an elongation variable e and a tensile force f. We are told that the potential energy in the spring has magnitude $f^2/2k$ whenever the force in the spring is f; i.e.,

$$\bar{V} = \frac{f^2}{2k},\tag{E–11}$$

where k is a given constant. Our problem is to deduce the constitutive relation for the spring from (E–11).

If we carry out the steps (E–9) and (E–10) for the given function (E–11), we obtain, in succession,

$$f = \frac{f}{k}\frac{df}{de}$$

$$df = k\,de$$

$$f = k(e - e_0),\tag{E–12}$$

where e_0 is an undetermined constant of integration. Thus, from (E–11), we can recover the constitutive relation only to within a constant of integration. The situation is illuminated in Fig. E–1. Several representative constitutive relations of the form (E–12) are sketched in (a). For each constitutive relation there is a distinct potential-energy function $V = \frac{1}{2}k(e - e_0)^2$, as indicated in (b). The magnitude of the potential energy, when considered as a function of the *force* in the spring, is, however, the *same* for all cases, as indicated in (c). Because of this loss of information inherent in the function $\bar{V}(f)$, we say that the potential energy does not depend naturally on the force f.

NATURAL DEPENDENCE OF KINETIC STATE-FUNCTIONS

The foregoing argument can be readily applied to determine the natural dependence of other state-functions. It can be extended to state-functions which have more than one independent variable, e.g., (4–45), the transducer state-functions of Chap. 6, and the state-functions of thermodynamics.[1] In every case the fundamental definition of the state-function implies a natural dependence on particular independent variables. The introduction of complementary independent variables by simple substitution always entails a loss of information. The concept of natural dependence is particularly important in an application such as thermodynamics, where major attention is focused on the state-functions and a secondary role is assigned to the constitutive relations. In applications where the principal emphasis is put directly on the constitutive relations

---

[1] *Ibid.*

themselves, it must be admitted that the natural-dependency argument is of secondary importance.   The treatment of the kinetic relations in classical dynamics has traditionally been in this latter category.

In newtonian dynamics the velocity-momentum relation $p = mv$ is customarily taken as fundamental, and the question of "recovering" it from the kinetic energy never arises.   If, however, one were to admit only the fundamental definition of kinetic energy T, as work done in increasing momentum, one can only recover the relation $p = m(v - v_0)$

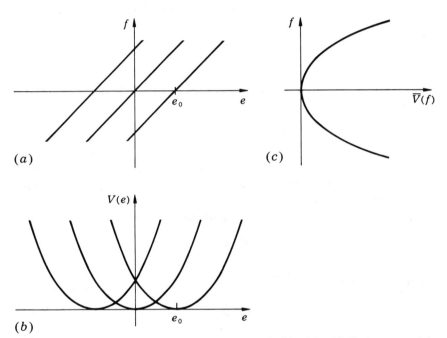

Fig. E-1.   Example E-1.   Family of constitutive relations (a), with distinct potential energies when considered as functions of elongation (b), but with common potential energy when considered as function of force (c).

if one is told that kinetic energy is $\tfrac{1}{2}mv^2$.   The detailed argument for this result is given in the next paragraph.   Since there is a loss of information in expressing the kinetic energy as a function of the velocity, we say that kinetic energy does not have a *natural* dependence on velocity.   Kinetic energy does have a *natural* dependence on momentum, because the complete velocity-momentum relation can be recovered when the kinetic energy is expressed as a function of momentum.   The complete velocity-momentum relation can also be recovered from the function $\tfrac{1}{2}mv^2$, provided that $\tfrac{1}{2}mv^2$ is taken to be the kinetic *coenergy* T^*, since the Legendre transform T^* does have a *natural* dependence on velocity.

To develop the result stated in the preceding paragraph, we begin with the basic definition of kinetic energy. Consider a uniaxial motion of a particle with momentum p and velocity v which is acted on by a force f. The instantaneous power delivered is vf, and in an infinitesimal time interval dt the increment of work done on the particle is

$$vf \, dt = v \, dp, \tag{E-13}$$

on using the linear-momentum principle. The kinetic energy $T(p)$ is defined as the work done in increasing the momentum, from the datum $p = 0$, to the present momentum p,

$$T = \int_0^p v \, dp. \tag{E-14}$$

If $T(p)$ is known, the velocity-momentum relation is implicitly contained since

$$\frac{dT}{dp} = v. \tag{E-15}$$

We therefore say that kinetic energy has a natural dependence on momentum. On the other hand, if we are told that when the particle has velocity v its kinetic energy is $\frac{1}{2}mv^2$, we cannot completely recover the velocity-momentum relation. From (E-15) we have

$$\frac{dT}{dv}\frac{dv}{dp} = v, \tag{E-16}$$

and on inserting $\frac{1}{2}mv^2$ for T we obtain

$$mv\frac{dv}{dp} = v. \tag{E-17}$$

When we integrate (E-17) we get

$$p = m(v - v_0), \tag{E-18}$$

where v_0 is an undetermined constant of integration. Thus kinetic energy does not depend naturally on velocity. A similar argument can be used to show that the kinetic *coenergy* T^* depends naturally on velocity, but not on momentum.

Answers to problems

Chapter 1

1-1. $\theta = \dfrac{e_1 + 2e_2}{6a}$

1-2. $\tau_A = \dfrac{2\tau_1 + \tau_2}{3}$

1-3. $V = \dfrac{ce^4}{4}, \; V^* = \dfrac{3}{4}\sqrt[3]{\dfrac{f^4}{c}}$

1-4. $f = k_1 x + k_2 x + cx^3$

1-5. $\dfrac{f}{k_2} = \dfrac{f_A}{k_1} + \dfrac{f_A}{k_2} + \sqrt[3]{\dfrac{f_A}{c}}$

1-6. $P = \dfrac{1}{2}\left(1 - \dfrac{r_2}{r_1}\right)$

1-7. $L = L_0 + \dfrac{n\pi d W}{k}$

1-8. $20 \text{ lb} < W < 100 \text{ lb}$

1-9. $(\cos \theta)^3 = \dfrac{b}{L}$

1-10. $R = \dfrac{Pa}{b}$

1-11. $H = \dfrac{2P}{3}$

1-14. $\dfrac{\partial F}{\partial T} = -S, \; \dfrac{\partial F}{\partial V} = -P$

1-17. $m_1\left(\dfrac{d^2 x_1}{dt^2}\right) + k_1 x_1 - k_2(x_2 - x_1) = 0; \; m_2\left(\dfrac{d^2 x_2}{dt^2}\right) + k_2(x_2 - x_1) = 0$

1-19. $\dfrac{m \, d^2 x/dt^2}{(1 - v^2/c^2)^{3/2}} + kx = f, \; v = \dfrac{dx}{dt}$

1-22. $A_{\text{nat}} = 0, \; A_{\text{adm}} = -\pi a^2 \sqrt{km}$

447

Chapter 2

2-1. $\mathbf{a}_B = L(\ddot{\theta}\cos\theta - \dot{\theta}^2\sin\theta)\mathbf{u}_x$

2-2. $\mathbf{a}_P = (\ddot{r} - r\dot{\theta}^2)\mathbf{u}_r + (r\ddot{\theta} + 2\dot{r}\dot{\theta})\mathbf{u}_\theta + \ddot{z}\mathbf{u}_z$

2-3. $\mathbf{v} = (v_0 - b\omega_0)\mathbf{u}_x + x\omega_0\mathbf{u}_y$

$\mathbf{a} = -x\omega_0^2\mathbf{u}_x + (2\omega_0 v_0 - b\omega_0^2)\mathbf{u}_y$

2-4. (a) $\frac{1}{2}(a\omega_0 + v_0)$ east, $\dfrac{\sqrt{3}}{2}v_0$ north

(b) $\dfrac{\sqrt{3}}{2}\omega_0(\frac{1}{2}a\omega_0 + v_0)$ north, $\frac{3}{2}\omega_0 v_0$ west, $\frac{1}{4}a\omega_0^2$ vertically downward

2-5. $\mathbf{a}_P = (a\dot{\omega}\cos\theta - b\ddot{\omega} - a\omega^2\sin\theta)\mathbf{u}_x + (a\omega^2\cos\theta + a\dot{\omega}\sin\theta)\mathbf{u}_y$

2-6. $a\omega = (b-a)\dot{\theta}$

2-7. 29.15 ft/sec

2-14.

$$\begin{bmatrix} \cos\theta_{z''}\cos\theta_{y'} & \sin\theta_{z''}\cos\theta_x + \cos\theta_{z''}\sin\theta_{y'}\sin\theta_x & \sin\theta_x\sin\theta_{z''} - \cos\theta_{z''}\sin\theta_{y'}\cos\theta_x \\ -\sin\theta_{z''}\cos\theta_{y'} & \cos\theta_{z''}\cos\theta_x - \sin\theta_{z''}\sin\theta_{y'}\sin\theta_x & \sin\theta_x\cos\theta_{z''}\sin\theta_x + \sin\theta_{z''}\sin\theta_{y'}\cos\theta_x \\ \sin\theta_{y'} & -\cos\theta_{y'}\sin\theta_x & \cos\theta_{y'}\cos\theta_x \end{bmatrix}$$

2-15. $(l,m,n) = \dfrac{1}{\sqrt{14}}(1,2,3); \ \theta = 63.9°$

2-19. $f(r) = \dfrac{\text{constant}}{r}$

2-24. $k = \dfrac{2Mg}{a}$

2-25. $mc^2\ddot{\theta} + (a^2 k_a + b^2 k_b)\theta = 0$

2-26. $m_1\ddot{x}_1 + b\dot{x}_1 + kx_1 - kx_2 = 0$

$-kx_1 + m_2\ddot{x}_2 + kx_2 = 0$

2-27. $(m_1 + 4m_3)\ddot{y}_1 + b\dot{y}_1 + 2m_3\ddot{y}_2 = (2m_3 - m_1)g$

$2m_3\ddot{y}_1 + (m_2 + m_3)\ddot{y}_2 = (m_3 - m_2)g$

2-29. $y = x\tan\alpha$, and $m\ddot{x} + \quad\quad \lambda\sin\alpha = 0$

$m\ddot{y} - mg - \lambda\cos\alpha = 0$

2-30. (b) 2

2-31. (a) 5

(b) 3

2-32. Constraint is piecewise holonomic.

2-33. $\Xi_r = f\sin\theta\sin\varphi; \ \Xi_\theta = fr\cos\theta\sin\varphi; \ \Xi_\varphi = fr\sin\theta\cos\varphi$

2-34. $(a^2 + b^2)\ddot{\theta} + gb\sin\theta = 0$

2-35. $m_1\ddot{x}_1 + (k_1 + k_2)x_1 - \quad\quad\quad k_2 x_2 \quad\quad\quad\quad\quad\quad\quad = 0$

$-k_2 x_1 \quad + (m_2 + m_3)\ddot{x}_2 + k_2 x_2 + m_3 L(\ddot{\theta}\cos\theta - \dot{\theta}^2\sin\theta) = 0$

$m_3 L\cos\theta\ddot{x}_2 + m_3 L^2\ddot{\theta} + m_3 gL\sin\theta = 0$

2-36. $(a^2 + b^2 - 2ab\cos\theta)\ddot{\theta} + ab\sin\theta\,\dot{\theta}^2 + ag\sin\theta = 0$

2-37.

(c) $\frac{1}{2}m(\dot{x}_1^2 + \dot{y}_1^2 + \dot{x}_2^2) - \frac{1}{2}k\left[\left(\frac{1}{2}x_1 + \frac{\sqrt{3}}{2}y_1\right)^2 + x_2^2 + \left(\frac{1}{2}x_2 - \frac{1}{2}x_1 + \frac{\sqrt{3}}{2}y_1\right)^2\right]$

2-38. (c) $\frac{1}{2}m(2\dot{x}^2 + \dot{y}_1^2 + \dot{y}_2^2) - \frac{1}{2}\dfrac{mg}{a}(2x^2 + y_1^2 + y_2^2)$

2-39. (c) $\frac{1}{2}m(\dot r^2 + r^2\dot\theta^2) - \frac{1}{2}k(r-a)^2$

2-40. (a) $m\ddot e + ke = mg$ (b) $m\ddot\xi + k\xi = 0$

2-41. $m(\ddot r - r\dot\theta^2) + k(r-a) + mg(1-\cos\theta) = 0$
$$m(r\ddot\theta + 2\dot r\dot\theta) + mg\sin\theta \qquad = 0$$

2-42.
$$\ddot\theta_1 + \left(\frac{k}{m} + \frac{g}{a}\right)\theta_1 - \frac{k}{m}\theta_2 = 0$$
$$-\frac{k}{m}\theta_1 + \ddot\theta_2 + \left(\frac{k}{m} + \frac{g}{a}\right)\theta_2 = 0$$

2-43. $m\ddot\xi_i + k\xi_i = 0;\ \xi_i = x,\ y,\ z,\ \theta_x,\ \theta_y,\ \theta_z$

2-44. $f = m(g\cos\theta + a\dot\theta^2)$

2-45. $\cos\theta_E = \frac{2}{3}$

2-51. (a) $(0,3L)$
   (b) $-2v_0\mathbf{u}_x$

2-52. $m\ddot x + kx = \dfrac{mv_0^2 L^2}{(L+x)^3}$

2-53. $\theta - \varphi$, where $\tan\theta = \dfrac{v_0^2}{gb}\left(1 + \dfrac{a}{b}\sin\theta\right)$

2-57. Net force is zero.

2-58. $V(r) = -\dfrac{KMm}{2b}\left(3 - \dfrac{r^2}{b^2}\right),\qquad r \le b$

2-59. 84.6 min

Chapter 3

3-1. $I_{zz} = \frac{2}{5}Ma^2$

3-2.
$$[I]_c = \frac{M}{12}\begin{bmatrix} L^2 + h^2 & 0 & 0 \\ 0 & h^2 + t^2 & 0 \\ 0 & 0 & t^2 + L^2 \end{bmatrix}$$

3-7. $T^* = 1,554\ \text{(translation)} + 1,504\ \text{(rotation)}$
    $= 3,058\ \text{ft-lbf}$

3-8. $T^* = 25,900\ \text{ft-lbf}$

3-12.
$$[I]_0 = \begin{bmatrix} \frac{19}{12} & 0 & 0 \\ 0 & \frac{7}{12} & -\frac{1}{2} \\ 0 & -\frac{1}{2} & \frac{3}{2} \end{bmatrix} ML^2$$

3-13.
$$[I]_0 = \begin{bmatrix} 49\frac{1}{3} & 6 & 6 \\ 6 & 27\frac{1}{3} & -18 \\ 6 & -18 & 27\frac{1}{3} \end{bmatrix} mL^2$$

3-14. (a) $I_1 = I_{yy} = M\dfrac{R^2}{2},\ I_2 = I_3 = I_x = I_z = M\left(\dfrac{R^2}{4} + \dfrac{h^2}{12} + L^2\right)$

   (b) $\{\omega\}^t = \left\{0, \dfrac{-\omega_0 L}{R}, \omega_0\right\}$

3-15. $\{H'\}_o^t = \{2.6, -1.8, -13.0\}$

3-18. $\{u_1\}^t = \{0.707, -0.500, 0.500\}$
    $\{u_2\}^t = \{-0.707, -0.500, 0.500\}$
    $\{u_3\}^t = \{0, 0.707, 0.707\}$

3-19. $I_1 = \frac{1}{6},\ I_2 = I_3 = \frac{11}{12}$

3–21. $I_1 = 0$, $I_2 = I_3 = 2mL^2$

3–23. $I_1 = 97,400$ lbm-ft^2, $I_2 = 150,000$ lbm-ft^2, $I_3 = 252,600$ lbm-ft^2

Chapter 4

4–1. $\omega_n^2 = \dfrac{g}{L(1 + I/ML^2)}$

4–2. $\dfrac{Mga^2}{L\omega_0^2}$

4–3. $T^* = \frac{1}{2}M(L^2 + 2aL + 2a^2)\dot{\theta}_1^2$
$\qquad\qquad + \frac{1}{2}m[(L + a)^2\dot{\theta}_1^2 + a^2\dot{\theta}_2^2 + 2a(L + a)\dot{\theta}_1\dot{\theta}_2 \cos(\theta_2 - \theta_1)]$
$\quad V = (M + m)g(L + a)(1 - \cos\theta_1) + mga(1 - \cos\theta_2)$

4–4. $(M + \frac{1}{2}m)b^2\ddot{\theta}_1 \qquad\qquad - \frac{1}{2}mb(b - a)\ddot{\theta}_2 \quad = 0$
$\quad -\frac{1}{2}mb(b - a)\ddot{\theta}_1 + \frac{3}{2}m(b - a)^2\ddot{\theta}_2 + mg(b - a)\sin\theta_2 = 0$

4–5. $y = \frac{1}{2}L \cos\theta$ and $m\ddot{y} + mg - \lambda = 0$
$\qquad\qquad\qquad\qquad\qquad m\ddot{x} = 0$
$\qquad\qquad\qquad I\ddot{\theta} - \frac{1}{2}L\lambda \sin\theta = 0$

4–6. (a) $\dfrac{2Mg \sin\theta}{2M + I/r^2}$

4–7. $\mu \geq \frac{1}{3} \tan\theta_0$

4–9. $\omega_n^2 = \dfrac{3}{4}\left(\dfrac{g}{a}\right)$

4–10. $s = \pm\left[\dfrac{g}{a}(3 \pm \sqrt{10})/8\right]^{\frac{1}{2}}$ Unstable because one root has positive real part

4–14. $\theta_0 = 0 \left(\text{stable for } \omega_0^2 \leq \dfrac{3}{2}\dfrac{g}{L}\right)$

$\qquad\quad \theta_0 = \cos^{-1}\dfrac{3}{2}\dfrac{g}{L\omega_0^2}\left(\text{stable for } \omega_0^2 \geq \dfrac{3}{2}\dfrac{g}{L}\right)$

4–15. (d) $I_3\dot{\varphi}\dot{\psi} + (I_1 + ML^2 - I_3)\dot{\varphi}^2 \cos\theta = MgL$

4–16. $T^* = \frac{1}{2}I\dot{\varphi}^2 + \frac{1}{2}M(a^2\dot{\theta}^2 + a^2 \sin^2\theta\dot{\varphi}^2) + \frac{1}{2}I_a(\dot{\psi} + \dot{\varphi}\sin\theta)^2 + \frac{1}{2}I_d(\dot{\theta}^2 + \dot{\varphi}^2 \cos^2\theta)$
$\quad V = Mga(1 - \cos\theta)$

4–17. $T^* = \frac{1}{2}(I + I_1)\dot{\varphi}^2 + \frac{1}{2}ML^2(\dot{\theta}^2 + \dot{\varphi}^2 \sin^2\theta) + \frac{1}{2}I_3\dot{\psi}^2$
$\quad V = MgL(1 - \cos\theta)$

4–18. $T^* = \frac{1}{2}M(\dot{r}^2 + r^2\dot{\theta}^2 + r^2\dot{\varphi}^2 \sin^2\theta) + \frac{1}{2}I_1(\dot{\theta}^2 + \dot{\varphi}^2 \sin^2\theta) + \frac{1}{2}I_3\dot{\varphi}^2 \cos^2\theta$
$\quad V = -Mgr \cos\theta + \frac{1}{2}k(r - L)^2$

4–19. $T^* = \frac{1}{2}M\dot{z}^2 + \frac{1}{2}I_{xx}\dot{\theta}_x^2 + \frac{1}{2}I_{yy}\dot{\theta}_y^2$
$\quad V = \frac{1}{2}k_{11}(4z^2 + \frac{1}{2}a^2\theta_x^2 + \frac{1}{2}b^2\theta_y^2) - k_{12}(a\theta_x^2 + b\theta_y^2) + (k_{22} + k_{33})(\theta_x^2 + \theta_y^2)$

4–20. $|\omega| = \Omega \cot\alpha$

4–21. (c) $(I_3 - I_1)\Omega \cos\theta + I_3\omega_3 = Mgh$

4–22. (c) $7\dot{x} + 2\Omega y = c_1$, $7\dot{y} - 2\Omega x = c_2$

4–25. (b) $\Omega = \dfrac{Mga}{I_3s}$

$\qquad$ (d) Down

4–26. $\omega_n^2 = \dfrac{k}{I_1} + \dfrac{1}{2}\dfrac{I_3^2\omega_3^2}{I_1^2} \pm \dfrac{I_3\omega_3}{I_1}\left(\dfrac{k}{I_1} + \dfrac{1}{4}\dfrac{I_3^2\omega_3^2}{I_1^2}\right)^{\frac{1}{2}}$

Chapter 5

5-1. $q = E_0 C(1 - e^{-t/RC})$

5-2. $e = I_0 R \dfrac{\omega RL \cos \omega t + \omega^2 L^2 \sin \omega t}{R^2 + \omega^2 L^2}$

5-3. $L\ddot{q} + R\dot{q} + \dfrac{1}{C} q = E(t)$

5-4. $\dfrac{I_0 \cos \omega t}{\omega C(\omega^2 LC - 2)}$

5-5. $L\ddot{q} + R\dot{q} + \dfrac{1}{C} q = \left(\dfrac{i_0 - \dot{q}}{a}\right)^{\frac{1}{2}}$

5-6. $L_1\ddot{q}_1 + \dfrac{q_1}{C_1} + \dfrac{q_1 - q_2}{C_2} = -E_1$

$\qquad L_2\ddot{q}_2 - \dfrac{q_1 - q_2}{C_2} = E_1 - E_2$

5-7. (b) or (c)

$C_1\ddot{\lambda}_1 + C_3(\ddot{\lambda}_1 - \ddot{\lambda}_2) + \dfrac{\lambda_1}{L} = 0$

$C_2\ddot{\lambda}_2 - C_3(\ddot{\lambda}_1 - \ddot{\lambda}_2) + \dfrac{\lambda_2}{L} = 0$

5-9. (b) $L_1\ddot{q}_1 + \dfrac{1}{C} q_1 - \dfrac{1}{C} q_2 = 0$

$\qquad -\dfrac{1}{C} q_1 + L_2\ddot{q}_2 - \dfrac{1}{C} q_2 = 0$

5-10. (b) $L\ddot{q}_1 + R\dot{q}_1 + \left(\dfrac{1}{C_1} + \dfrac{1}{C_2}\right) q_1 = -R\dot{q}_0 - \dfrac{1}{C_2} q_0,$ where $\dot{q}_0 = I(t)$

5-13. (a) $C_1\ddot{\lambda}_1 + \dfrac{1}{L} \lambda_1 \qquad\qquad - C_1\ddot{\lambda}_2 \qquad\qquad\qquad\qquad = \dfrac{1}{L} \lambda_0$

$\qquad\qquad - C_1\ddot{\lambda}_1 + (C_1 + C_2)\ddot{\lambda}_2 + \dfrac{1}{R_1} \dot{\lambda}_2 - C_2\ddot{\lambda}_3 \qquad\quad = 0$

$\qquad\qquad\qquad\qquad\qquad - C_2\ddot{\lambda}_2 \qquad\qquad + C_2\ddot{\lambda}_3 + \dfrac{1}{R_2} \dot{\lambda}_3 = 0$

where $\dot{\lambda}_0 = E(t)$

5-14. (b) $L\ddot{q}_1 + \dfrac{1}{C} q_1 \qquad\quad - \dfrac{1}{C} q_2 \qquad\qquad\qquad = E(t)$

$\qquad -\dfrac{1}{C} q_1 + L\ddot{q}_2 + \dfrac{2}{C} q_2 \qquad\quad - \dfrac{1}{C} q_3 = 0$

$\qquad\qquad\qquad - \dfrac{1}{C} q_2 + L\ddot{q}_3 + \dfrac{2}{C} q_3 = 0$

5-15. $C_1\ddot{\lambda}_1 + \dfrac{1}{R_2} \dot{\lambda}_1 + \dfrac{1}{L_2} \lambda_1 \qquad\qquad\qquad - \dfrac{1}{L_2} \lambda_2 \qquad\qquad - \dfrac{1}{R_2} \dot{\lambda}_3 \qquad\qquad = 0$

$\qquad\qquad - \dfrac{1}{L_2} \lambda_1 + C_2\ddot{\lambda}_2 + \dfrac{1}{R_1} \dot{\lambda}_2 + \dfrac{1}{L_2} \lambda_2 - C_2\ddot{\lambda}_3 \qquad\qquad\qquad = 0$

$\qquad - \dfrac{1}{R_2} \dot{\lambda}_1 \qquad\qquad - C_2\ddot{\lambda}_2 \qquad\qquad\qquad + C_2\ddot{\lambda}_3 + \dfrac{1}{R_2} \dot{\lambda}_3 + \dfrac{1}{L_1} \lambda_3 = 0$

5-16. $\omega_n^2 = (n^2 LC)^{-1}$

5-17. $\omega_n^2 = (a^2 C_1 C_2)^{-1}$

5-18. Since conductor does not accelerate, external agent must supply $-df$ to balance df from field. Therefore power delivered *by* mechanical agent equals power flow *into* external electric circuit.

Chapter 6

6-1. (a) $\theta = 0$ (stable), $\dfrac{\pi}{2}$ (unstable), π (stable), $\dfrac{3\pi}{2}$ (unstable)

(b) $\omega_n^2 = \dfrac{E_0^2 C_1}{J}$

(c) 7.38×10^{-6} ft-lbf

(d) $\mu = 0.007$

6-2. With $d = a - x - L \sin \theta$,

$$M\ddot{x} + kx = \frac{q^2}{16\pi\epsilon d^2}$$

$$mL^2\ddot{\theta} + mgL \sin \theta = \frac{q^2 L \cos \theta}{16\pi\epsilon d^2}$$

6-3. Stable equilibrium for $x = -x_0$, where x_0 is smallest root of

$$\frac{x_0}{d_0} = \frac{C_0 E_0^2}{2k(d_0 - x_0)^2}$$

Natural frequency ω_n given by

$$\omega_n^2 = \frac{k}{m}\left(1 - \frac{2x_0}{d_0 - x_0}\right)$$

6-5. $\omega_0^2 = \dfrac{x_0}{L_0 C_0 d_0}$

6-6. $m_1\ddot{x}_1 + kx_1 - kx_2 = \frac{1}{2}L'\dot{q}^2$

$m_2\ddot{x}_2 + kx_2 - kx_1 = \dfrac{q^2 C'}{2C^2}$

$$\frac{d}{dt}(L\dot{q}) + \frac{q}{C} = E_0$$

6-7. $m\ddot{x} + b\dot{x} + \dfrac{q^2}{2C_0 d_0} = 0$

$$R\dot{q} + \frac{q}{C_1}\left(1 + \frac{x}{d_0}\right) = RI(t)$$

6-9. 12.9 kv

6-10. $G(\omega) = \dfrac{j\omega b + k}{-\omega^2 m + j\omega b + k};$ $H(\omega) = \dfrac{E_0 C_0}{d_0}\dfrac{G(\omega)}{C_0 + C_1 + 1/j\omega R}$

6-13. $C(x) = C_0\left[1 + \left(\dfrac{x}{d_0}\right)^2\right]^{-1}$ $k(q) = \dfrac{q^2}{C_0 d_0^2}$

6-14. $L(x) = L_0\left[1 + \left(\dfrac{x}{x_0}\right)^2\right]^{-1}$ $k(\lambda) = \dfrac{\lambda^2}{L_0 x_0^2}$

6-17. 75.4 rad/sec (720 rpm)

6-18. 0.0885 lbf/amp

Chapter 7

7-1. $\dfrac{\partial^2\varphi}{\partial x^2} = \dfrac{\rho}{G}\dfrac{\partial^2\varphi}{\partial t^2}$ $\qquad 0 < x < L$

$GI_\rho\dfrac{\partial\varphi}{\partial x} = I_1\dfrac{\partial^2\varphi}{\partial t^2} + k\varphi$ $\qquad$ at $x = 0$

$-GI_\rho\dfrac{\partial\varphi}{\partial x} = I_2\dfrac{\partial^2\varphi}{\partial t^2}$ $\qquad$ at $x = L$

7-2. $\dfrac{\partial^2\xi}{\partial x^2} = \dfrac{\rho}{E}\dfrac{\partial^2\xi}{\partial t^2}$ $\qquad 0 < x < L$

$EA\dfrac{\partial\xi}{\partial x} = M\dfrac{\partial^2\xi}{\partial t^2} + k\xi$ $\qquad$ at $x = 0$

$-EA\dfrac{\partial\xi}{\partial x} = b\dfrac{\partial\xi}{\partial t}$ $\qquad$ at $x = L$

7-3. Solution given in Figs. 7–15 to 7–17 applies if $\epsilon_0 L \to a$, $L \to \dfrac{L}{2}$, $c^2 \to \dfrac{P}{\rho A}$

7-5. $\omega_0^2 = \dfrac{\pi^2 P}{\rho A L^2}$

7-6.
$$\left.\begin{array}{l} \dfrac{\partial\eta}{\partial x} = 0 \\[2mm] \dfrac{\partial}{\partial x}\left(EI\dfrac{\partial^2\eta}{\partial x^2}\right) + M\dfrac{\partial^2\eta}{\partial t^2} = 0 \end{array}\right\} \text{ at } x = 0 \qquad \left.\begin{array}{l} \dfrac{\partial}{\partial x}\left(EI\dfrac{\partial^2\eta}{\partial x^2}\right) - k\eta = 0 \\[2mm] \dfrac{\partial^2\eta}{\partial x^2} = 0 \end{array}\right\} \text{ at } x = L$$

7-7. $\omega_n^2 = \dfrac{EI}{\rho A}\left(\dfrac{n\pi}{L}\right)^4$

7-8. $\omega^2 = \dfrac{EI}{\rho A}k^4 + \dfrac{P}{\rho A}k^2$

7-9. $\omega^2 = \dfrac{EI}{\rho A}k^4 + \dfrac{K}{\rho A}$

7-11. $C\dfrac{\partial^2\lambda}{\partial t^2} + G\dfrac{\partial\lambda}{\partial t} = \dfrac{\partial}{\partial x}\left(\dfrac{1}{L}\dfrac{\partial\lambda}{\partial x}\right)$ $\qquad 0 < x < l$

$\dfrac{1}{L}\dfrac{\partial\lambda}{\partial x} = -I(t)$ $\qquad$ at $x = 0$

$-\dfrac{1}{L}\dfrac{\partial\lambda}{\partial x} = \dfrac{1}{R_1}\dfrac{\partial\lambda}{\partial t}$ $\qquad$ at $x = l$

7-12. Let $i = \partial q_1/\partial t$ be series current, $(\partial q_2/\partial x)\,dx$ be charge on shunt capacitance $C\,dx$, and $(\partial^2 q_3/\partial x\,\partial t)\,dx$ be current through shunt conductance $G\,dx$; then $q_1 + q_2 + q_3 = 0$, and

$$L\dfrac{\partial i}{\partial t} + iR + \dfrac{\partial}{\partial x}\left(\dfrac{1}{C}\dfrac{\partial q_2}{\partial x}\right) = 0$$

$$L\dfrac{\partial i}{\partial t} + iR + \dfrac{\partial}{\partial x}\left(\dfrac{1}{G}\dfrac{\partial^2 q_3}{\partial x\,\partial t}\right) = 0.$$

When L, C, R, and G are constants, these combine to yield

$$LC\dfrac{\partial^2 i}{\partial t^2} + (RC + LG)\dfrac{\partial i}{\partial t} + RGi - \dfrac{\partial^2 i}{\partial x^2} = 0.$$

7-13. $\frac{1}{2}mv_\infty^2 = \frac{q_0^2}{2C_0}$

7-15. $\rho A \to L; b \to R; P \to \frac{1}{C}$

7-16. $\rho h \frac{\partial^2 \zeta}{\partial t^2} - S\left(\frac{\partial^2 \zeta}{\partial x^2} + \frac{\partial^2 \zeta}{\partial y^2}\right) = f$

7-20. $\dfrac{2\sqrt{\rho AP}}{b}$

7-21. $\omega^2 = \dfrac{EI}{\rho A}k^4 \qquad V_g = 2c = 2\sqrt{\dfrac{EI}{\rho A}}\,k$

7-22. $\omega^2 = \dfrac{EIk^4}{\rho A + \rho Ik^2}$

7-23. With notation of (7-70), $\omega^2 = \dfrac{a_1^2 a_2^2 k^4}{a_1^2 k_0^2 + a_2^2 k^2}$

7-24. $C_1(\cosh k_1 x - \cos k_1 x - \alpha_1 \sinh k_1 x + \alpha_1 \sin k_1 x)\sin(\omega_1 t + \varphi_1)$, where $k_1 L = 1.87510$; $\alpha_1 = 0.73410$; $\omega_1 = \dfrac{3.51602}{L^2}\sqrt{\dfrac{EI}{\rho A}}$

7-25. $\rho A \dfrac{\partial^2 \eta}{\partial t^2} + EI \dfrac{\partial^4 \eta}{\partial x^4} + P \dfrac{\partial^2 \eta}{\partial x^2} = 0 \qquad 0 < x < L$

$$\left.\begin{array}{c} \eta = 0 \\[2mm] \dfrac{\partial \eta}{\partial x} = 0 \end{array}\right\} \quad \text{at } x = 0$$

$$\left.\begin{array}{c} \eta = 0 \\[2mm] EI \dfrac{\partial^2 \eta}{\partial x^2} = 0 \end{array}\right\} \quad \text{at } x = L$$

7-26. $\omega_n = (n - \tfrac{1}{2})\dfrac{\pi}{l\sqrt{LC}} \qquad n = 1, 2, \ldots$

7-27. $C_2 = 0$ for characteristic impedance, i.e., no reflected wave

Chapter 8

8-1. $\dfrac{1}{\gamma - 1}\dfrac{p_0}{\rho_0}\left[\left(\dfrac{\rho}{\rho_0}\right)^{\gamma - 1} - 1\right]$

8-2. $\dfrac{\partial}{\partial Z}\left(\rho^2 \dfrac{dU}{d\rho}\right) + \rho_0 g = 0 \qquad 0 < Z < h$

$$\rho^2 \dfrac{dU}{d\rho} = p_0 \qquad \text{at } Z = 0$$

$$\rho^2 \dfrac{dU}{d\rho} = 0 \qquad \text{at } Z = h$$

8-4. $\rho \dfrac{Du}{Dt} = -\dfrac{1}{A}\dfrac{\partial}{\partial x}(pA)$

8-5. $\rho_0 R^2 \dfrac{\partial^2 \chi}{\partial t^2} = -\chi^2 \dfrac{\partial p}{\partial R}$ (momentum)

$\chi^2 \rho \dfrac{\partial \chi}{\partial R} = R^2 \rho_0$ (continuity)

8-6. $\rho \dfrac{Du}{Dt} = -\dfrac{1}{r^2} \dfrac{\partial}{\partial r} (r^2 p)$ (momentum)

$\dfrac{\partial \rho}{\partial t} + \dfrac{\partial}{\partial r} (\rho u) + \dfrac{2}{r} \rho u = 0$ (continuity)

8-7.

$\rho_0 \ddot{\chi} + \dfrac{\partial}{\partial X} \left(\rho^2 \dfrac{dU}{d\rho} \right) = 0$ $0 < X < L$

$\rho^2 \dfrac{dU}{d\rho} A + k\chi + m \dfrac{\partial^2 \chi}{\partial t^2} = 0$ at $X = 0$

$\dfrac{\partial \chi}{\partial t} = V$ at $X = L$

for $t > 0$, and with initially

$\chi(X,0) = X - \dfrac{p_0 A}{k}$, $0 < X < L$

8-8. $\rho_0 c A \dfrac{\sin (\omega L/c) - [(m\omega^2 + k)/\rho_0 A c\omega] \cos (\omega L/c)}{\cos (\omega L/c) + [(m\omega^2 + k)/\rho_0 A c\omega] \sin (\omega L/c)}$ Re $(ju_0 e^{j\omega t})$

8-21. 53 ft/sec

Chapter 9

9-1. (b) $\pi^2 = 2mL^2[E - mgL(1 - \cos \xi)]$

9-2. (a) $\mathcal{K} = \dfrac{\pi_1^2}{2m} + \dfrac{\pi_2^2}{2m\xi_1^2} + V(\xi_1)$

9-3. (a) $\mathcal{K} = \dfrac{\pi_1^2}{2M} + \dfrac{\pi_2^2}{2I} + k(\xi_1^2 + a^2\xi_2^2)$

9-4. (a) $\mathcal{K} = \dfrac{\pi^2}{2m} - \pi \dfrac{dg}{dt} + \tfrac{1}{2}k\xi^2$

    (b) $\dfrac{\pi^2}{2m} + \tfrac{1}{2}k\xi^2$

9-5. $p = \pm \sqrt{2mE}$ in free travel, and total travel is $L + 2\sqrt{2E/k}$.

9-7. (a) $\mathcal{K} = \dfrac{\pi^2}{2mL^2} + \tfrac{1}{2}kL^2\xi^2 - mgL(1 - \cos \xi)$

    (c) $\xi = 0$, or root of $\dfrac{\sin \xi}{\xi} = \dfrac{kL}{mg}$

    (d) $E = 0$, or $-\tfrac{1}{2}kL^2 \left[2 \dfrac{\xi(1 - \cos \xi)}{\sin \xi} - \xi^2 \right]$, when ξ is one of the above roots

Index

Index

Acceleration, 43, 66
  angular, 67
  centripetal, 67, 70, 72
  Coriolis, 67, 70, 72
Acoustic approximation, 377–385
Action, 32
  principle of stationary, 30n., 31–35
Active variable, 16
Admissible force state, 10
Admissible geometric state, 4
Admissible motion, 22, 24
Admissible variations, 5, 110–114, 123
Algebra, of matrices, 50, 55, 58, 425–429
  of vectors, 44, 50, 58, 430
Alternator, single-phase, 323–326
Ampere, 258
Analogy, dual electrical, 274
  dual mechanical, 288, 436
  electrical-mechanical, 273–275
Angle, Euler, 224, 225
Angular acceleration, 67
Angular displacement, 48, 49, 53–61
Angular momentum, of particle, 85
  of rigid body, 168–171, 178
  of system, 145
Angular momentum principle, 84–87, 144–146, 178, 226–228
Angular velocity, 49–51, 61–63
Ashley, H., 395n.
Avogadro's number, 406

Axis, of inertial symmetry, 193
  instantaneous, 50, 61
  principal, 190
Axle of gyroscope, 242

Balance of rotors, 221–222
Beam model, Bernoulli-Euler, 346, 365
  Rayleigh, 368
  with shear compliance, 368
  Timoshenko, 343–348
Benjamin, T. B., 395n.
Bergman, P. G., 18n.
Biot, M. A., 360n.
Body coordinates, 167, 224
Boundary conditions, natural, 340, 343, 348, 350, 377, 382, 393, 421

Calculus of variations, 22, 23, 28, 417–424
Callen, H. B., 17n., 441n., 444n.
Canonical equations, 403
Capacitance, 262
Capacitor, 260–262
  coenergy of, 262, 293
  energy of, 261, 293
  movable plate, 292–294
Center of gravity, 163
Center of mass, 144, 167, 168

Centripetal acceleration, 67, 70, 72
Centroid, 144, 167, 168
Cherry, E. C., 257n., 275n.
Chu, L. J., 258n.
Coenergy, electrical, 262, 293
    kinetic, of bar, 337
        of beam, 346
        of particles, 11, 87, 445–446
        of rigid body, 175–177
        of string, 341
    magnetic, 265, 295, 299, 320, 321,
        350
    potential, 11, 433, 440–443
Commutativity, lack of, in rotational
        transformations, 50, 51, 58
    of variation and differentiation, 23,
        28, 418, 419
Complementary principle, in dynamics,
        31, 433–440
    in statics, 10–13
Complementary state function, 14
Completeness, of coordinates, 115
    of variations, 116
Compliance, coefficients of, 219
    matrix of, 219
Condensation, 377
Condenser, 260–262
    (*See also* Capacitor)
Conservation, of energy, 136–137
    of momentum, 139
Conservative element, 97
Conservative field, 92
Conservative force, 8
Conservative system, 137
Constitutive relations, 3, 17, 23, 24,
        102, 269
    for elastic beam, 345
    for electrical elements, 261, 263,
        264
    for a gas, 17, 395
    for isentropic flow, 373, 374
    for two-force element, 96, 97
Constraints, geometric, 2, 117–122
    holonomic, 119, 423
    nonholonomic, 119–122, 156, 424
    piecewise holonomic, 122
    rolling, 117

Constraints, time-varying, 118
    workless, 99–101
Contact force, 81, 82
Contact interaction, 218
Continuity equation, 387, 411
Control volume, 386
Coordinates, body, 167, 224
    complete, 115
    configurational, 407
    cyclic, 139
    eulerian, 385
    generalized, 115
    ignorable, 139
    independent, 116
    lagrangian, 372
    material, 372
    momentum, 407
    spatial, 385
    spherical, 70
Coriolis acceleration, 67, 70, 72
Cosine, direction, 53
Coulomb, 258
Coupling, gyroscopic, 243
Cowper, G. R., 346n.
Crandall, S. H., 2n., 21n., 31n., 33n.,
        215n., 218n., 333n., 345n., 429n.
Curl, 431, 432
Cyclic coordinate, 139
Cycloid, 44

Dahl, N. C., 2n., 218n., 333n., 345n.
D'Alembert's principle, 204
Dashpot, ideal, 97
Degree of freedom, 117
Dependence, natural, 22, 441–446
Deresiewicz, H., 346n.
Direct method, 2, 4, 31, 102, 104, 107–
        109, 300
Direction, invariable, 231
    principal, 190, 211
Direction cosine, 53
Dispersion, 358
Dispersion relation, 357
Displacement, 43
    angular, 48, 49, 53–61

Distribution function, 407, 410
Divergence, 431
Divergence theorem, 432
Dreyfus, S. E., 33n.
Dual analogy, electrical, 274
  mechanical, 288, 436
Dual variational principle, in dynamics, 31, 433–440
  in statics, 10
Duality, 10
Dummy variable, 5n.
Dynamic balance, 222
Dynamic programming, 33n.

Eigenvalue, 191
Eigenvector, 191
Element, electrical, 260–265
  two-force, 96–98
Energy, conservation of, 136, 137
  electrical, 261, 293, 350
  free, Gibbs', 17
    Helmholtz, 40
  internal, 17, 373
  kinetic, 18, 87, 137, 445, 446
  magnetic, 264, 295, 299, 319, 321, 323
  potential, 90–101, 440–443
    of bar, 337
    of beam, 220, 346
    of element, 98
    of field, 92–96
    of spring, 6, 98
    of string, 341
    of system, 98, 99
  rest, 21
  total, 21
Energy ellipsoid, 237
Ensemble, 408
Enthalpy, 17
Entropy, 17, 373
Equation of state, 373
Equations of motion, formulation of, 101–110
  linearization of, 134, 135, 313–318
Equilibrium, requirements for, 2, 142, 143, 161
  stability of, 315, 326

Eulerian coordinate, 385
Euler's angles, 224, 225
Euler's equation, 420
Euler's equations, 226–228
Extremal, 420

Fano, R. M., 258n.
Farad, 262
Faraday's law, 264–265
Feynman, R. P., 19n.
Field, conservative, 92
  electromagnetic, 81, 90
  electrostatic, 81, 259
  of force, 80, 81, 90–96
  gravitational, 81, 93–96, 163
  magnetic, 81, 90, 263, 265, 266
Flügge, S., 32n., 390n.
Flügge, W., 215n.
Flux linkage, 263, 264
Flyball governor, 88–90, 100, 101, 107–110
Force, 79–82
  central, 80
  conservative, 8, 92, 97
  contact, 81, 82
  electromagnetic, 81, 90
  electrostatic, 81, 259
  external, 141
  field of, 80, 81, 90, 96
  generalized, 124, 125, 127
  gravity, 81, 93–96, 163
  inertia, 27, 40
  internal, 141–143
  Lorentz, 81, 266
Frame, inertial reference, 83
  intermediate, 63–78
  relative motion of, 46–53
  rotation of, 46–63
Free energy, Gibbs', 17
  Helmholtz, 40
Frequency, 356
Frequency response, of acoustic transmission system, 382–385
  of loudspeaker, 309–311
  of microphone, 317, 318

g (unit of acceleration), 43
Generalized coordinates, 115
  completeness of, 115
  for electrical networks, 280–285
  for gyroscopes, 240–242
  independence of, 116
  for transducers, 302
Generalized forces, 124, 125, 127
Generalized momentum, 125, 128
Generalized variables, 115–128
Generalized velocities, 124
Geometric admissibility, 4, 24, 115
Geometric requirements, 23, 24
Gradient, 431
Gravitational field, 93–96
  uniform, 95, 162, 163
Gravity, center of, 163
Group velocity, 358–360
Gyrator, ideal, 256, 290, 318*n*.
Gyroscope, 239–246
  generalized coordinates for, 240–242
Gyroscopic coupling, 243

Hamilton, W. R., 30*n*.
Hamiltonian, 403
Hamilton's canonical equations, 403
Hamilton's principle, 30, 128–135
  complementary, 31, 433–440
  for continuous systems, 336–351
  for electrical networks, 276
  for fluid flow, 376, 389
  for holonomic system, 131
  for transducers, 301
Haus, H. A., 22*n*., 390*n*.
Haviland, G., 395*n*.
Henry, 265
Hermann, G., 395*n*.
Hildebrand, F. B., 17*n*., 44*n*.
Holonomic constraint, 119, 423
  piecewise, 122
Holonomic system, generalized, 122–128
Hooke's law, 333
Housner, G. W., 395*n*.
Hunt, F. V., 316*n*.
Huntingdon, H. B., 143*n*.
Hysteresis, 264

Ignorable coordinate, 139
Impedance, 310
  characteristic, 370, 384, 397
Impulse, 434
Independence, of coordinates, 116
  of variations, 116
Indicator, variational, 5, 25–27
Indirect method (*see* Variational method)
Inductance, 264
  mutual, 319–321
    coefficient of, 320
    variable, 321–323
  self, 264
    coefficient of, 321
Inductor, 263–265
  coenergy of, 265, 295
  energy of, 264, 295
  movable core, 294–296
Inertia, moment of, 172
  principal, 190
  product of, 172
Inertia force, 27, 40
Inertia matrix, 173
  for composite bodies, 182–185
  for parallel axes, 180–182
  for rotated axes, 186–189
  symmetry of, 173
Inertia tensor, 171–175
Inertial frame, 83, 224
Inertial symmetry, 193
Initial conditions, 104, 340
Integrability requirement, 16, 92, 121, 319
Internal energy, 17, 373
Inverse, of matrix, 428
  of matrix product, 429
Inverse square law, 80, 93, 259
Invertibility requirement, 16, 17
Ipsen, D. C., 258*n*.
Isentropic flow, 373

Jammer, M., 79*n*.
Joule, 258

Karnopp, D. C., 439*n*.
Kilogram, 83, 84

Kinematics, 42–78
Kinetic coenergy, of bar, 337
  of beam, 346
  of particles, 11, 87, 445, 446
  of rigid body, 175–177
  of string, 341
Kinetic energy, 18, 87, 137, 445, 446
Kirchhoff's laws, 268

Lagrange multiplier, 112–114, 422, 424
  interpretation of, 113, 114
Lagrange's equations, 132, 280, 281,
    302
Lagrangian, 30, 131
Lagrangian coordinate, 372
Lanczos, C., 30n.
Legendre transformation, 13–22
  for electrical state functions, 262, 293
  for kinetic state functions, 18–22, 87
  between lagrangian and hamiltonian,
    403
  for magnetic state functions, 265,
    295, 320, 323
  for potential state functions, 15, 442
  in thermodynamics, 17, 18, 40
Leighton, R. B., 19n.
Lin, C. C., 390n.
Linearity, electrical, of capacitors, 261,
    262, 293, 294, 330
  of inductors, 264, 265, 295, 296,
    330, 331
Linearization, of equations of motion,
    134, 135, 313–318
  of gravitational potential, 94–96
Liouville's theorem, 411
Lorentz-force law, 81, 266
Loudspeaker, 308
  impedance of, 309–311

McLean, L., 8n.
Magnetic coenergy, 265, 295, 299, 320,
    321, 350
Magnetic energy, 264, 295, 299, 321,
    323
Magnetic field, 81, 90, 263, 265, 266

Mandelstam, S., 21n.
Mass, center of, 144, 163, 167, 168
  continuous distribution of, 165
  discrete distribution of, 165
  relativistic, 21
  rest, 19
Material coordinate, 372
Material derivative, 386, 389
Matrix, compliance, 219
  inertia, 173
  rotation, 55
  skew-symmetric, 60, 62
  stiffness, 219
Matrix algebra, 425–429
Matrix notation, 55, 172, 425–429
Maxwell, J. C., 221n., 257n.
Meisel, J., 21n.
Merriam, C. W., III, 31n.
Mesh, 268
Microphone, 312
  frequency response of, 317–318
Microstate, 407
Mindlin, R. D., 346n.
Mohr's circle, 198
Moment, 85
Momentum, 18
  angular, of particle, 85
    of rigid body, 168–171
  linear, of particle, 18, 79
    of rigid body, 168
Momentum principle, angular, for par-
    ticle, 84–87
  for rigid body, 178, 226–228
  for system, 144–146
  linear, for particle, 82
  for rigid body, 178
  for system, 141–144
Motion, admissible, 22, 24
  of beam, 343–348, 360–363
  conservative, 137
  of elastic bar, 337–340, 352–356
  equations of, 101–110
  gyroscopic, 239–246
  natural, 30
  plane, 208–221
  relative, 46–53, 63–78
  stability of, 238

Motion, of string, 340–343
  torque-free, 228–239

Natural boundary conditions, 340, 343, 348, 350, 377, 382, 393, 421
Natural dependence, 22, 441–446
Natural motion, 30
Natural state, 5
Nemat-Nasser, S., 295n.
Newton, 83
Newtonian mechanics, 18–22, 79–87
Newton's second law, 82
Nodal line, 225
Node, 268
Nutation, 254

Ohm, 262
Ohm's law, 262
Oravas, G., Æ., 8n.
Order notation, 59

Parallel-axes theorem, 182
Particle, 79
  angular momentum principle for, 84–87
  linear momentum principle for, 82
Passive variable, 16
Pendulum, bifilar, 157
  compound, 157, 159, 160, 246
  double, 132–135
  inverted, 414
  simple, 214, 246
  spherical, 126–128, 139, 140
Penfield, P., Jr., 22n., 390n.
Period, 357
Phase fluid, 411
Phase space, 404
Phase velocity, 356
Poisson's ratio, 143
Port, power-flow, 318
Potential, electric, 259
Potential coenergy, 11, 433, 440–443
Potential energy, 90–101, 440–443
  of bar, 337

Potential energy, of beam, 220, 346
  of element, 98
  of field, 92–96
  of spring, 6, 98
  of string, 341
  of system, 98, 99
Pound, 83
Poundal, 83
Precession, 232
Principal axis, 190
Principal direction, 190
Principal moment of inertia, 190
Principle, D'Alembert's, 204
  of stationary action, 30n., 31–35
  of stationary coenergy, 13
  of stationary energy, 8
  of virtual work, 8
  (*See also* Hamilton's principle; Momentum principle)
Probability-density function, 410
Product, cross, 430
  dot, 427, 430
  of inertia, 172
  inner, 427, 430
  inverse of, 429
  matrix, 426, 427
  scalar triple, 428
  transpose of, 429
  vector, 430
  vector triple, 429
Programming, dynamic, 33n.
Pseudo-potential function, 153

Rayleigh beam model, 368
Reciprocity, 221
Relativity, special, 18–21
Requirements, for currents, 268, 269
  for forces, 23, 24, 102
  geometric, 23, 24, 102
  for integrability, 16, 92, 121, 319
  for invertibility, 16, 17
  for voltages, 269
Resistor, 262
Right-hand rule, 49
Rigid body, 165–168
  angular momentum of, 168–171

Rigid body, dynamic principles for, 177–180, 226–228
  gyroscopic motion of, 239–246
  kinetic coenergy of, 175–177
  linear momentum of, 168
  plane motion of, 208–221
  torque-free motion of, 228–239
Rigid-body variations, 143, 161
Rolling-cone representation, 233
Rotation, 48, 66
  matrix representation of, 53–63
Rotors, balancing of, 221, 222

Sands, M., 19n.
Seitz, F., 143n.
Serrin, J., 390n.
Slug, 84
Solenoid, 294
  door-chime, 303–308
Sound, velocity of, 378
Source, current, 267
  voltage, 267
Space, phase, 404
  state, 404n.
Spin, 232
Spring, ideal, 96, 97
Stability, of equilibrium, 315, 326
  of torque-free motion, 238, 239
State, equation of, 373
State function, 14, 18, 87–90, 92, 93, 98, 293, 295, 299, 300, 320, 323
State space, 404n.
Static balance, 222
Stiffness matrix, 219
Stokes' theorem, 92, 432
Stokesian derivative, 386
Substantive derivative, 386
Synge, J. L., 32n.

Tensor, of inertia, 171–175
  invariant nature of, 173
Thermodynamics, 17, 18, 40
  first law of, 373
  natural dependence in, 444
Time, universal, 43

Timoshenko, S., 343n.
Timoshenko beam model, 343–348
  shear coefficient for, 346, 367, 368
  wave propagation along, 360–363
Torque, 85
Torque–angular-momentum relation, for particle, 84–87
  for rigid body, 178, 226–228
  for system, 144–146
Torque-free motion, 228–239
  stability of, 238–239
Tou, J. T., 33n.
Toupin, R. A., 21n., 433n.
Transducer, energy-storage, 291, 293, 295
  energy-transfer, 292, 299
  frequency response of, 308–318
  movable-core, 294–296
  movable-plate, 292–294
  moving-coil, 296–300
  mutual-inductance, 321–326
Transformation, improper orthogonal, 56
  rotational, 55
  of tensor components, 187
  of vector components, 57
Transformer, ideal, 289, 318n.
Translation, 47, 66
Transmission line, 348–350
Transpose, of matrix, 426
  of matrix product, 429
Truesdell, C., 390n.
Turnbull, D., 143n.
Two-force element, 96–98

Units, electrical, 258–265
  mechanical, 42, 43, 83, 84, 258

Variable, active, 16
  dummy, 5n.
  passive, 16
Variational calculus, 22, 23, 28, 417–424
Variational indicator, 5, 25–27

Variational method, 2, 4, 31, 102, 104–106, 109, 110, 132–134, 213, 216, 220, 221, 279, 281–285, 301, 302, 306, 313, 324, 325, 336–351, 379–382, 391–393
Variational principle, dynamic, 24–34, 128–135
   complementary, 31, 433–440
   electrical, 276
   electromechanical, 301
   static, 2–10
   complementary, 10–13
Variations, 5, 116, 417, 418
   admissible, 5, 110–114, 123
   completeness of, 116
   independence of, 116
   rigid-body, 143, 161
   virtual, 5
Vector analysis, 430–432
Vectors, algebra of, 44, 50, 80, 430
   invariant nature of, 173
Vehicle point, 66, 67
Velocity, 43
   angular, 49–51, 61–63
   of energy transport, 360
   generalized, 124, 127
   group, 358–360
   phase, 356
   of propagation, 352
   of sound, 378
Velocity-momentum relation, linear, 19–21
   nonlinear, 19–21

Velocity-momentum relation, for particle, 19–21, 23, 24, 102
   for rigid body, 178
Vibration, of cantilever beam, 216–221
   of rotating machinery, 215
Virtual variation, 5
Virtual work, 8, 38, 143
Volt, 259

Watt, 260
Wave, deep-water, 399
   in elastic bar, 352–356
   shallow-water, 399
   along Timoshenko beam, 360–363
Wave equation, 351–356
Wavelength, 357
Wave number, 356
Weber, 264
White, D. C., $21n.$, $257n.$
Wilkinson, J. H., $194n.$
Woodson, H. H., $21n.$, $257n.$
Work, 5, 90–101
   complementary, 11
   virtual, 8, 38, 143
Work expression, 105, 130, 276–279, 302, 350
Work increment, 5, 7, 8, 25, 29, 90, 98, 124, 125, 130, 277
Workless constraint, 99–101

Yourgrau, W., $21n.$